J. Seemann Y. I. Chirkov
J. Lomas B. Primault

Agrometeorology

With 89 Figures

Springer-Verlag
Berlin Heidelberg New York 1979

Professor Dr. J. SEEMANN, Karlsbader Str. 11, D-6056 Heusenstamm

Professor Dr. Y. I. CHIRKOV, Moscow Agricultural Academy, USSR-Moscow 125008

Professor Dr. J. LOMAS, Meteorological Service, POB 25, IL-Bet-Dagan

Dr. B. PRIMAULT, Institut Suisse de Météorologie, Krähbühlstr. 58, CH-8044 Zürich

ISBN 3-540-09331-1 Springer-Verlag Berlin Heidelberg New York
ISBN 0-387-09331-1 Springer-Verlag New York Heidelberg Berlin

Library of Congress Cataloging in Publication Data. Main entry under title: Agrometeorology. Bibliography: p. Includes index. 1. Meteorology, Agricultural. I. Seemann, J. S600.5.A37.630′.25′15.79–9757.

© by Springer-Verlag Berlin Heidelberg 1979

Printed in Germany

Typesetting, printing, and bookbinding: Brühlsche Universitätsdruckerei, Lahn-Gießen, 2122/2131-543210

Preface

Agrometeorology is a comparatively young science. The beginnings of agrometeorological work came in the 20's of this century, when agrometeorology was a working branch of climatology. In the years following 1950 it then developed widely to an independent science. In this process, agrometeorology has not only gained a vast knowledge of the influence of meteorological conditions on plants and livestock in agriculture and damage prevention, but additionally evolved new advisory methods which are of great practical use in agriculture.

Up to the present time there has been practically no specific training for an agrometeorologist. Agrometeorologists are drawn, according to their training, from the ranks of general meteorology or from agriculture and its related biological disciplines. They must, therefore, themselves gather the knowledge for their agrometeorological work and combine for themselves the complex of agrometeorology from biological and meteorological information. This is usually far from easy, as the relevant literature is scattered among the most widely differing journals, partly in little-known foreign languages, and is thus very difficult of access. Comprehensive writings are to be found only in very few partial fields of agrometeorology. The subject of training problems has thus been treated as of utmost importance at the meetings of the Commission for Agrometeorology (CAgM) of the World Meteorological Organization (WMO), especially as agrometeorology has won such great significance and usefulness not only in the so-called underdeveloped countries in advancing a more productive agriculture, but also in countries whose agricultural standard is already high.

In order to encourage the necessary training or further education in agrometeorology, the proposal was made at one of the latest CAgM meetings to write a book, which should provide an introduction to the present-day problems of agrometeorology. The four authors who agreed to undertake the task now present this book. It is neither a text book nor a handbook. It consists simply of a series of accounts, each complete in itself of agrometeorological problems. The authors hope that in this way, useful knowledge can be imparted to all those who wish to concern themselves with the subject of agrometeorology.

May 1979   J. SEEMANN

Contents

Part I   Introduction

The Atmosphere From an Ecological Viewpoint

Y. I. CHIRKOV

Agriculture interfaces with a complex dynamic system of natural conditions, among which meteorological factors are the most prevalent and the most changeable. The effect of these factors on agricultural processes and subjects governs to a considerable extent harvest size, product volume, cost, and operating efficiency. For this reason, the qualified calculation and measurement of meteorological components, as well as the use of information concerning weather phenomena and climate resources, is needed for the efficient solution of many agricultural problems.

In a "soil-plant-atmosphere" system, soil and atmosphere are the physical media through which plants exist, thereby taking on an ecological aspect. The physical bases of the phenomena and processes which occur in the atmosphere and in the region of interaction between the atmosphere and the subjacent surface are studied by means of conventional meteorology. The atmosphere of the near-earth air layer as a habitation medium for plant and animal organisms, which affects developmental processes and productivity formation in these organisms, is studied by means of agricultural meteorology.

The study of the effects of weather and climate on agricultural installations is a basic problem in agrometeorology. In order to establish quantitative expressions for the dependence of the formation, growth, and development of crop productivity on meteorological factors, it is necessary to study the basic meteorological components and process regularities which occur in a free atmosphere and in a near-earth air layer as it interacts with the subjacent surface.

Agrometeorological factors act upon living organisms in association; however, the basic factors which determine the vitality of plant organisms – air, light, heat, and moisture – are singled out as the most importance. None of these vital factors can replace the others. A specific combination of these factors determines the development, growth, and productivity of organisms. Other meteorological components only correct the action of the basic factors, either intensifying or diminishing them. For example, evaporation, and consequently, moisture consumption, is either intensified or diminished as a function of wind strength in the near-earth air layer; cloudiness varies both the quantity and quality of the sunlight which falls on young crops, etc.

The corrective components take on autonomous importance only when their intensity is great, i.e., when a direct harmful action develops with respect to the organisms of a given component. For this reason, the basic factors are studied first when formulating models of the dependence of growth and productivity formation in agrometeorology.

The changeability of meteorological factors in time and space is diverse. During clear weather, the influx of sunlight to the Earth's surface varies by territory to a considerably lesser extent than the temperature in the near-earth air layer. The distribution of precipitation during the warmer seasons is characterized by a great variegation, and consequently, a great variability is observed in productive moisture reserves in the soil. Therefore, data from a fixed network of meteorological stations is insufficient for the assessment of meteorological conditions in cultivated fields, gardens, and meadows. Such data provides only an approximate representation. One must know the regularities of the variation of meteorological conditions in the near-earth air layer as influenced by relief and by the properties of the subjacent surface. In other words, it is necessary to take the microclimate of farm areas and the phytoclimate of young crops into consideration. This is especially important with respect to the agroclimatic bases of an agricultural technique for the cultivation of plants and as concerns reclamative measures for the purpose of optimizing the microclimate of fields, meadows, and gardens.

Examining the near-earth air layer from an ecological point of view when deriving numerical expressions for the relationship between plant development and meteorological factors, it is necessary to lend ecological import to this relationship; in other words, to proceed on the basis of the biological features of a plant and of its reactions to various systems of meteorological conditions. Therefore, a necessary condition for the development of meteorological models is the derivation of physically and biologically based numerical expressions for the relationship of plant growth and development to meteorological factors, as well as research on the manner in which these factors are manifested in various soil-climate regions, with allowance for the ability of the climate to meet the light, heat, and moisture requirements of plants in individual years.

At present, relative to the accumulation of materials concerning interrelated plant and weather observations, taking the ever-increasing efficiency of computer technology into consideration, multifactorial dependences are being developed which more fully express the complex effect of meteorological conditions on plant productivity.

Basic agrometeorological information is presented in this textbook, providing a look at the state of contemporary research, as well as its application for the purpose of the more efficient use of climate resources in increasing agricultural productivity throughout the world.

Part II Physical and Meteorological Principles of Agrometeorology

1 Solar Radiation

J. SEEMANN

1.1 Influence of the Atmosphere

Agricultural meteorology concerns itself primarily with the meteorological conditions in the lowest layers of the atmosphere, in which the energy conversion from solar radiation on the surface of the soil and of the plants most directly influences the meteorological occurrences. For this reason, solar radiation is, as a meteorological parameter, of special importance to agricultural meteorology.

Of the total solar radiation that amounts to an average of $1.94 \, cal \cdot cm^{-2} \cdot min^{-1}$ (solar constant) in vertical impingement on the outer limits of the atmosphere, the soil surface receives approximately only 43%. Forty two percent are directly reflected into space from the edge of the atmosphere. About 15% are absorbed or scattered in the atmosphere by air molecules, water droplets, and dust particles. The sum of absorption and scattering is called the extinction of solar radiation in the atmosphere. The loss of radiated energy through absorption and scattering is dependent on the path lengths of the solar rays in the atmosphere, on the condition of the atmosphere, its moisture content, and its dust-turbidity. The applicable permeability of the atmosphere can be determined according to Lambert's Law which is valid for all radiation filters. Accordingly, the intensity of sunshine on vertical irradiation is

$$I_m = I_0 \cdot q^m$$

at the earth's surface, where I_0 represents the solar constant, q, the transmission factor for the layer thickness 1 (solar angle 90°) and m represents the thickness of the amount of traversed air. If, instead of the transmission factor q, one introduces the extinction coefficient $a = \ln q$, the equation is transformed into

$$I_m = I_0 \cdot e^{-a \cdot m}.$$

The thickness of the amount of air transradiated is given approximately by the air pressure b and the solar height h, as follows:

$$m = \frac{b}{760} \cdot \frac{1}{\sin h}.$$

Tables 1 and 2 present the values of m at normal air pressure (b = 760 mm) for several solar angles.

The following gases are involved in a special measure in the absorption of solar radiation: water vapor, O_3 and O_2. One introduces that height w (in cm) of a liquid water column that would fictitiously be present after condensation of the

Table 1. Thickness of the air amount (m) in its dependence on solar height (solar angle) (h)

h	5°	10°	15°	20°	30°	45°	60°	90°
Elevation above sea level = 0 m	10.40	5.60	3.82	2.90	2.00	1.41	1.15	1.00
Elevation above sea level = 500 m	9.80	5.28	3.60	2.73	1.88	1.34	1.08	0.94

Table 2. Transmission factors and extinction coefficients for total radiation in their dependence on the absolute air-mass for pure, dry air

Absolute air mass Mp	0.5	1	2	3	4	6	8	10
Transmission factor q	0.900	0.905	0.914	0.921	0.927	0.935	0.941	0.946
Extinction efficient $a = \ln q$	0.105	0.099	0.089	0.082	0.076	0.0671	0.0600	0.055

water vapor (precipitable water), as the measure for the amount of the water vapor. The following corresponds to a partial pressure e mm Hg of water vapor:

(mm)	2.0	5.5	7.3	8.5	9.4	11.0	12.0
w (cm)	0.1	0.5	1.0	1.5	2.0	3.0	4.0.

The weakening of direct solar radiation by water vapor, ozone, and oxygen at various solar angles is:

Solar angles:	90°	60°	30°	10°	5°
H_2O (w = 1 cm)	6.8 %	7.3 %	8.7 %	12.4 %	14.8 %
O_3 (2 mm) and O_2	2.1 %	2.1 %	2.8 %	4.5 %	5.7 %.

The turbidity caused by dust and other impurities in the atmosphere is designated as aerosol. Aerosols lead to a visible turbidity and, additionally, weaken solar radiation in its penetration of the atmosphere. Linke's turbidity factor is primarily used as the measure for turbidity. The value of the turbidity factor is conditioned to the countryside. It is higher in summer than in winter. On the average, the turbidity factor is given by Linke (1953) with the following values:

High mountains	T = 1.90
Flat land	T = 2.75
Major city	T = 3.75
Industrial area	T = 5.00.

Because of the influence of the atmosphere, solar radiation reaches the soil surface both as parallel radiation (also called direct radiation), and as diffuse radiation (also called sky radiation).

1.2 Measuring Systems

Generally, the solar energy that is radiated per unit time to unit surface is measured as gram-calories per square centimeter per minute ($cal \cdot cm^{-2} \cdot min^{-1}$). In recent years Joule has been introduced as a measuring unit instead of calories in radiation measurements. In some cases, especially in publications with technical content, Watt is also being used. Table 3 juxtaposes these three measurement units. For further conversions, it is, among others, useful to remember that 860 kcal correspond to a kilowatt hour.

Table 3. Measuring units of solar energy

	$cal \cdot cm^{-2} \cdot min$	$J \cdot cm^{-2} \cdot min$	$W \cdot cm^{-2}$
$cal \cdot cm^{-2} \cdot min^{-1}$	1	4.1868	0.069
$J \cdot cm^{-2} \cdot min^{-1}$	0.238	1	0.00165
$W \cdot cm^{-2}$	14.3	60.6	1

1.3 Global Radiation

Global radiation is the sum of solar- and sky-radiation on a horizontal surface. The total earth surface receives an annual 860×10^{18} kcal from global radiation. This is a million times the amount of energy than that represented by the total electricity production of industry on earth (Schulze, 1970). The values of Table 4 are to present a survey on the magnitude of global radiation in its dependence on the solar angle. This shows quite clearly the effect of the varying aerosol concentrations on the intensity of the global radiation, as it is received over the soil. An example for the annual course of global radiation and sky radiation, according to values from Kinshasha and Vienna (Dirmhirn, 1964) under clear and average cloudy sky in Fig. 1, permits the visualization of geographical differences of radiation consumption. Further extended orientations on the distribution of global radiation on earth can be derived from Ashbel's (1979) maps, among others. These contain the global radiation measurements from the geophysical years.

The predominance of the proportions of direct- and sky-radiations is determined by the atmospheric conditions. Direct solar radiation is weakened by the lengthening of the path-length in the atmosphere (low solar angle, low elevation above sea level), by stronger turbidity and higher degree of clouding. These, however, increase diffuse radiation (see also Table 5).

A comprehensive presentation of sky radiation in its geographical distribution will remain problematic because of the considerable regional fluctuation of the number of factors upon which its intensity is dependent. This becomes especially difficult when clouding effects are added. According to Schulze (1970), global radiation is reduced to about 64% with an average clouding of 6/10.

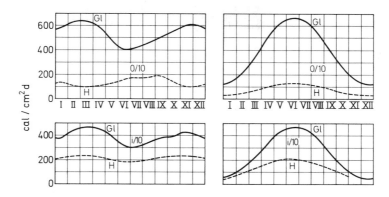

Fig. 1. Annual course of daily sums of global- and sky-radiation for cloudless skies (*upper*) and for average cloud counditions (*lower*), for Kinshasa (*left*) and Vienna (*right*). (After Dirmhirn, 1964)

Nicolet and Doginaux (1951) have investigated the relationships between duration of sunshine and sums of radiation for average latitudes and established the following formula:

$$S = S_0 \cdot \frac{s^n}{s_0},$$

where S = total radiation with corresponding clouding; S_0 = sum of radiation under cloud-free sky; s = duration of sunshine with corresponding clouding; s_0 = duration of sunshine under cloud-free sky; n = exponent: $1 + 0.8 \dfrac{s}{s_0}$.

With the aid of this formula and the values for global radiation from Table 4, we obtain the least order of magnitude determination of actual radiation sums on cloudy days when radiation measurements are lacking.

Table 4. Global radiation (a) and sky radiation (b) in their dependence on solar angle h, the turbidity factor T under cloudless skies in $\text{mcal} \cdot \text{cm}^{-2} \cdot \text{min}^{-1}$. (After Collmann, 1964)

T/h	5°	10°	15°	20°	25°	30°	35°	40°	45°	50°	55°	60°
a												
1	127	278	440	599	756	906	1053	1191	1318	1438	1544	1640
2.72	73	176	302	438	577	716	850	976	1100	1211	1311	1399
3.75	61	147	257	383	514	647	774	897	1013	1121	1218	1306
5.00	56	129	228	344	468	599	726	850	963	1071	1165	1252
b												
1	29	45	58	67	73	78	84	90	94	99	103	103
2.75	36	57	74	88	100	110	118	124	131	135	138	140
3.75	40	66	87	108	125	139	150	159	167	172	176	180
5.00	45	79	111	140	168	193	216	233	248	259	266	273

1.4 Radiation on Inclined Surfaces

The previous treatment of solar radiation concerned a flat, horizontal surface as the receiving surface. In topoclimatological work, the radiation conditions on horizontal surfaces will play a role only to a very limited extend, all the more, however, in respect of inclined surfaces, i.e., slopes of varying inclination and exposure. For these, radiation measurements are very rarely available, resulting in voluminous calculations. However since, especially in topoclimatology, the effects of radiation are of special importance, this problem will be treated thoroughly here

Table 5. Contribution of sky radiation to global radiation on cloud-free days (at 1 cm H_2O; 2 mm O_3 and O_2) in percent

Solar angle	90°	30°	10°
$T = 2.75$	8	16	37
$T = 3.73$	12	24	49

(Table 5). The knowledge of some basics from spherical trigonometry is required for the determination of irradiation of vertical or inclined surfaces. These basics have been treated very thoroughly by Kaempfert (1942) in respect of practical application to agricultural meteorology.

If the intensity I_m of solar radiation, i.e., its surface strength on vertical incidence at a given time in a location on earth is known, then the strength of irradiation I on any desired surface that forms the angle y (incidence angle) with the solar radiation is calculated

$$I = I_m \sin y.$$

Here, I_m refers to the plane. The angle of incidence that is formed by the incident solar ray with the surface upon which it impinges then precisely corresponds to the solar angle. Thus it is dependent only on the geographic latitude φ, the solar declination δ corresponding to the calendar date and the hour-angle t that results from the time of day. Then:

$$\sin y = \sin h = \cos \varphi \cos \delta \cos t + \sin \varphi \sin \delta.$$

φ represents the geographic latitude of the location. δ can be derived from the astronomical annuals. The hour-angle t results from the intersection of the solar vertical with the horizon. Uniformly, the hour-angle amounts to 15° for each full hour of the day, and is counted with the course of the sun from the daily solar culmination in the South (noon $= 0°$) positively through West ($18^{00} = 90°$) to North ($24^{00} = 180°$).

Kaempfert (1942) has derived the following formula for the angle of incidence on a vertical wall (y_w):

$$\sin y_w = \sin a' \cos h,$$

Table 6. Azimuth table (according to Kaempfert). Solar azimuth a : South $=0°$, wall azimuth a^1 : course of wall $=0°$

Sun position azimuth		a^1 for walls and slopes								
Direction	a	N	NE	E	SE	S	SW	W	NW	N
N	− 180	90	45	0	−	−	−	180	135	90
NE	− 135	135	90	45	0	−	−	−	180	135
E	− 90	189	135	90	45	0	−	−	−	180
SE	− 45	−	180	135	90	45	0	−	−	−
S	− 0	−	−	180	135	90	45	0	−	−
SW	+ 45	−	−	−	180	135	90	45	0	−
W	+ 90	0	−	−	−	180	135	90	45	0
NW	+135	45	0	−	−	−	180	135	90	45
N	+180	90	45	0	−	−	−	180	135	90

where a′ is the "wall azimuth", namely the angle between the wall that would be produced by raising the slope to the vertical, and the solar vertical, counted positively from the South $=0°$ through West towards North.

cos h is obtained from the relationship for the determination of the solar azimuth : a

$$\sin a = \frac{\cos \delta}{\cos h} \sin t$$

$$\cos h = \frac{\cos \delta}{\sin a} \sin t .$$

The solar azimuth a is also counted positively with the course of the sun from the South $=0°$, through West toward North.

Table 6 serves the conversion of the solar azimuth a (South $=0°$) into the wall azimuths a′ (wall course $=0°$) for the various wall- and slope directions.

The conditions become considerably more difficult when the radiation, as in the practical application of this calculation to topoclimatology, impinges upon a randomly oriented surface.

If one considers the inclination of the surface with the angle v as being the rotation of a vertical wall by its horizontal central axis (v $=90°$ would then correspond to the extreme case of the vertical wall, v $=0°$ to the opposite extreme case of the horizontal plane), then the wall azimuth a′ of the vertical surface changes into the slope azimuth $\alpha(=90° - a')$. The angle of incidence of the sloped surface, i.e., a slope, is then obtained through the general relation:

$$\sin y = (\sin \varphi \cos v - \cos \varphi \sin v \cos \alpha) \sin \delta$$
$$+ (\cos \varphi \cos v + \sin \varphi \sin v \cos \alpha) \cos \delta \cos t$$
$$+ \sin v \sin \alpha \cos \delta \sin t .$$

The extended Lambert formula is suitably used for the determination of the intensity of the sunshine I_m. Then,

$$I_m = I_0 \cdot e^{-a_m \cdot T \cdot m} .$$

Table 7. Daily course of solar radiation on May 15 under cloudless sky on a horizontal surface with completely free horizon at 200 m a.s.l. in mcal·cm^{-2}·min^{-1}

Time	05:00/ 19:00	06:00/ 18:00	07:00/ 17:00	08:00/ 16:00	09:00/ 15:00	10:00/ 14:00	11:00/ 13:00	12:00
Measured	17	149	366	600	804	980	1076	1110
Calculated	14	145	350	580	794	965	1074	1110

Where: I_0 represents the solar constant. e = 2.718, the basis of the natural logarithms, a_m is the extinction coefficient (relative to the air amount m) T is the turbidity factor that was introduced by Linke and m the amount of air transradiated.

The strength of irradiation I on a randomly sloped surface that forms the angle y with the solar rays is then:

$$I = I_0 \cdot e^{-a_m \cdot Tm} \cdot \sin y.$$

With this formula and the previous equation for sin y, the conditions are now given for the calculation of the strength of irradiation from direct solar radiation. Since such calculations are connected with a considerable effort, one will, today, simply use the computer. It is appropriate for the determination of the daily values of the strength of irradiation of slopes, to use short-period, perhaps 5-min steps for the computer calculations, and then to integrate these values numerically for the entire day.

As can be seen from the formula that was given for the calculation of the solar radiation intensity, this process determines only direct solar radiation under cloudless sky. It would, of course, be possible to utilize the global radiation for the calculation, perhaps as it can be obtained from Collmann's Table 4.

Since the differences in the albedo are strongly involved in the values of the sky radiation, difficulties are still encountered, for example, in the calculation of topographically very different areas. However, the determination of the strength of irradiation from direct solar radiation should be completely sufficient for topoclimatological purposes, since differences worth mentioning are to be expected only in the meteorological conditions in the air layers close to the ground and in the heat conversion on the surfaces, when direct solar radiation is present.

The degree to which calculated values of solar radiation actually agree with the measured conditions, when calculated according to the present extended Lambert formula, can be seen from a juxtaposition with values that were obtained by Sauberer and Dirmhirn (1958) at about $\varphi = 47°$ latitude North (Table 7).

Gräfe (1956) has compared the calculation of the solar radiation intensity on a south-exposed slope with v = 45°, according to Kaempfert's formulas with direct measurements (measuring instrument at 45° angle exposed towards the south!). The calculated values are, as was to be expected, below those measured for the maximally possible global radiation on a cloudless day. Gräfe found that the individual daily totals of the maximal possible global radiation onto the 45° southern slope are approximately 16% higher than those of solar irradiation.

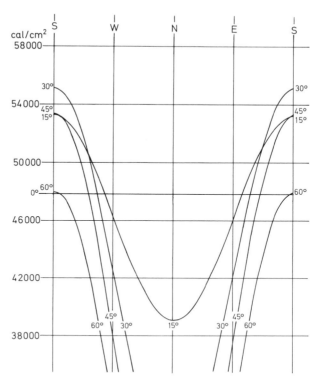

Fig. 2. Flat-surface energy availability from direct solar radiation ($\varphi = 50°$, 50′ N) during the vegetation period (April to October) under consideration of the a verage duration of sunshine

   The method for the calculation of the strength of irradiation that is described here naturally only leads to maximally possible values of solar radiation. The influence of clouds is, initially, not considered. Thus, for example, a slope, the irradiation maximum of which is in a time of day, when strong clouding is generally present, will obtain only slight radiation intensities, despite theoretically high solation values. The influence of clouding is, however, difficult to consider since – especially for extended periods – no adequate observations are available. An indirect conclusion concerning the clouding conditions and the consequent reduction of radiation can, however, be obtained from registrations of duration of sunshine.
   If one considers that the maximal radiative energy of a day-time hour can be registered only during unlimited sunshine, then a survey on the actual hourly energy application of a slope or of a horizontal surface that is sufficient for practical use can be obtained by multiplying the hourly sum of the calculated irradiation that was calculated, with the time-wise appropriate value of sunshine duration (in 1/10 h).
   Figure 2 presents, according to these considerations, the surface-oriented strength of irradiation of direct solar radiation during the entire vegetative period

Table 8. Sunrise (A) and sunset (U) in valleys with 10°, 20°, 30°, and 45° horizon superelevation for northern latitudes (NL) 25° and 50°

Horizon superelevation		10°		20°		30°		45°	
		A	U	A	U	A	U	A	U
N–S-Tal									
NL 25°	21 June	5^{57}	18^{05}	6^{55}	17^{19}	7^{55}	16^{30}	8^{43}	15^{23}
	21 March 23 Sept.	6^{46}	17^{12}	7^{25}	16^{32}	8^{08}	15^{10}	9^{10}	14^{50}
	21 Dec.	7^{35}	16^{27}	8^{13}	15^{50}	8^{50}	15^{10}	9^{40}	14^{18}
NL 50°	21 June	5^{00}	19^{00}	6^{00}	17^{52}	7^{21}	10^{40}	8^{45}	15^{22}
	21 March 23 Sept.	7^{05}	16^{55}	7^{55}	16^{00}	8^{50}	15^{05}	9^{51}	14^{10}
	21 Dec.	9^{05}	15^{00}	9^{53}	14^{15}	10^{30}	13^{40}	10^{55}	13^{05}
O–W-Tal									
NL 25°	21 June	5^{28}	18^{38}	5^{50}	18^{15}	6^{10}	17^{52}	6^{35}	17^{25}
	21 March 23 Sept.	6^{00}	18^{00}	6^{00}	18^{00}	6^{00}	18^{00}	6^{00}	18^{00}
	21 Dec.	7^{19}	16^{45}	7^{50}	16^{15}	8^{37}	15^{27}	–	–
NL 50°	21 June	4^{30}	19^{25}	5^{05}	18^{58}	5^{30}	18^{35}	5^{52}	18^{08}
	21 March 23 Sept.	6^{00}	18^{00}	6^{00}	18^{00}	6^{00}	18^{00}	6^{00}	18^{00}
	21 Dec.	9^{30}	14^{30}	–	–	–	–	–	–

between April and October under consideration of the average duration of sunshine. The values apply to a location of $\varphi = 50°$ latitude North. The calculation takes place for surfaces of all directions and slopes from 0° to 60° with increments of 15°.

1.5 The Influence of Superelevated Horizons

In determinations of radiation conditions in orographically pronouncedly organized terrane or in areas near structures, the superelevated horizon often plays an important role. The duration of sunshine and, therefore, direct irradiation is abbreviated by corresponding obstacles. Such an abbreviation of the ecliptic can be readily determined.

The energy reduction that is produced by the limitation of the horizon must be deducted from the strength of irradiation of the corresponding surface.

Table 8 compiles the sunrise and sunset times for a North–South and for an East–West valley with varyingly high horizon superelevation to show the effect of horizon limitation for general orientation. Horizon superelevation limits the intensity as well of the solar radiation as of the sky radiation, i.e., global radiation is being reduced.

A horizon superelevation not only has an effect on the energy intake during the daylight hours, but it also has an important role in radiation conditions during the night. In sloped surfaces that practically have their own horizon superelevation, as

Table 9. Effective radiation in dells with varying horizon super-elevation in per mil. (After Lauscher, 1928)

Horizon superelevation	0°	5°	10°	20°	30°	60°
	1000	996	984	915	793	282

Table 10. Spectral distribution of global radiation

	Wavelength range	Strength of irradiation in cal cm^{-2}	Percent of the total strength of irradiation	
Ultraviolet		0.1	6.3	
UV–B	0.28–0.315			0.35
UV–A	0.315–0.40			5.9
Visible ligth		0.8	52.1	
	0.40–0.48			11.8
	0.48–0.60			18.2
	0.60–0.78			22.1
Infrared		0.6	41,6	
	0.78–1.4			30.6
	1.4 –3.0			10.8

well as in dells and valleys, there occurs an increase of the reflective radiation and thus a reduction of the balance of long-wavelength radiation, or of the effective radiation. Süssenberger (1935) has shown the effects of sloped surfaces in relation to zenith distance. Lauscher (1928) calculated the effect of horizon superelevation on effective radiation Table 9 presents an excerpt from his calculations.

1.6 Spectral Distribution of Global Radiation

It is known that solar radiation is not homogeneous; it is composed, rather, of radiation of avrying wavelengths. We separate this radiation into the ultraviolet part (UV), the range of the visible radiation, i.e., the light and the infrared (see Table 10).

In ultraviolet radiation, one differentiates between two ranges, that are also different in their effects. UV – A produces direct pigmentation on human skin, UV – B causes erythema.

The light is of special concern to the agricultural meteorologist. It involves the largest portion of solar radiation and has strongly different biological effects in the various wavelength ranges. The following color ranges are differentiated in the composition of light:

Violet, 360–424 nm; blue, 424–492 nm; green, 492–535 nm; yellow, 535–586 nm; orange, 586–647 nm, and red, 647–760 nm.

Table 11. Global illumination of the horizontal surface in klx under cloud-free sky

Solar angle	5°	10°	15°	20°	30°	40°	50°	60°	70°	80°
		6.5	14.5	24.5	35.5	57.7	78.0	96.0	110.9	121.5	127.8

Table 12. Non-day intensity of global illumination in various geographic latitudes under cloudless sky in klx

Northern latitude	0°	10°	20°	30°	40°	50°	60°	70°
March–Sept.	128	126	117	105	92	87	59	36
June	117	126	130	128	122	113	101	84
Dec.	117	105	90	71	50	28	2	–

Spectral composition of light changes with the ratio between direct and diffuse radiation. Diffuse radiation, percentually, contains more short-wavelength radiation (i.e., green and blue) than does direct radiation.

The human eye senses the spectral distribution of sunlight as "white", unless individual spectral ranges are missing. If, for example, a loss occurs in the red spectral range, the object appears to be green, in the case of losses in the green, it will appear as red.

The effectiveness curve of the human eye, also called spectral sensitivity, peaks at about 555 nm. The product of this effectiveness curve and the spectral distribution of solar radiation lead to the strength of solar radiation, expressed in lux (lx) or in lumen/m^2 (lm/m^2). The unit lux always refers to the total spectral range of light sensitivity of the human eye, based upon its spectral sensitivity distribution. In referring to the measurement of individual spectral ranges, one may, therefore, not apply lux as a measurement unit.

The extraterrestrial strength of illumination amounts to an average of 140,000 lx. The measurements of Siedentopf and Reger (1944) in Table 11 can provide a certain overview about the strength of illumination on earth in its dependence on the solar angle, as can the data on noon-time intensities in various latitudes, given in Table 11.

Clouding naturally influences both the strength of illumination as well as global radiation very considerably. Under a cloud-covered sky, the illumination will average only about 27% of that under a clear sky. However, in the presence of scattered clouds, it is possible that reflection effects may produce an increase of the strength of illumination in comparison to a clear sky (Table 12).

Literature on p. 30.

2 The Biological Effects of Light

J. SEEMANN

Generally considered, great importance must be assigned to the optical radiation in the biosphere of earth. The specific biological effects can be grouped in the following manner:

1) Construction of organic substance
   a) photosynthesis
   b) formation of vitamin D
   c) formation of anthocyanines.
2) Transformation of matter
   a) pigmentation
   b) erythema formation
   c) bactericidal effects
3) Irritating effects
   a) photoperiodism
   b) phototropism
   c) phototaxis
   d) photonastic movement
   e) light- and darkness-germination
   f) photomorphosis
   g) stimulation of nerves and glands.

The effects of light in its importance to plant production are to be primarily discussed here. In this, of course, photosynthesis naturally stands in the foreground. Photoperiodism, which is of importance to plant propagation and culture of ornamental plants, will be discussed in connection with artificial illumination in chapter on greenhouse climates.

The importance of the energy output of solar light to plant production becomes obvious when one determines that energy for photosynthesis of the order of about $8.6 \cdot 10^{12}$ kcal per day are required for the nutrition of mankind on earth alone. This enormous amount of energy is, in a manner of speaking, biochemically stored solar radiation.

In photosynthesis, glucose is produced from carbon dioxide and water under bonding of energy from the light, whereby oxygen is being released:

$$6CO_2 + 6H_2O + 675\,kcal = C_6H_{12}O_6 + 6O_2.$$

This primary assimilation then produces sugar, starch, protein, and fats. CO_2-assimilation can be carried out only by plants that contain chloroplasts in their organs (generally the leaves). Absorption of energy from light takes place through chlorophyll.

Carbon dioxide assimilation rises at constant temperature and sufficient CO_2-supply from the air with increasing light intensity. This does not occur proportionally, however, but on equal percentual increase, more rapidly in the lower

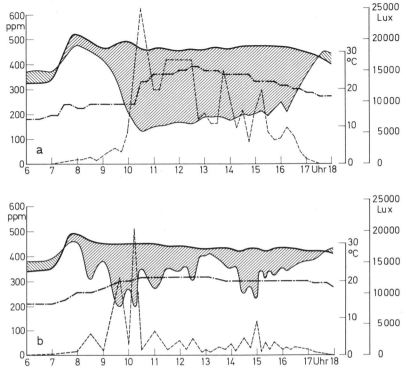

Fig. 1. *a* CO_2 assimilation of cyclamen at relatively high light intensity. —— CO_2 content of the air entering into the container where photosynthesis takes place, —— CO_2 content of the air leaving this container, ––– light, –·–·–· air temperature. *b* CO_2 assimilation of cyclamen at low light intensity (Seemann, 1965)

light ranges than at higher intensities. This becomes clear from Fig. 2 in chapter on greenhouse climates. Beyond a certain light intensity, no further increase of photosynthesis takes place. The excess radiation is partially converted into heat. A remnant radiates through the leaf. Only 0.6–7.7% of "white" light is converted into chemical energy.

The asymptotic decrease of the assimilation curve takes place more rapidly in typical shade-leaves than in sun-leaves. Shade-leaves often assimilate somewhat more strongly than sun-leaves in weak light. The latter are, however, definitely superior at higher light intensities.

Alongside of the CO_2-assimilation which is known to take place only in the presence of light, the opposite process of respiration continues simultaneously day and night. While matter is being formed from CO_2-assimilation, respiration consumes matter. If the light intensity drops below a certain level, respiration dominates during the day. The point where production and simultaneous consumption of assimilates are equal is called the compensation point.

Figure 1 shows examples for the daily course of CO_2-assimilation of a cyclamen plant (by Seemann, 1965). The upper part represents an example of relatively high light intensity, the lower at a lesser illumination intensity. The extent of the CO_2-consumption is represented by the hatched area. It is limited by the CO_2-content of the air flowing into and out of the assimilation vessel, from which a part of the carbon dioxide has been removed during positive photosynthesis. It can be clearly seen that during the dark phase in the morning and the evening, and at times of insufficiently high light intensity, the course of this curve is inverted. During these periods the plant releases carbon dioxide from its respiration into the atmosphere. Starting with a certain illumination intensity, one can determine that the compensation point has been exceeded. While the assimilation rate reacted very strongly to any change of the light intensity in the example of generally low light intensity, the other example shows that a "light saturation" had already occurred during various times. The assimilation output practically no longer increases, although the light intensity rises further, beyond a certain level.

Individual ranges of the total light spectrum more or less energetically promote CO_2-assimilation. The effect of light on carbon dioxide assimilation is therefore not only dependent on light intensity, but also on its spectral composition. The assimilation peak is located in the red between 660 and 680 nm. A secondary peak is in the blue, between 470 and 550 nm. The wavelengths between 470 and 550 nm are of only very limited effectiveness. This spectral distribution also agrees with the absorptive ability of chlorophyll. Red and blue are absorbed considerably more strongly than green-yellow. In addition the blue spectral range supports the synthesis of bio-growth-materials and inhibits the etiol element. The section on artificial illumination shows by examples in Fig. 2 the effects of light of various spectral composition on photosynthesis.

The course, as well as the extent, of CO_2-assimilation is controlled not only by light. Temperature, along with the availability of carbon dioxide, also has a considerable influence on the course of this photochemical process. Added to this is the fact that respiration that proceeds simultaneously with assimilation is solely temperature-controlled. The respiration of assimilants increases quite considerably with increasing temperature. Thus, photosynthesis, for example, can drop below the compensation point with only low light intensities and high temperatures, as the CO_2-assimilation is low because of the small energy supply from light, and respiration is too high as a consequence of high temperatures. How different the light yield can be at variously high temperatures, can readily be seen from Figure 2. For example, at 10 °C 15,000 lx produce an assimilation rate of about 14 mm³/h. With the same light intensity, but a temperature of 30 °C, the assimilation output rises to 35 mm³/h.

The influence of carbon dioxide will, generally, only become noticeable in assimilation, when its concentration is greatly reduced. This will frequently be the case when, under otherwise good conditions, the carbon dioxide in the proximity of the plant is consumed and the resupply from the more remote atmosphere does not take place rapidly enough because of low air circulation. Such a carbon dioxide impoverishment can take place in closed greenhouses as well as, under certain conditions, in plants of the open country. Such a lack of carbon dioxide is often eliminated in greenhouses by artificial increases of the concentration (Seemann, 1965). Such a procedure can only meet with limited success in open

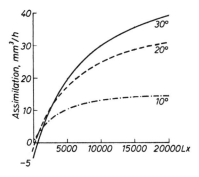

Fig. 2. Depandence of assimilation on the light availability at various temperatures

country, because the carbon dioxide molecules are rapidly exchanged in an increased partial pressure drop in the atmosphere.

Under certain conditions, an increase in carbon dioxide can produce an increase in assimilation beyond the normal measure. The reason for this is either that all other factors (light and temperature) are available to maximum assimilation in such an advantageous ratio that more carbon dioxide can be consumed than is generally available at normal exchange in the atmosphere, or that the diffusion resistance for the entrance of carbon dioxide with normal partial pressure is too great because of physiological conditions (closure or constriction of stomata or permeability limitations).

In all of these considerations, however, the fact cannot be altered that with advantageous temperature conditions, as well as a sufficient availability of carbon dioxide, light as energy source must be understood to be the primarily limiting factor in CO_2-assimilation. Light cannot be replaced in this by any other factor.

In order to provide an approximate concept of the photosynthetic efforts of some agricultural plants, Table 1 presents a compilation according to Lundegårdh (1957).

Table 1. Intensity of assimilation of some plants at 18–20 °C, with maximal illumination (1/1–1/4 light), CO_2 about 0.03 %

Assimilation in mg CO_2 per 50 cm² leaf surface		$C_6H_{12}O_6$ per 1 cm² leaf surface in g
Potato	9.57	1.30
Tomato	8.42	1.15
Sugar beet	9.26	1.26
Spinach	9.78	1.33
Vicia faba	8.83	1.20
Phaseolus vulgaris	9.27	1.26
Oats max.	20 mg per 1 g fresh weight	

	Maximum assimilation (at 1.22 % CO_2) in mg CO_2 per 50 cm² leaf surface and hour	
Potato	40.7	(at 30.2 °C)
Tomato	30.3	(at 35.5 °C)
Cucumber leaves	29.9	(at 36.1 °C)

These data can only be considered to be general indicative values. The estimate of the assimilation-output of entire plantings is naturally very difficult. The assimilation of entire plantings does not only depend on the types of plant, but also, perhaps in a stronger measure, on the developmental condition, the density, and the height of the planting. After all, all leaves participate to a certain ratio in the total assimilation output. The degree to which lower-position leaves, for example, assimilate, will depend on the degree to which light can penetrate into the planting with sufficient intensity.

Literature on p. 30.

3 Net Radiation

J. Seemann

It is always the absolute energy that is of importance for the radiation conversion of a surface. It alone can be converted into another form of energy, e.g., heat-energy. It is known that a part of the solar energy that is radiated to earth is reflected and is no longer available for energy conversions. Energy is also reflected from the surface of the ground or from plants, and subsequently reradiated (reflected radiation) by the atmosphere to the same surface to a certain measure.

We call the difference between global radiation (S = H) and reflection (R) (also called albedo), the short-wave and the difference of radiation between long wavelength radiation (E + r) and of reflection (A), the long wavelength radiation balance (net radiation). r is the reflection of atmospheric reflection at the ground. The absolute radiative energy that can be converted at the surface, e.g., into heat, is the total radiative balance or, more briefly, the net radiation (Q).

$$Q = (S + H) - R + A - E - r$$
$$= (S + H - R) - (E - A + r).$$

Net radiation is, therefore = short wavelength less long wavelength balance. Basically, the long wavelength radiation balance is effective throughout the day, while the short wavelength balance is limited to that period of time during which the sun is located above the astronomical (or also the orographical) horizon. The short wave balance is subject to far stronger fluctuations than is the long wavelength balance. Although the long-wavelength balance produces higher negative values during the day (the surface temperature is higher during the day!), it is considerably exceeded in moderate latitudes during Spring and Summer by the short wavelength balance. Under these conditions, the net balance is therefore positive in the course of the day during daylight, and negative at night. This is not to say, that negative radiation balances do not occur after sunrise or before sundown. Figure 1 shows a typical example for the course of the day in respect of the individual net radiation under a clear sky.

The value of the individual radiation components is, naturally, dependent upon various factors. The reflection value is primarily influenced by the surface conditions. In this respect, one differentiates between diffuse and mirrored reflection. The former occurs mostly on rough surfaces, the mirrored reflection on smooth surfaces – water, wax-coatings on leaves etc. The value of reflection also depends to a certain measure on the solar angle. In addition, reflection is a function of wavelength. Table 1 provides a survey on the dependence of reflection on the surface for over-all radiation (by Kirnov, 1953).

Long wavelength radiation of individual formations is a function of the surface temperature (T_s). Stefan-Bolzmann's law

$$E = \varepsilon \sigma T^4 \, \text{cal} \cdot \text{cm}^{-2} \cdot \text{min}^{-1}$$

applies to these.

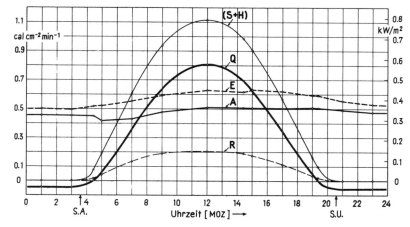

Fig. 1. Radiation components and net radiation on June 5, 1954 above meadow ground at Hamburg-Fuhlsbüttel. *A*, reflected radiation; *S+H*, global radiation; *E*, radiation; *R*, ground-reflected global radiation; *Q*, net radiation. (After Fleischer, 1953, 1954)

Dependent upon the surface, ε is subject to minor fluctuations (ε for lawns = 0.97, for sand = 0.98, and for water = 0.95). Since the sum E + r (long wave radiation plus reflection of the atmospheric reradiation) is regularly measured in practice, it corresponds to measurement accuracy when ε = 1. The intensity of long wavelength radiative intensity (back body radiation) of surfaces in the semi-infinite body can be determined from the daily courses of surface temperatures with the aid of Table 2.

Since the measurement of surface temperature is usually connected with difficulties, Schulze (1970) has used the temperature from within the weather observation hut (2 m) for the determination of E + r. He found the following relationship:

$$E + r = 0.99 \, \sigma \, T_{Air}^4 \, cal \cdot cm^{-2} \cdot min^{-1}.$$

This formula should be useful for order-of-magnitude estimates.

Table 1. Reflection on the surface in percent (from Krinov, 1953)

		Total	Light
Meadow	Dry	15–35	5–12
Grain	Green	15–25	7–15
Clover		17–32	
Fallow soil	Dry	8–15	
Sand	White, yellow	34–40	15–35
Loam	Dry	23	
Grey soil	Dry	25–30	11–20
Black	Dry	8–16	4–7
Needle-wood forest		10–14	3–7
Water surfaces		5–12	6–16

Table 2. Long wavelength radiation into the hemisphere

$E+r=\sigma T^4 (\sigma=0.826\cdot10^{-10})$ in mcal·cm^{-2}·min^{-1}						
Temp. in °C	− 30	− 25	− 20	− 15	− 10	− 5	0
E+r	288	312	339	366	395	426	459
Temp. in °C	5	10	15	20	25	30	35
	493	530	568	609	651	696	743

Atmospheric re-radiation is primarily dependent on the water vapor content of the atmosphere and cloud conditions. Its intensity is almost always below that of the long wavelength radiation of the soil. Ångström established the following experimentally supported formula for the calculation of atmospheric counter radiation:

$$A=\sigma T_s^4(0.82-0.25+10^{-0.12e}).$$

In respect of the temperature in the weather observation hut (T_{Air}) the reflection can be described in simple manner as follows:

$$A=0.82\sigma T_{Air}^4 \, cal\cdot cm^{-2}\cdot min^{-1} \quad \text{for clear sky (daily average)}$$
$$A=0.94\sigma T_{Air}^4 \, cal\cdot cm^{-2}\cdot min^{-1} \quad \text{for cloudy sky (daily average)}.$$

The average of the daily course of reflected radiation is not very great. Table 3 can approximately show the influence of clouding upon reflected radiation. It is based upon values obtained in Austria.

The night-time long wavelength net radiation can be approximately calculated with the aid of the above-mentioned equation. Where there are:

$$-Q=(E-r)-A$$
$$=0.162\sigma T_{Air}^4 \quad \text{for clear sky}$$
$$=0.05 \; \sigma T_{Air}^4 \quad \text{for overcast sky}.$$

One must clearly understand that, in the utilization of these formulas, one can only expect to obtain general values for a nightly radiation thus calculated. Bolz and Fritz (1950) have determined that the calculated re-radiation can deviate from the measured by ±5% and, in extreme cases by ±10%. In addition to this, air temperatures are being used here, while, in relation to radiative laws, on radiation, it is really a matter of the temperature of the surface.

Table 3. Daily totals of reflected radiation in 200 m above NN in the Austrian Alps in cal·cm^{-2}

Month	January	May	September
Clouding 0/10	450	627	631
5/10	505	678	682
10/10	600	771	776

The momentary net radiation values generally fluctuate very strongly, even in the most narrow space, caused primarily by the various albedo and the surface temperatures. Sauberer (1937) for example, determined the following differing net radiation values on a fine September day around 13.30h at Linz (Austria):

Vegetable and potato fields	$0.530 \, \text{cal} \cdot \text{cm}^{-2} \cdot \text{min}^{-1}$
Meadows	$0.445 \, \text{cal} \cdot \text{cm}^{-2} \cdot \text{min}^{-1}$
Paths (solid)	$0.395 \, \text{cal} \cdot \text{cm}^{-2} \cdot \text{min}^{-1}$

Literature on p. 30.

4 Radiation Measurement Technology

J. SEEMANN

Agricultural meteorology generally requires the following measurements of solar radiation:
Global and reflected radiation (albedo)
Net radiation,
Light
Infrared radiation.
Pyranometers, i.e., measuring instruments with an angle of 180° are used almost exclusively for measurements of global radiation. Their sensitivity reaches approximately 3μ. The most frequently used instruments are the solarimeter according to Moll-Gorczynski, the Eppley pyranometer, the stellar pyranometer of Linke and the actinograph of Robitzsch.

The Moll-Gorczynski is a black surface instrument with a built-in thermo-column that produces a thermal voltage proportional to the irradiation intensity. The thermal column is covered with two concentrically arranged glass culottes. In this manner the zero points are reduced only very slightly. The Eppley pyrano-meter is a black-and-white instrument, in which the sensor surfaces are arranged in a circular pattern. Linke's stellar pyranometer is also a black-and-white instru-ment. The sensor surfaces in it are arranged – black and white – in a stellar arrangement. Both Eppley- and stellar pyranometers also work with thermo-columns. The bi-metal actinograph of Robitzsch has proven itself in measure-ments in which no specially high requirements for accuracy are established. In contrast to the afore-mentioned pyranometers, this instrument is a mechanical one. Three bi-metal strips form a flat surface as radiation receivers. The outer two bimetals are white, the central one is black. These are counter-circuited to each other, so that temperature changes that are based on causes other than radiation are compensated. Only the effect of irradiation, that, by increased absorption heats, the black strip more strongly than the white, leads to a curvature of the surface. It is transferred to the recorder by means of lever systems. In this manner, any change of the radiation is registered in the form of a curve on a recording strip attached to a drum. A glass callot is also located above the measuring surface. With average maintenance, registration with this instrument permits good sum-mary of the radiation course in a location. The daily values of the actinograph are in good agreement with each other.

Along with zero point stability, the stability of the standardization constant and, especially, the best possible agreement with the "cosine law", i.e., that the radiation incident under various angles of incidence is as effective as possible in its correspondence to the cosine of the angle of incidence. These are the points that are determinant for the quality of a pyranometer. At the same time, pyranometer should have equal sensitivity for all wavelengths within the range of global radiation.

With the exception of the actinograph according to Robitzsch, all other pyranometers mentioned here, can, in addition to global radiation measurements, also be used for albedo measurements.

Net radiation meters could be built in useful form only after the introduction of polyethylene (Lupolen H) for covering caps (Schulze, 1953). The net radiation meter consists, in a manner of speaking, of two pyranometers, of which one is directed upwards, and the other downward. The upper measuring device measures solar and sky radiation (J + H), as well as the atmospheric reflection. The downward measuring device accepts the reflected global radiation (R), the reflected atmospheric reflection (r) and the reflection from the surface. The polyethylene caps that are mounted over the pyranometers, in contrast to instruments with glass callots, permit the measurement of the long wavelength range (0.3–60 μ) in global radiation measurements. The net radiation meter, therefore, is an instrument that can measure all wavelength ranges that result from solar- and heat-radiation on earth.

Net radiation meters are produced, today, in various forms. The best-known and technically most expensive is the net radiation meter according to Schulze. In almost all net radiation meters, the upper and the lower thermoelectric piles are arranged in a differentiation circuit, so that the value of the net radiation appears directly on the registering apparatus or a corresponding galvanometer. With Schulze's net radiation meter, it is also possible to measure short wavelength- and long wavelength-radiation separately.

For practical purposes, only photoelectric instruments are applicable to light measurements. In this regard, one differentiates between cells with outer or interior photoeffects. Photocells are among the former, the latter are represented by photoresistors and photoelements.

Vaccum cells with alkali cathodes are used as photocells (Görlich, 1951). The principle is as follows: electrons are emitted from metal surfaces upon exposure to light. If one connects the negative pole of a voltage source to the metal surface, and the positive pole to a second electrode (alkali), the electrons are removed towards the anode. A current flows that has been caused by the light exposure. It is important that proportionality exist between photo-current and radiation intensity for the measurement to be reproducible. This requirement is satisfied by vacuum cells in respect of monochromatic light, provided that the illumination intensity is not too high. Cadmium, potassium, rubidium and cesium are primarily used for alkali cathodes. Cadmium cells are used for UV-measurements. The spectral sensitivity of potassium reaches from about 0.3 to 0.67 μ with a peak at 0.4 μ. Cesium covers the wavelength ranges between 0.3 and 0.73 μ with a minimum at 0.4 and a peak at 0.56 μ. The sensitivity of rubidium ranges between 0.4 and 0.6 μ with a peak at 0.48 μ. Temporary changes can occur in photocells by aging and fatigue. While aging represents a permanent change that is irreversible, fatigue represents a temporary change that is caused by strong or extended loads and that is reduced again when the cell is not being used.

While photoresistors are hardly useful for agriculture-meteorological applications because of very limited applicability in the outdoors, photoelements find extensive application. The photoelement (Lange, 1939, 1940), consists of a metallic base plate with an applied layer of selenium or silicon and a thin light-permeable front-electrode. The base plate is the positive cell-pole, while the front electrode represents the negative, from which the current is taken by means of a sprayed-on metallic ring. When light is incident on the cell, it produces electrons in the selenium or in the silicon, which then flow off by way of the front electrode. The photocurrent rises proportionally to the strength of the light. When using silicon, one will obtain about ten times the current than is obtained from selenium. The spectral sensitivity of the selenium cell ranges from 350 to 750 nm with a peak at 550 nm. In this, it coincides essentially with the sensitivity of the human eye. In the silicon cell, the sensitivity extends farther into the long wavelength range (1170 nm) and the peak is located between 800 and 900 nm. Silicon cells have a relatively high dark resistance (about 500 kOhms). For this reason they are not as well suited to agricultural meteorological light measurements as are the selenium cells.

The measuring values of instruments with selenium cells are generally standardized in lux for light measurements and are frequently offered by the manufacturers as "luxmeters". The electrical stability of selenium photocells, after a certain time, shows a reduction of sensitivity. Investigations on this have been carried out by Wörner (1955), among others. More frequent standardization is therefore recommended. A certain control in this is offered by simultaneous measurement with a pyranometer in solar radiation. An exact conversion from lux to absolute units (cal) is possible with the knowledge of the color temperature of the radiator. In the case of a color temperature of about 5800 K (sun), 1 lux corresponds to $1 \text{ lx} = 1.146 \cdot 10^{-5} \text{ cal} \cdot \text{cm}^{-2} \cdot \text{min}^{-1}$.

As has already been mentioned, the spectral sensitivity of the selenium photocell coincides with that of the human eye, and is therefore well suited as an instrument for the measurement of the strength of illumination. Some difficulties arise for agricultural meteorological purposes in respect to the spectral sensitivity. Comparisons of the light intensity of artificial light sources with varying spectral composition are not readily possible. In outdoor measurements, also, the measurement values can in practice only be considered as relative values, when one considers that the photosynthetically active radiation within the spectrum has a totally different intensity distribution than that measured by the selenium cell. For this reason, special measurement methods have been developed for photosynthetic purposes. Schulze (1949) compiles a filter combination (Schott-filters) that permits measurements with selenium cells in the range of chlorophyll absorption. Lambda Instruments Corp. (Lincoln, Neb., USA) has developed a measuring instrument that much more suitably measures the wavelength range between 380 and 740 nm, than does the selenium cell (Fig. 1). This sensor is especially suited for light-dependent investigations in connection with photosynthesis. With this instrument it is possible to determine photosynthetically active radiation directly. Measurements are in microeinsteins (1 microeinstein $= 6.023 \cdot 10^{17}$ photons).

The basic concept of this measurement system is, briefly, as follows: one must first consider that only absorbed radiation currents can become energetically effective. Reflection and transmission based on differing leaf positions and solar angles make reference more difficult. Secondly, photosynthesis represents a physicochemical process, i.e., the transition of the physical form of energy into the biochemical. Thirdly, one can consider electromagnetic radiation as being corpuscular, i.e., as consisting of discrete quanta. The mental prerequisite for the solution

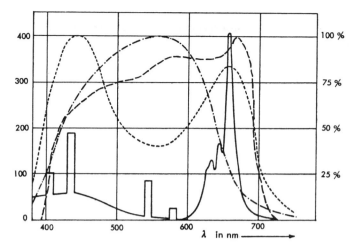

Fig. 1. Relative spectral radiation density distribution of the sensitivity of a selenium photocell (–·–·–) in comparison to the absorption of a green leaf (–––), the relative spectral radiation density of a fluorescent lamp (Osram L-Fluora) (———) and the sensitivity of a lambda sensor instrument (– – –)

of the problem of the conversion of energy for this case would be provided with the aid of Einstein's Law on photochemical equivalence. According to this, there exists a simple integral relationship between the number of the molecules that are photochemically changed (chemical energy acceptance) and the number of the absorbed quanta in the range between 0.4 and 0.7 µ. The latter are presently called photons. In this, it is decisive that the energy content of a photon, as concerns photosynthesis, is unimportant, i.e., that each photon contributes equally to the production of matter in the mentioned spectral range. It is known that the relative energy content of a photon in the range of 0.7 µ is only about 4/7 of the energy amount of a photon in the 0.4 µ range. However, since only constant amounts are required for electrochemical molecular changes, photons of shorter wavelength ranges contribute more strongly to heat dissipation than do photons of higher wavelength ranges, because energy amounts that are not required for photo-synthesis are converted into heat. For the photosynthetic matter correlation, it is only necessary to count the number of photons in the visible wavelength range. The photon stream in the range between 0.4 and 0.7 µ is counted as the photosynthetically active solar radiation that is being used exclusively for photosynthesis. Energy that is not utilized photosynthetically is dissipated as heat. This definition is, among others, in agreement with McCree (1972, 1973) and Björkmann (1968), who also found a quasiconstant yield in the spectral range between 0.4 and 0.7 µ in using photon numbers. Gabrielson (1960) had made a suggestion along these lines. An international consensus on this definition is still lacking.

When discussing radiation-measuring devices for agricultural meteorological purposes, one may not overlook the fact that infrared radiation measurements have been gaining in importance for some time. In agricultural meteorology, these are being used almost exclusively for contact-free temperature determination of

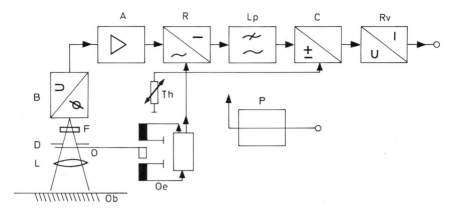

Fig. 2. Block schematic of an infrared radiation instrument. *Ob*, Object; *L*, Lins; *O*, Oscillator; *Oe*, Oscillator electronic; *D*, Diaphragm; *F*, Filter; *B*, Bolometer; *A*, Amplifier; *R*, Rectifier; *Lp*, Low pass-filter; *C*, Compensation-stage; *Rv*, Reverse of the voltage; *T*, Thermometer

surfaces. A block schematic (Fig. 2) can aid in the visualization of the function and construction of such an infrared radiation measuring device. The radiation emitted from a surface (measuring object) goes through an infrared transparent lens and an edge filter to the radiation receiver (bolometer). The object radiation is periodically interrupted with a frequency of 35 cycles by means of a vibrating modulator that is fastened in front of the diaphragm opening of the bolometer. On each interruption, the bolometer receives the radiation of the modulator, which is a function of the modulator temperature. The output voltage is an alternating voltage, whose value depends only on the difference between the two radiation fluxes as a function of the object- and the modulator-temperature. After amplification and phase-dependent rectification and smoothing, a compensation step adds a voltage that is modulator temperature-dependent, so that the output voltage becomes independent of the modulator temperature, i.e., is dependent only on the object temperature. In order that the additional reflected solar radiation from the object can be essentially eliminated, the sensitivity is set for the wavelength range above 8 mμ. As a result, such an instrument is especially suited to contact-free temperature measurements in the outdoors. Temperature resolution is at approximately ±0.25 °C in the range between −20 and +60 °C.

Literature (Chaps. 1–4)

Ashbel, D.: Global solar radiation world maps during international quiet sun years 1964–1967. Jerusalem: Hebrew University 1970

Bjorkmann, D.: Further studies on differentiation of photosynthetic properties in sun and shade ecotypes of solidago virgurea. Physiol. Plant. *21* (1968)

Bolz, H.M., Fritz, H.: Tables and diagrams for the calculation of reflected radiation and re-radiation. Z. Meteorol. *4* (1950)

Collmann, W.: Measurement on the relation between sky radiation and turbidity. Meteorol. Rundsch. *5* (1964)

Decoster, M.W., Schuepp, W., van der Elst, N.: The radiation on vertical planes at Leopoldville. Mem. Acad. Roy. Sci. Col. Sci. Tech. Nouv. Ser. II, fasc. 1 (1955)

Dirmhirn, J.: The radiation field in the ecosphere. Frankfurt: Akad. Verlagsgesellschaft 1964

Gabrielsen, E.K.: Light wavelengths and photosynthesis. In: Ruhland, Handbook of plant physiology. Ruhland (ed.). Berlin, Heidelberg, New York: Springer 1960

Görlich, P.: The photocells. Leipzig 1951

Gräfe, K.: Radiation reception of plane vertical surfaces, Global radiation of Hamburg. Ber. d. DWD (DWD = Deutscher Wetterdienst – German Meteorological Service) 29 (1956)

Kaempfert, W.: Solar radiation on plane, wall, and slope. Reich Office for Weather Service: Wiss. Abh. B, IX (1942)

Kirnov, E.L.: Spectral reflection properties of natural formations (Russ., 1974). Transl.: Tech. Trans. Natl. Counc. Can., Ottawa (1953)

Lange, B.: The photocells and their application, Vols. I, II. Leipzig: 1939–40

Lauscher, F.: Reports on measurements of nighttime radiation on the Stolzalpe. Meteorol. Z. 45 (1928)

Linke, F., Baur, F.: Meteorological pocketbook. Leipzig 1953

Lundegårdh, H.: Climate and soil. Jena 1957

McCree, K.J.: The action spectrum absorptance and quantum yield of photosynthesis in crop plants. Agric. Meteorol. 9 (1972)

McCree, K.J.: The measurement of photosynthetically active radiation. Sol. Energy 15 (1973)

Nicolet, M., Dogniaux, R.: Study of the solar global radiation. Mem. Inst. R. Meteorol. Belg. XLV, II (1951)

Sauberer, F.: Measurements of the radiation household of horizontal planes in pleasant weather. Meteorol. Z. 54 (1937)

Sauberer, F., Dirmhirn, J.: The radiation climate. In: Climatology of Austria. Steinhauser, Eckel, and Lauscher (ed.). Vienna 1958

Schulze, R.: Effect-related radiation in meteorology. Ber. d. DWD, Special Ed. of Meteorol. Rundsch. (1949)

Schulze, R.: On a radiation measuring device with UV transmitting wind protection cap at the meteorological Observatory Hamburg. Geofis. Pura Appl. 24 (1953)

Schulze, R.: Radiation climate of the earth. Darmstadt: D. Steinkopf 1970

Seemann, J.: A contribution of the carbon dioxide problem in greenhouses. Gartenbauwissenschaft 4 (1965)

Siedentopf, H., Reger, E.: The illumination by the sun. Meteorol. Z. 61 (1944)

Süssenberger, E.: New investigations on the nighttime effective radiation. Gerlands Beitr. Geophys. 45 (1935)

Wörner, H.: The stability of selenium photocells. Z. Meteorol. 9 (1955)

5 Heat Balance

J. SEEMANN

The processes involved in heat conversion on the soil are most easily made understandable by means of a model. If one assumes a vegetation-free partial section of the earth's surface that can be generally considered to be flat, then the positive net radiation that is converted into heat on the surface is led away in varying manner. A portion of the heat is led off into the ground and leads to warming the ground. A further portion serves the heating of the air. This portion is called the sensible heat stream. A third portion of the heat conversion is, in the presence of water at the surface, used for vaporization. This heat is latently preserved in the vapor. Since the energy conversion from the net radiation takes place on a surface, heat cannot be stored, i.e., the total of the energy is divided between the three heat streams. Expressed in a simple equation, this means that net radiation (Q) less heat flux into the soil (B), less heat transfer to the air (L), less the heat that is consumed in vaporization, (V), will have to be equal to zero

$$Q - B - L - V = 0.$$

This energy equation is called the heat household equation.

If the assumed earth surface is covered with vegetation, then an additional heat flux into the "plant bodies" is added. An additional part of the energy is also consumed in photosynthesis. This portion is, however, in comparison with the other energy fluxes, so small that it can be neglected.

In the case of a negative net radiation, the heat fluxes are reversed. The soil and the air release heat to the surface. In addition, the latent heat present in the water vapor is also released to the surface by way of condensation (formation of dew). The heat released to the surface is then again converted into radiative energy. The heat household equation at night, therefore is

$$-Q + B + L + V = 0.$$

In the case of negative net radiation, the heat fluxes are, in other words, directed towards the surface, while in the case of positive net radiation, they point in the opposite direction. That is the rule. However, occasionally, the soil heat flux can have a positive sign during the day and lead heat to the surface.

The model concept on energy conversion, as explained by the heat household equation, is of especially great practical importance in agricultural meteorology. It permits the experienced agricultural meteorologist to make estimates of the orders of magnitude of temperature conditions and water conversion in the most varying situations. This model is, however, of special importance in the determination of the actual evapotranspiration, which will be discussed elsewhere.

Here it is to be shown on hand of some examples in what relationships the individual energy streams can stand to each other when heat conversion takes place over various soils with partially different vegetation. Three different topological

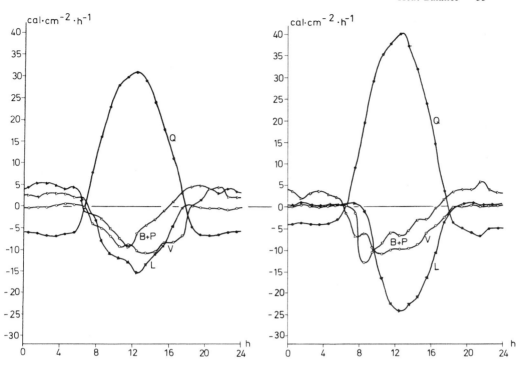

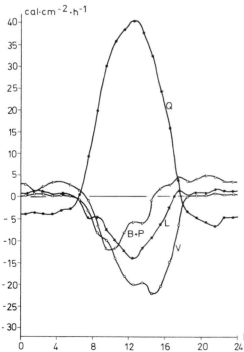

Fig. 1. Components of energy balance (average of 4 cloud-free September days) on three different surfaces (*from left to right:* fine-grain loamy deposits, high moor and low moor). (From Miess, 1968)

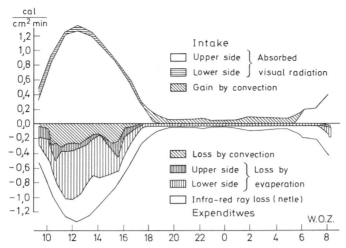

Fig. 2. Daily course of heat householf of a leaf of *Alocasia indica*. (After research by Raschke)

areas are to be treated: sandy soils of fine grain and loamy deposits, high moor, and low moor. The investigations were carried out by Miess (1968) during a 4-day fine-weather period in September 1964 in Northern Germany. Figure 1 presents the average course of energy conversions during 4 days. The regular course of the net radiation that is symmetrical to the sun is a characteristic of a cloud-free sky. This is also reflected in a correspondingly symmetrical course of the current of sensible heat (L). Not corresponding, however, is the course of vaporization (V). In the low moor, the largest portion of the energy from the net radiation is utilized for vaporization, while, in the high moor, the largest part of the energy is applied to the heating of the air. This can be explained by the differing water content of the uppermost soil layers. Wherever there is sufficient water available, the major portion of the energy will be applied to vaporization. In the case of wet surfaces, this portion can be designated as being 70–80 % of the energy from the net radiation.

Heat household investigations are, today, available from the most varied climatic regions. These deal, primarily, with publications of measurement results that are very hard to compare with one another. Geiger (1960) has published a compilation of such results in his book *The Climate of the Ground-Proximate Air Layer*.

Special importance must be assigned to the heat conversion directly on the plants, especially the leaves. By means of this heat conversion, the water household and, therefore, indirectly also photosynthesis are controlled to a large measure. Raschke (1956) has investigated the heat household on leaves. The daily course of the heat household of a horizontally standing leaf of *Alocasia indica* is given as an example for this in Fig. 2. The measurements were carried out on a cloud-free day at Poona (India). In the curves half-hour average values are connected to each other so as to exclude short-period fluctuations. The differing energy portions for vaporization on the underside, vs. the upper side of the leaves should be especially noted.

Literature on p. 44.

6 Heat Flux in the Soil

J. SEEMANN

Energy conversions on the soil surface and the temperature conditions that result from them in the ground-proximate air only become entirely clear when one knows more concerning the heat movement in soil and in the air. In order to understand heat movement in the soil, one first needs to know something concerning the physical properties of soils. Every natural soil consists of three components that differ in character, the actual substance of the soil of mineral and organic nature, the mobile, not chemically bonded water, and the pore-filling air. The density of natural soil ϱ_m and its specific heat (C_s) can be calculated from these three individual parts. If one considers that the density of the water $=1$, that of the air can be neglected because of its small mass, then:

$$\varrho_m = 0.01(v_s \varrho_s + v_w) \quad (g \cdot cm^{-3}),$$

where v_s is the volume portion of the actual substance and v_w, that of the water in percent with ϱ_s as the density of the substance.

The volume heat can be calculated in similar manner (heat capacity per unit volume) $(\varrho_0)m$. It indicates the amount of heat (in cal) that must be supplied to $1 \, cm^3$ of soil in order to raise its temperature by $1 \, °C$. With the application of the above simplification, and considering the fact that the specific heat of water is $c_w = 1 \, cal \cdot g^{-1} \cdot deg^{-1}$, there results:

$$(\varrho_c)m = 0.01(v_s \varrho_s c_s + v_w) \quad (cal \cdot cm^{-3} \cdot deg^{-1}).$$

For water-free soils, the volume heat is with great approximation a soil constant, but the value increases with increasing water content.

In order to determine heat movement in the soil, one initially assumes that the soil is homogeneous. In that case, all points of the same depth x (cm) will have the same temperature t (°C). The heat flux B that flows toward the surface with increasing temperature with depth is proportional to the temperature gradient $\frac{dt}{dx} (deg \cdot cm^{-1})$

$$B = \lambda \frac{dt}{dx} \quad (cal \cdot cm^{-2} \cdot s^{-1}).$$

Thermal conductivity is the proportionality factor λ, its dimension is $cal \cdot cm^{-1} \cdot s^{-1} \cdot deg^{-1}$. It is, therefore, that amount of heat that flows through a cube of $1 \, cm$ edge length of the corresponding substance in $1 \, s$, when the temperature difference of the opposing surfaces is $1°$, and no other temperature gradient contributes.

Table 1 presents the most important constants for the heat household of some soils.

The effect of the water household on heat conductivity λ, on the density and the volume heat, can be seen from Table 2.

Table 1. Some constants for the heat household of naturally deposited soil

Soil Type	Density ϱ_s $\mathrm{g\,cm^{-3}}$	Volume heat $(\varrho c)\,m$ $\mathrm{cal\cdot cm^{-3}\cdot deg^{-1}}$	Heat conductivity $1000\cdot\lambda$ $\mathrm{cal\cdot cm^{-1}\cdot s^{-1}\cdot deg^{-1}}$	Temperature conductivity $1000\cdot a$ $\mathrm{cm^2\cdot s^{-1}}$
Sand, wet	–	0.2 –0.6	2 – 6	4 – 10
dry	1.4–1.7	0.1 –0.4	0.4 – 0.7	2 – 5
Clay, wet	1.7–2.2	0.3 –0.4	2 – 5	6 – 14
dry	–	0.1 –0.4	0.2 – 1.5	0.5– 2.0
Moor soil wet	0.8–1.0	0.6 –0.8	0.7 – 1.0	0.9– 1.5
dry	0.3–0.6	0.1 –0.2	0.1 – 0.3	1 – 3
Rock	2.5–2.9	0.43–0.58	4 –10	6 – 23
For comparison				
Still air	0.001–0.0014	0.00024–0.00034	0.05–0.06	150 –250
Still water	1.0	1.0	1.3 – 1.5	1.3– 1.5

Table 2. Heat conductivity, density, and volume heat of the soil, dependent on water household

Water content v_w	0	10	20	30	40	Vol.-%
λ	0.0006	0.0024	0.0036	0.0040	0.0043	$\mathrm{cal\cdot cm^{-1}\,s^{-1}\,deg^{-1}}$
ϱm	1.50	1.64	1.70	1.80	1.90	$\mathrm{g\cdot cm^{-3}}$
$(\varrho c)\,m$	0.30	0.40	0.50	0.60	0.70	$\mathrm{cal\cdot cm^{-3}\,deg^{-1}}$

The heating of soil is not only dependent on the amount of transported heat, but also on the ability of the soil to absorb heat, i.e., the volume heat $(\varrho\cdot c)m$. The temperature change of a soil particle is caused by the change of the heat flux B with the soil depth x (cm). If, for instance, B is designated as being positive, the heat flux flows towards the surface, and if B increases with growing depth downward, the heat flux becomes smaller with the approach to the surface. Therefore, heat remains in the soil, increasing the temperature of the soil particles (heat storage in the soil).

If $\dfrac{dB}{dx} = (\varrho\cdot c)\,m\cdot\dfrac{dt}{dz}(\mathrm{cal\cdot cm^{-3}\cdot s^{-1}})$ and place for $B\lambda\dfrac{dt}{dx}$ there results

$$\frac{dt}{dz} = \frac{\lambda d^2 t}{(\varrho\cdot c)m\,dx^2} \quad (\mathrm{deg\cdot s^{-1}}).$$

By means of this equation, the spatial and temporal temperature change is described in a homogeneous soil. The factor $\dfrac{\lambda}{(\varrho c)m}$ is called the thermal diffusivity k.

The importance of thermal diffusivity in the determination of heat flux in the soil will be discussed elsewhere.

Heat flux within the soil has a great influence on the temperature conditions in the ground-proximate air layer. With stable energy requirements for vaporization, the air will warm up less strongly over soils with good heat conductivity than over

Table 3. Average temperatures (a) and annual amplitudes (b) of three different soils (From W. Krenz, 1943)

Depth in cm	Clay		Sand		Humus	
	a	b	a	b	a	b
5	9.0	19.8	9.3	21.0	10.1	22.0
10	9.5	19.8	9.3	20.6	10.4	22.0
20	9.4	19.0	9.6	20.0	10.4	21.0
50	10.3	16.4	10.7	16.8	10.7	18.4
100	10.6	14.4	11.1	14.2	11.3	14.2

soils with poor heat conductivity. Since heat conductivity has a strongly varying importance to heat flux, the warming of the soils is also dependent upon it. It appears useful, therefore, to show the influence of the soil type on the warming of the soil, briefly.

In agriculture the production capability of a soil is determined by its grain-size composition, its lime content, and the humus portion. One speaks in terms of sandy soils when coarser grains predominate and a maximum of 25 % of washable particles (diameter less than 0.01 mm) are present; of loamy soils when between 25 and 65 % and of clay soils, when more than 65 % of washable particles are present.

Kreutz (1943) has measured continuously throughout a three-year period, the soil temperatures in loam, sand, and humus (Table 3). In the table the average temperatures are compiled, as well as the annual amplitudes (difference between maximum and minimum). One recognizes that the average annual temperature increases with depth. The soils that were under equal weather conditions show different temperatures, in the individual layers as well as in the average. The average air temperature (2 m height) was 8.2° during the measuring period. It was, therefore, lower than the soil temperature in any location.

Literature on p. 44.

7 Heat Transport in the Air

J. Seemann

Heat transport in the air takes place in two ways, by molecular conduction, and by mass exchange (generally also only called "exchange"). The heat stream (L) in the case of molecular conduction is proportional to the heat gradient and is therefore

$$L = \lambda \frac{dt}{dx} \quad (cal \cdot cm^{-2} \cdot s^{-1}).$$

The proportionality factor λ ($cal \cdot cm^{-1} \cdot s^{-1} \cdot deg^{-1}$) is, as was already mentioned in connection with heat transport in the soil, the thermal conductivity. Heat transport in air by molecular conduction takes place only to a very limited extent. It is primary in the junction area of surfaces. If the total heat transport in the air were to take place only by thermal conductivity, then the daily temperature fluctuation would be determinable in the atmosphere up to about 3 m in height. However it is known that a marked difference exists in the air temperature between day and night up to at least 1000 m aboveground. The physical heat conduction, therefore, has only an insignificant part in the actual heat movement in the air. Decisive for the heat transport in air is mass exchange. In mass exchange, "air packets" of differing size are exchanged for each other, where the contents (heat, water vapor, gas, dust particles) are carried along. Schmidt (1925) already published basic knowledge on this exchange.

The exchange process takes place in an entirely irregular manner. The mixing processes that take place in the atmosphere are consequences of disordered movements (turbulence) that take place in the atmosphere. According to Sutton (1953) "the turbulent flow is a movement that contains random fluctuations of finite size, and which causes irregularities in particle flow, whose order of magnitude is comparable to that which determines the kinematics of the average movement, such as e.g., the shape of the boundary layer". In exchange processes one must consider, foor heat transport, that the temperature is height-dependent because of the reduction of air pressure. If, in other words, an air quantum is transported from elevation x_1 to elevation x_2, the temperature t_1 changes into temperature t_2.

The speed with which the transport of an air quantum takes place is expressed by the exchange coefficient A. For the heat transport L of 1 g air ($cal \cdot g^{-1}$), $c_p \cdot T$, where c_p is the specific heat at constant pressure ($0.241 \cdot cal \cdot g^{-1} \cdot deg^{-1}$) and T is the potential temperature (K). One understands that temperature, under the term potential temperature, that the air would have if it were brought to the pressure level of 1000 mbars. The potential temperature, therefore, satisfies the condition of being independent of external conditions. Within the ground-proximate stratum – and only there – the difference between potential and actual temperature is so small, that one can consider T to be the same as the direct temperature t, without producing any mentionable error. The transport of sensible heat in the air by means of exchange is therefore described by the following equation:

$$L = A \cdot c_p \frac{dt}{dx} \quad (cal \cdot cm^{-2} \cdot s^{-1}).$$

This equation completely corresponds to the heat conduction equation. The exchange coefficient A here replaces the heat conductivity λ. Heat movements in the air, therefore, obey the same regularity as in the soil, but, because of the differences of λ and A c_p, in different orders of magnitude. Therefore, however, A is exposed to the greatest fluctuations, both spatially and temporally. In the boundary layer, where molecular heat conduction is normally effective, A has the order of magnitude of $10^{-4} g \cdot cm^{-1} \cdot s^{-1}$. This value rises very rapidly with the distance from the ground. At altitudes between 1 and 10 m, it fluctuates between about 0.1 and $10 g \cdot cm^{-1}$. Other orders of magnitude are possible in the total atmosphere.

Two causes must be differentiated in respect to mass exchange, friction exchange, and convection exchange. Friction exchange (forced convection) is variable by the change of wind velocity with elevation (normally speed increase) and with the relative roughness of the ground surface. Therefore, the exchange coefficient increases with increasing wind velocity, as well as with elevation above ground.

Forced convection exists especially at night. During the day, when the soil is heated by positive net radiation, convection exchange is added to forced convection. Convection exchange occurs especially in the presence of weak stratification of the air (temperature reduction per 100 m elevation of more than 1 °C). Such over-adiabatic gradients are the rule in ground proximity at noon. Superheated air particles attempt to release themselves from the soil and to rise, cooler air particles drop into their place. In the same way as with forced convection, this vertical movement – designated as free convection – also produces irregular movements. However, the elements of turbulence of convection exchange are not as random as they are in forced convection. They have the tendency to resolve into rising and sinking flow threads.

The science of exchange has, today, become a speciality. Because of the limited extent to which this problem can be treated here, some important publications are pointed out, such as the basic thoughts of Schmidt (1925), the discussions of Lettau (1943, 1949), Sutton (1953), and Swinbank (1958) as well as the experimental investigations of Priestley (1956) and of Frankenberger (1958).

All theoretical considerations are based upon the concept that the exchange movement satisfies the random laws of probability, i.e., in a similar manner to that in which a molecular movement of greater dimension can be determined mathematically. However, even the calculation of A at the same time and in the same place (depending on which element is used for it), shows that, in addition to random, also nonrandom processes can play a role.

Literature on p. 44.

8 Measuring Technology

J. SEEMANN

In order to determine heat conversion by means of measuring techniques, a relatively large instrumental effort is required. The instruments for measuring individual energy flows are best mounted on a special instrument carrier. The individual measurement instruments are then placed onto so-called outriggers that are movably attached to a main mast in such a manner that they always stand free above the ground. There is no need to point out that such measurement installations are, today, equipped only with electrical measuring instruments that guarantee high measuring accuracy. The following instruments are parts of a complete set of equipment: energy uptake is measured with a net radiation meter. It should be mounted on the instrument carrier at a height of about 2 m on an outrigger, pointing to the South in horizontal position in such a manner that the typical underground is observed without shadow cast by the mast or instruments arranged above. Next, thermometers are mounted in two, often three measurement heights for the measurement of the sensible heat (heat transport in the air). In the same locations are instruments for the measurement of water vapor for the determination of water vapor transport. It is often usual to use, for the measurement of temperature and water vapor, aspiration psychrometers with electrically driven aspirators. Measuring heights over vegetation-free ground or over low vegetation, will be about 20 cm for the lowest instruments and the next lowest about 100 cm above ground level. A third measuring set can then be mounted at 180 or 200 cm elevation. In the case of measurements above higher vegetation, a vegetation surface must be determined, above which approximately the same heights are selected as over plain ground. In growing vegetation, the measurement heights must be adjusted to the longitudinal growth of the vegetation. Wind velocities are measured at the same elevations as the instruments for temperature and humidity measurements, by means of anemometers. In the selection of these instruments it is important that these produce measurements even with very low wind velocities. For heat household measurements over vegetation, a third anemometer should be placed into the vegetation at approximately 50 cm in height.

Certain difficulties arise in the measurement of heat flux in the soil. In principle, here, too, a temperature measurement profile (electrical thermometers at depths of about 5, 10, and 20 cm) would be sufficient, if, in each case, the thermal conductivity (λ) of the soil were known. This simple method is, however, not sufficient since λ also changes with changing soil moisture conditions. For this reason it is necessary to determine on a continuous basis, the heat conductivity along with the temperature differences at various soil depths. Various mathematical processes have been developed for this purpose that work essentially with the "temperature wave" in the soil. Berénji (1967) has reported on this.

The thermal conductivity can be calculated from four temperature values in the following manner according to a formula developed by Komolgorov (1967). In order to be able to carry out the calculation, one must measure the temperature at two depths Z_1 and Z_2, during the course of the day at four points-in-time that have the same distance from each other

$$k = 0.000274 \left(\frac{Z_2 - Z_1}{\ln s^2} \right)^2 (cm^2 \cdot s^{-1}),$$

$$s^2 = \frac{(U_1 - U_3)^2 + (U_2 - U_4)^2}{(U'_1 - U'_3)^2 + (U'_2 - U'_4)^2}.$$

U_1, U_2, U_3, and U_4 are the soil temperatures at depth Z_1 for established points in time, while U'_1, U'_2, U'_3, and U'_4 are the corresponding values at depth Z_2.

Schwerdtfeger (1970) has developed heat flux plates for the continuous measurement of heat flux in the soil, these are suitable for practical measurements. They are relatively simple and inexpensive instruments that produce a thermocurrent as output. Manufacturer in TND–TH at Delft, Holland. The plates are placed directly under the soil surface and several are placed in series. Since the soil heat flux is given as $B = \lambda \dfrac{dt}{dz}$, the thermal voltage is a direct function of the temperature difference between the upper and the lower side of the plates, and, hence, proportional to heat flux of the soil, with the assumption of constant heat conductivity. This condition is, however, not completely satisfied in practice. Especially in sandy soils the heat conductivity can fluctuate, corresponding to the moisture content, between about 0.001 and 0.004 $cal \cdot cm^{-2} \cdot s^{-1} \cdot deg^{-1}$. For this reason, the use of measuring instruments is recommended for direct measurement of heat conductivity. So-called heat needles are available for this purpose (producer: Stichting, Technical Service, Wageningen, Holland). These are 200 mm long and 1 mm thick steel needles that have, in their interior, a thermoelement as well as a heating coil. The needles are heated for only a few minutes with a constant electrical output of 0.1 W/cm. According to Büttner (1955) and Koitzsch (1960), the temperature T of the constant-heated sensor rises linearly after a start-up time that is dependent on the instrument itself and upon its soil contact, with the logarithm of time t so that

$$\lambda = \frac{\ln t_2 - \ln t_1}{T_2 - T_1}.$$

λ can, therefore, be derived directly from the slope of the curve.

If one now applies the individual parameters that have been obtained in the indicated measuring methods to the heat household equation, the various heat fluxes can be determined. Net radiation Q and the heat flux in the soil B can be measured directly. The flow of sensible heat l in the air and the proportion of latent heat for evaporation V must be calculated with the aid of appropriate measurement values. Initially, there is, according to the heat household equation

$$(Q + B) = -(L + V)$$

with the utilization of the transport laws, there is

$$L = A_L cp \frac{dT}{dz} (cal \cdot cm^{-2} \cdot min^{-1})$$

and in similar manner

$$V = A_V r \frac{dq}{dz} (cal \cdot cm^{-2} \cdot min^{-1}),$$

where T is the potential temperature, q, the specific moisture, cp the specific heat of air at constant pressure, r the heat of evaporation of water and z the height.

Starting with the transport equation for sensible heat and water vapor, Sverdrup (1936) under the condition of the assumption made by Schmidt (1925) that the exchange coefficient for water vapor and sensible heat is the same, i.e., $A_L = A_V$, has produced the separation of L and V with the aid of the quotient

$$\frac{L}{V} = c_p \frac{\dfrac{dT}{dz}}{r \dfrac{dq}{dz}}.$$

A simplification of the equation is obtained by introducing the actual temperature in place of the potential temperature T (actual temperature [t °C]) and, instead of the specific moisture, q, one introduced the vapor pressure [mmHg] into the equation. This is permissible in the ground-proximate range. In this, the connection of q with e must be considered by the following equation:

$$q = \frac{0.622 \cdot e}{P - 0.378 \cdot e} = c \cdot e$$

with $c = 0.836 \cdot 10^{-3}$ Torr^{-1}, $r = 592$ [cal·g^{-1}] and cp = 0.241 [cal·g^{-1}·°C^{-1}].
   The quotient then is:

$$\frac{L}{V} = 0.49 \cdot \frac{\dfrac{dt}{dz}}{\dfrac{de}{dx}},$$

where t must be introduced in [°C] and e in [mmHg]. The transition from the differential quotient to the difference quotient is permissible when the quotient A_L/A_V is independent of the height aboveground. If temperature and water vapor gradients are measured at the same height, the air mass is not entered into the equation, and one obtains for the quotient

$$\frac{L}{V} = 0.49 \cdot \frac{dt}{de}.$$

By the re-formulation of the heat household equation, one obtains

$$L = - \frac{0.49 \cdot \dfrac{dt}{de}(Q + B)}{1 + 0.49 \dfrac{dt}{de}},$$

$$V = - \frac{Q + B}{1 + 0.49 \cdot \dfrac{dt}{de}}.$$

All members of the heat household can therefore be determined by measuring the net radiation, the measurement of the heat flux in the soil, and the measurement of temperature and water vapor at two different elevations.

Certain difficulties will arise with the measuring and calculating methods given here, with higher wind velocities, or with weaker temperature gradients, because, in those cases, the accuracy of measurements is then often not sufficient. In these cases, it is recommended to derive the values for L and V from the turbulent mass transport. For this, the determination of the exchange coefficient A is needed with the aid of wind velocity measurements. It is for this reason that the initially described arrangement is equipped with anamometers.

The exchange coefficient A can be determined, again under the assumption that $A_L = A_V$, under the presumption of a logarithmic wind profile.

$$A = \frac{\varrho k_0^2 (\bar{u}_2 - \bar{u}_1)}{\ln Z_2/Z_1}.$$

ϱ is the air density, k_0 Karman's constant (0.492), Z is the elevation and $\bar{u}_1$ and $\bar{u}_2$ are the average horizontal wind velocity at elevations Z_1 and Z_2. This formula, that is valid for the adiabatic condition, was provided by Lettau (1949) with the correction factor

$$A_{Le} = A \cdot \frac{1}{(1 + Ri_L)^2}$$

corresponding to thermal stratification.

In this, Ri is Richardson's number,

$$Ri = \frac{\frac{g}{T} \cdot \frac{dQ}{dz}}{\left(\frac{du}{dz}\right)^2},$$

where g is the earth acceleration, $\frac{dQ}{dz}$ is the gradient of the potential temperature, which can be replaced without mentionable error by the temperature gradient $d/T/d/Z$.

If one introduces the exchange coefficients that were determined by the foreging method into the mass transport equation, one obtains (upon simultaneous integration between the limits of Z_1 and Z_2)

$$L = \frac{c_p \cdot \varrho k_0^2 (T_1 - T_2)(\bar{u}_2 - \bar{u}_1)}{(\ln Z_2/Z_1)^2}$$

and, with the use of the vapor pressure e, one obtains for V:

$$V = \frac{0.623 \cdot \varrho k_0^2 (e_1 - e_2)(\bar{u}_2 - \bar{u}_1)}{p \cdot (\ln Z_2/Z_1)^2}$$

(V in g/cm^2, if e and p in torr and u in cm/s).

If this process is used over taller vegetation, the geometric heights (Z_1 and Z_2 must be converted into effective heights by inclusion of the zero point displacement and the height of the roughness Z_0. In this case, d in no way corresponds to the height of the vegetation. The zero point (height d) is dependent upon the condition of the vegetation (leaf direction, leaf mass, flexibility in wind, etc.). The logarithmic wind law applies only above height d. Below this is the actual vegetation atmosphere. Thornthwaite and Holzman (1942) provide an approximation procedure for the calculation of d, in which the sought-for value is changed until both sides are equal in the following formula

$$\frac{\bar{u}_3 - \bar{u}_2}{\bar{u}_2 - \bar{u}_1} = \frac{\ln(Z_3 - d) - \ln(Z_1 - d)}{\ln(Z_2 - d) - \ln(Z_1 - d)}.$$

The roughness height Z_0 is given by the same authors as being:

$$Z_0 = \left[\frac{(Z_1 - d)^{\bar{u}_2}}{(Z_2 - d)^{\bar{u}_1}} \right]^{\frac{1}{\bar{u}_2 - \bar{u}_1}}.$$

That such tedious calculations are being carried out by means of computers these days need not be pointed out. The determination of the exceedingly great mass of measurement data will also be carried out by means of electronic instruments in order to even be able to carry out an extended measurement program, and the data will then be transferred directly to computercompatible data carriers. A discriminator will aid in the determination in the computer program, of whether the energy conversions are to be determined according to the heat household equation or according to the "transport formula".

Literature (Chaps. 5–8)

Berényi, D.: Microclimatology. Stuttgart: G. Fischer 1967

Büttner, K.: Evaluation of soil heat conductivity with cylindrical test bodies. Trans. Am. Geophys. Union 36 (1955)

Frankenberger, E.: The exchange coefficient over ground. Beitr. Phys. (1958)

Geiger, R.: The climate near the ground. Cambridge: Blue Hill Met. Observ. Harvard University 1960

Koitzsch, R.: Experiments on the determination of moisture content on thermal basis. Abh. Met-Hydrol. Dienst, DDR 8 (1960)

Kolmogorov, A. N.: In: Microclimatology. Berényi (Ed.) Stuttgart: G. Fischer 1967

Kreuz, W.: The annual course of the temperature in various soils under equal weather conditions. Z. Angew. Meteorol. 60 (1943)

Lettau, H.: On the nightly course of the vertical exchange coefficient in connection with meteorological occurrences. Gerlands Beitr. Geophys. 59 (1943)

Lettau, H.: Isotropic and non-isotropic turbulence in the atmospheric surface layer. Geophys. Res. Pap. No. 1, Cambridge, Mass. (1949)

Miess, M.: Comparative presentation of meteorological measurement results and heat household investigations on three different locations in Northern Germany. Ber. Inst. Meteorol. Klimat. TU Hannover, No. 2 (1968)

Priestley, C. H. B.: Free and forced convection in the atmosphere near the ground. Q. J. 81 (1955); 82 (1956)

Raschke, K.: Micrometeorologically measured energy conversions of an alocasia leaf. Wrach Met. (B) *7* (1956)

Schmidt, W.: The mass exchange in free air and related occurrances. Hamburg: H. Grand 1925

Schwerdtfeger, P.: The measurement of heat flow in the ground and the theory of heat flux meters. Tech. Rep. Cold Regions Engr. Lab., Hannover, N.H. (1970)

Sutton, D. G.: Atmospheric turbulence. London: Methuen and Co. 1953

Sverdrup, H. U.: The maritime vaporization problem. Ann. Hydrograph. Marit. Meteorol. *32* (1936)

Swinbank, W. C.: Turbulent transfer in the lower atmosphere. Proc. Canberra Symp. 1956. UNESCO, Paris 1958

Thornthwaite, C. W., Holzmann, B.: Measurement of evaporation from land and water surfaces. U.S. Dept. Agric. Tech. Bull. No. 817. Washington 1942

9 Transfer of Quantities of Air Masses
(Momentum Transfer)

B. PRIMAULT

9.1 Basic Physical Principles

9.1.1 The Air Which Surrounds Us

The planet Earth, third in order of distance from the sun, is surrounded by a gaseous envelope called the "atmosphere". This film of air is made up of a mixture of various gases, most of which consists of nitrogen (chemical symbol N), accounting for about 4/5 while oxygen (O) comprises the final fifth. Certain rare gases are also to be found in the atmosphere, such as argon (Ar), xenon (X), neon (Ne), krypton (Kr), hydrogen (H), helium (He), and so on.

These main gases are in almost constant proportion with one another up to about 90 km above the earth. Beyond this altitude, the mixture undergoes a slow modification.

In addition to the main constituent gases of the atmosphere, there are other gases found in proportions which vary widely depending on the altitude, time of year and/or meteorological conditions.

Primary among them is *ozone* (O_3) which, at a certain altitude, becomes extremely important because of its repercussions on physicochemical processes. Even at the altitude where it is densest, however, it is found in a proportion of only 7 ppm (parts per million). Near the earth's surface, the proportion of ozone is minimal, fortunately, since this is a highly toxic gas which originates under the effect of the sun's ultraviolet rays on the upper layers of the atmosphere.

Carbon Dioxide (CO_2). This gas originates as the result of the combustion of carbon under the influence of fire or of the vital metabolic processes of plants or animals. It is found mainly in the lower layers of our atmosphere, due to the fact that it originates in the immediate vicinity of the earth's surface (either just above it or just below it). It cannot be a matter of natural sedimentation caused by its specific weight, which is, nonetheless, greater than that of air, since other rare gases are much heavier (e.g. xeonon, krypton). The proportion of carbon dioxide varies tremendously from place to place and from season to season in accordance with vital activities (vegetation period, flock movements) or fires which may occur. Furthermore, this proportion increases slowly but constantly as a result of numerous combustion phenomena, whether from industry, the heating of households or commercial buildings, or traffic, especially automobile traffic. Thermal plants which produce electricity also make a perceptible contribution to this increasing proportion of carbon dioxide.

Water Vapor (H_2O). This gas is present everywhere in the atmosphere and is extremely important, but is found in proportions varying greatly in time and space. Most of this water vapor originates from evaporation, not only over the oceans, but over all water surfaces. This evaporation depends on the difference in temperature between the water and the air, but most of all on the absorption capability of the air mass reposing on the water. Apart from the water vapor created by evaporation from the oceans, lakes, and waterways, or from the sublimation of ice and firn, respiration and evapotranspiration contribute a large quantity of water vapor to the atmosphere. In fact, all beings living in an emerged state (plants or animals) need oxygen to survive. For the osmotic gaseous exchanges to occur in their respiratory organs, the air must be saturated with water when it contacts the walls of these organs, otherwise their cells would dry out and the consistency of their walls would be unsuited for these osmotic exchanges. Thus, the living organism must add to the water vapor breathed in from the atmosphere until saturation is reached. The air expired thereafter contains the maximum amount of water for its pressure and temperature. The simply respiratory process is thus responsible for the release of tons of water, in the form of vapor, into the atmosphere every minute. As we have indicated, those processes valid for animal respiration are likewise valid for plant respiration. By covering vast expanses of our planet, plants therefore contribute substantial quantities of water vapor to the atmosphere.

In the same way that the rare gases complement the basic atmospheric mixture, other occasional gases have relative degrees of importance in the mixture surrounding a certain area. The origins of these gaseous pollutants are varied, and may be either natural or industrial. Included among them are methane (NH_4), sulphur dioxide (SO_2), hydrogen sulphide (H_2S) and carbon monoxide (CO).

Based on the proportion of each of these gases in the atmospheric mixture of an area and on the thickness of the layer of air over this area, each of these components exerts its own pressure on it. The total pressure, therefore, can be considered as the sum of the partial pressures of the oxygen and nitrogen and of each of the other gases present. In practice, however, a distinction is often made between the partial pressure exerted by the water vapor and the total pressure, so that it can be considered for its own sake since it gives a precise indication of humidity conditions at a given moment.

9.1.2 The Standard Atmosphere

As we have seen above, the pressure which the air exerts on any solid or liquid body depends first on its makeup (mixture of gases, especially humidity, temperature, etc.) and then on the height of the column of air surrounding the object in question. This pressure, then, decreases as the altitude increases according to a determinant function:

$$p = 1013.25 \left(\frac{288.16 - 0.0065 \cdot h}{288.16} \right)_{5.2544807} \tag{1}$$

in which p is the total pressure expressed in mbar and h is the altitude in meters.

Since Newton's law states that phenomena of universal attraction decrease with the square of the distance, for the sake of precision an additional function should be introduced in the above formula. This function would take gravitational variations into account and would thus account for the decrease in this attraction with altitude. The same would apply to the latitude, which also acts on the

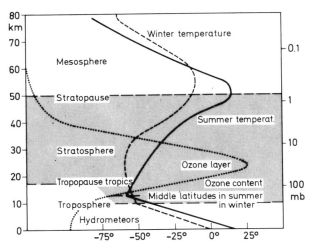

Fig. 1. Composition of the atmosphere

acceleration of gravity. The results obtained with this formula, therefore, are merely an initial approximation, but one which is excellent because it differs from the standard atmosphere by only 1 mbar at an altitude of 12,000 m.

In addition, the average temperature of the layer of air decreases as it moves away from the equator (see Sect. 9.2.1), with a resulting increase in its density. This means that the pressure exerted by a column of air of the same height will be higher at the pole than at the equator. Because of centrifugal phenomena caused by the rotation of the earth, however, the atmosphere is thinner at the pole than along the equator, resulting in an equalization of the pressure on the ground. As the latitude gradually increases, however, these phenomena diminish the accuracy of the above formula.

As the altitude increases, not only does the pressure decrease, but the temperature decreases as well. Yet, while the first of these phenomena is infinite, the second does not occur until about 210 K. This initial stratus, called the troposphere, is the point of origin of hydrometeors (clouds, precipitation, etc.), with a narrow transitional zone over it known as the tropopause.

Above, the pressure continues to drop as the altitude increases, but the temperature remains constant. This is a second layer termed the stratosphere, which is separated from the mesosphere by the stratopause. The low temperatures account for the very small amount of water vapor in the air circulating in the stratosphere, and thus the absence of clouds (see Fig. 1).

Working from the principle that pressure decreases with altitude, instruments have been developed for measuring the altitude based on pressure, or for measuring pressure based on the altitude. These instruments are called altimeters. For them to be calibrated accurately, however, (i.e., so that readings will be comparable on comparable instruments), a predetermined atmosphere had to be selected, known as the "standard atmosphere". All of the altimeters used in aviation are calibrated in accordance with this standard atmosphere, which is calculated using the laws governing perfect gases (cf. Sect. 9.1.4 below and Fig. 2).

The ICAO (International Civil Aviation Organization) defines the standard atmosphere as follows:

initial pressure at sea level: 1013.25 mbar
temperature at sea level: 15 °C = 288.15 K;
the atmosphere consists of dry air, so there are no condensation phenomena;
uniform thermal gradients of −6.5°/km between 0 and 11,000 m
0° /km between 11,000 and 20,000 m
+1.0°/km between 20,000 and 32,000 m.

9.1.3 Compression and Decompression

If a particle of air is displaced vertically from one altitude to another by any degree of mechanical force whatsoever, its pressure will vary according to the difference in height of the column of air above it. If the column of air is higher after the displacement than before, the pressure will be higher, and vice versa, if the column is lower, the pressure will decrease.

For example, if a quantity of air is displaced from sea level to an altitude of 1000 m, the column of air surmounting it will have decreased in height by 1 km. Consequently, the pressure which this same particle of air exerts in its new position will be lower than the pressure it exerted previously. On the other hand, the volume which it occupies will have increased proportionally, since the number of molecules composing the particle will have remained the same. This change in pressure with respect to volume is accompanied by thermal phenomena: the mass of air cools.

Similarly, if this same particle of air is displaced from an altitude of 1000 m to sea level, the pressure which it exerts will be higher because of the greater height of the column of air above it. The molecules composing it will occupy a smaller volume, and this decrease in volume will be accompanied by a release of heat energy expressed by an increase in the particle's temperature.

9.1.4 Dry Adiabatic Transformation

The processes which we have just described obey strict physical laws. The term "adiabatic phenomenon" is used to indicate the change in the state of a gas when the gas neither gives up heat to the ambient atmosphere nor takes heat from it. Its pressure, temperature, and density, therefore, are related by means of a simple formula:

$$\varrho = \frac{p}{R \cdot T} \qquad (2)$$

in which ϱ is the density expressed in $kg\,m^{-3}$, p is the pressure in Pa, T is the temperature in K and R is the specific constant of the dry air. The value of this constant is $2.8704 \cdot J\,kg^{-1}\,K^{-1}$. This relationship is based on the following hypothesis: despite their heterogeneous composition (a mixture of gases rather than a pure gas), both dry air and moist air behave as perfect gases.

Furthermore, based on the fact that in adiabatic processes there is no heat exchange at all between the particle of air and the ambient atmosphere, formulas can be established for calculating the temperature, pressure, or volume of the air mass every time, while the two other quantities remain constant.

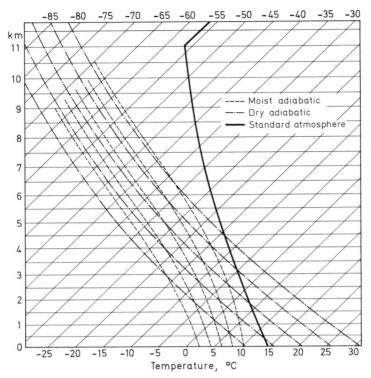

Fig. 2. Thermodynamic diagram

Based on heating phenomena caused by compression or cooling phenomena caused by expansion, temperature variation curves can be established as a function of pressure (with respect to altitude, since these two quantities go hand in hand); these are referred to as dry adiabatic curves (see Fig. 2).

The temperature of any moving particle at a given altitude can easily be found. If its temperature at a reference altitude is known, then by merely following the corresponding curve either upwards or downwards, its temperature at any other altitude can be found.

9.1.5 The Mixing Ratio

We saw above that water vapor is one of the important constituents of the atmosphere covering our planet. This water vapor is not simply one of the constituent elements of the gaseous layer, but is found in various forms in a good many of the meteorological phenomena which are normally measured. Not only does the humidity of the air depend on it (specific, absolute, and relative humidity), but clouds and precipitation as well. It can readily be seen why various methods have been developed to determine its rate.

A given volume of air, i.e., whose temperature is known, can only contain a certain quantity of water vapor. This is known as saturation. The ratio between the mass represented by water vapor alone and the mass of dry air is termed the mixing ratio. Calculating the ratio between the mass of water vapor and the mass of humid air gives the specific humidity. Thus, this relative number cannot exceed a certain maximum, representing saturation. As Chapter 11 discusses the special features of the humidity of air, we will not dwell on this subject any further at this point.

9.1.6 Changes in the State of Water

As we noted above, those quantities of water vapor which could be kept in suspension in the atmosphere were solely dependent upon the temperature of the atmosphere.

The amount of water vapor in every air mass does not always reach this maximum. It is much more often less than the maximum, i.e., saturation is not reached. The ratio between the quantity of vapor actually in suspension in the mass and the quantity corresponding to the saturation of this mass is termed the "relative humidity".

We have also seen that temperature decreases with pressure, in other words, a mass of air borne aloft expands, thereby decreasing in temperature. Consequently, any elevation of an air mass will, ipso facto, bring with it an increase in its relative humidity until saturation is reached. If the ascending movement continues and the air mass cannot hold any more water vapor, the state of this water changes, condensing to liquid form; clouds are produced (and possibly fog).

However, this change in the state of the water from the gaseous phase to the liquid phase is accompanied by thermal phenomena. Heat is released, and this latent heat passes to the mass of air and warms it. There follows a reverse phenomenon whereby the mass can – at least theoretically – hold more water and part of the water in liquid form returns to the gaseous phase, restoring the equilibrium. The drop in temperature caused by the elevation, however, will compensate for the rise in temperature inherent in the condensation process, so that in nature, the phenomenon occurs only in one direction.

In a descending movement, analogous but opposite phenomena will be observed: the air is warmed by compression and can hold more water vapor. The drops in suspension (clouds or fog) consequently evaporate, and the heat required for this change in state is taken from the mass of air.

9.1.7 Moist Adiabatic Processes

The evaporation or condensation heat mentioned above is an integral part of the primary air mass. Thus, when evaporation or condensation occurs, there is no input or transfer of heat with the ambient air. Adiabatic phenomena can thus be spoken of in accordance with the definition given in Sect. 9.1.4. Since, however, one of the constituents of the air mass undergoes a transformation (water changing from a gas to a liquid or vice versa), what we really have is a direct application of the definition of an adiabatic phenomenon.

However, since these transformations are accompanied by thermal phenomena, the curves resulting from these phenomena when a saturated air mass is displaced vertically are quite different from those which we saw in Sect. 9.1.4. The condensation or evaporation heat released, as well as required, by the process in which the state of the water is changed, is such that the decrease in heat with altitude is clearly lesser in the moist adiabatic process than in the dry adiabatic process (see Fig. 2).

If, then, an air mass rises, it will first follow the dry adiabatic process until saturation occurs. If the ascending movement continues, the laws of the moist adiabatic process will thereupon govern the mass. The transfer point from one to the other is thus of prime importance in meteorology, since it determines the altitude at which clouds are formed. Further on, we shall observe impact of this point on general meteorology as well as on wind formation.

9.2 The Reasons Behind Large-Scale Transfers of Air Masses

9.2.1 Pressure Belts

At the equinox, when the sun beams its rays toward the earth, the earth's temperature is not uniformly affected. The equator and a relatively narrow zone on either side of it receive the rays of the sun perpendicularly, while the poles receive them tangentially. This gives maximum heating to the equator and virtually none at all to the poles.

As a result of this intense heating of the land on the equator, the air above it is also heated by contact. It then rises and undergoes the effects of convection (see Sect. 9.1.4). This results in the formation, all along the equator, of storm systems which, when juxtaposed, create an almost continuous low-pressure belt on land and cause an intake of air on land which is compensated for by constant winds from subtropical regions to the equator: the trade winds.

At high altitude, however, the ascending movement described above cannot continue indefinitely. It is stopped by the tropopause and then spreads out toward the poles. While a certain quantity of air actually does flow as far as the upper latitudes, most of it comes back down toward the earth in the subtropical regions. This generates belts of relative high pressure to the north and south of the tropical low-pressure belt and extending on both sides of the Tropic of Cancer and the Tropic of Capricorn, where air flows either toward the equator or the poles. At high altitude, the air mass undergoes a general collapse, causing the disintegration of the clouds as well as the rarefaction or even disappearance of the rains. The presence of these high-pressure belts is responsible for the formation of the major deserts such as the Sahara, the Arabian, the Indian and the Mexican deserts in the northern hemisphere, the Kalahari desert, the Australian interior and the Argentinian steppes in the southern hemisphere.

Beyond the subtropical high-pressure belts there is a zone of relatively low pressure, but this belt lacks the consistency of the previous two. Anticylonic circulation and moving depressions alternate with each other in this zone, yet in general, pressure is lower than in the subtropical belts. This is called the temperate zone.

Toward the poles, with their long cold periods, once again zones of higher pressure are encountered, driving the air along the earth's surface toward the temperate regions. Over the earth, they are supplied by the remaining air that had been driven high above by the equatorial heating.

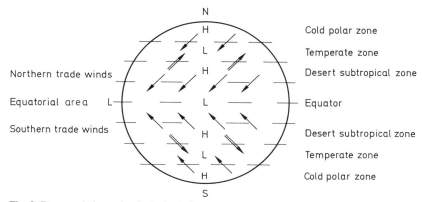

Fig. 3. Pressure belts and principal winds

The presence of these alternating high- and low-pressure belts causes gradient currents in the low layers of the atmosphere whose intensity, evenness, and direction also depend on local and momentary conditions. The belts, however, appear in striking fashion when long barometric pressure means are established. Average-gradient winds can then be determined, revealing the exchanges which exist between tropical air masses and polar air masses, each of which have their own special features (Fig. 3).

9.2.2 Exchanges of Air Masses on Land

"Nature abhors a vacuum" was once a common belief. For air masses, however, there is no need of a total vacuum to provoke a transfer from one place to another. All that is required is a difference in pressure, even slight. As soon as a pressure gradient is formed, the air tends to restore the equilibrium through a horizontal movement of translation. The lack of equilibrium generally originates from a difference in pressure, but may also result from a change in the density of two neighboring air mass caused by differences in temperature and/or humidity.

As we observed above, the oval shape of the earth prevents the rays of the sun from reaching the earth's surface at the same angle everywhere, resulting in various degrees of heating. Furthermore, the earth's surface is neither inclined nor oriented everywhere alike, so that orographic differences also cause heating which varies often over small surface areas. The opposition presented by large water surfaces (seas, lakes, etc), and emerged lands or surfaces covered with vegetation in proximity to lands which are bare or almost bare, causes appreciable variations in the local humidity of the air from one point to the next. This situation alone provides us with a permanent source of gaseous (i.e., air) exchanges between places which are close to each other (orographic variations) or very far away (oceans/land masses, poles/equator, etc.).

Thus, there will be two main sources of energy permanently at work on the surface of our planet. The first is twofold: variations in radiance and variations in evaporation and condensation. The second counteracts the pressure imbalance caused by the first. What is involved are movements of air masses called winds.

9.2.3 Seasonal Fluctuations

Our demonstration in Sect. 9.2.1 refers only to a temporary state of the sun's radiance, the equinoxes occurring on March 21 and September 23. Because of the angle made by the earth's axis of rotation with the plane of the ecliptic (23°27′), this situation is gradually modified from the spring equinox to the summer solstice. It then returns to normal from the summer solstice to the autumn equinox, then once again is modified, this time in reverse, from the autumn equinox to the winter solstice, finally returning to normal between the winter solstice and the spring equinox.

If we follow this line of thinking, we see that the parallel along which the rays of the sun shine at noon, local time, perpendicular to the earth's surface, moves northward between March 21 and June 22. This movement naturally causes a shift in the equatorial low-pressure belt and consequently, a similar shift in the other pressure belts described. Thus, little by little the high-pressure belt of the north pole loses in breadth to the high-pressure belt of the south pole. The width of the other belts does not change to any appreciable degree; they simply move along the earth's surface.

From the summer solstice to the winter solstice, a reverse movement occurs. Directly over the Tropic of Cancer, the sun gradually passes perpendicular over the equator (September 23, or autumn equinox), then over the Tropic of Capricorn on December 21. After this date, it rises once more toward the equator, reaching a position directly over it on March 21 (spring equinox), thus completing a yearly cycle.

This pendulum-type movement is what causes the formation of the seasons.

In the tropics, this movement is responsible for alternating rainy and dry seasons, varying in length according to the amount of time the sun is perpendicular to the area. Near the equator there are even zones which have two annual rainy seasons corresponding to each passage of the sun.

In most regions of the equatorial zone, however, this pendular motion is not sufficient to cause a dry season per se. The almost daily rain storms merely become less intense at the approach of the solstices, then reach their maximum at the equinox. In the polar regions, however, this pendular motion, accompanied by a lengthening of the days, then of the nights until they last for 24 h, has considerable effects on temperature, on cloud development and especially on the humidity content of the air (absolute humidity).

To conclude, we should add that the respective maximum and minimum temperatures in the temperate zones do not correspond to the longest and shortest days. There is a shift caused by the fact that the earth serves as a type of thermal steering wheel. It becomes warm slowly and its temperature reaches its maximum only about one month after that angle of incidence of the sun's rays has reached its highest point. The minimum temperature occurs one month and a half to two months after the sun has begun to rise on the horizon.

9.2.4 Coriolis Force

The exchanges of air masses (winds) which tend to equalize differences in pressure should occur on the surface of the earth along the line of the highest gradient, in the same way that water flows down a slope. This would be a direct, rectilinear motion along the large circle joining the center of a high-pressure zone to the center of a low-pressure zone. In actual fact, the wind never follows the line of the highest gradient. On the contrary, the flow of air is always modified because of the fact that our planet rotates.

Let us imagine that in the middle latitudes, an air mass under relatively high pressure is situated closer to the equator than another, neighboring air mass under lower pressure. In theory, the air would flow along the large circle linking the two centers. If this large circle is a meridian, the current should than head from the equator toward the pole, along a south-north path in the northern hemisphere and a north-south path in the southern hemisphere. Since, however, the molecules of air composing the moving mass are animated by a west-to-east horizontal movement of translation because of the rotation of the earth, this kinetic energy is maintained in the equator-to-pole movement. The further away from the equator, the more the length of the parallels decreases, i.e., a given distance (for example in km) will represent a larger angle in the center than on the preceding parallel. The lateral (west-east) displacement imparted to an object by the rotational movement of the earth, while remaining the same when expressed in distance per unit of time (km/h), gradually increases, moving away from the equator, in angular velocity longitudinally, this time expressed as an angle per unit of time (° longitude per h).

A similar, but opposite, approach can be used for movements of air displaced from one of the poles toward the equator. The lateral movement of the air mass entrained at its place of origin by the earth's rotation decreases in angular velocity as it gradually nears the equator.

Consequently, a gradient wind displaced parallel to a meridian will be deflected to the right in the northern hemisphere and to the left in the southern hemisphere.

The vector indicating the Coriolis force is always parallel to the equator or, in other words, perpendicular to the axis of the poles. Its magnitude corresponds to the following formula:

$$F_C = 2 \times m \times V_r \times U_F$$

in which F_C, the Coriolis force; m, the mass in motion; V_r, the relative velocity of this mass; U_F, the angular velocity of translation.

Coriolis force applies not only to moving air masses, but to any body moving along the earth's surface (e.g., a rock falling into a well or a drop of water falling from a cloud to the earth). In fact, the demonstration given for an air mass moving on the earth's surface is also valid vertically: the top of a mountain travels a greater distance per unit of time (day, hour, etc.) than a point located on the seashore at the same latitude.

The paths of all bodies, whatever their direction, are therefore affected by a vector answering to this formula.

9.2.5 The Effects of Air Friction

A moving mass exerts friction on the surface over which it moves, i.e., part of its kinetic energy is transformed into heat. Consequently, if no other force is at work, this kinetic energy gradually decreases and the translational velocity of the mass is reduced proportionally.

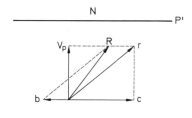

Fig. 4. Parallelogram of forces

Fluids are no exception. They are subject to the same laws as the air which moves upon contract with the ground, losing part of its kinetic energy, which is transformed into heat, and decreasing in velocity. This, then, results in an appreciable difference between wind velocity on the ground and wind velocity in the air, both of which would be occasioned by the same pressure gradient (see Sect. 9.4 and Chap. 10.4).

This new force acts in reverse of the Coriolis force. Coriolis force results in a rotation of the current to the right in the northern hemisphere and to the left in the southern hemisphere, but the force caused by friction reduces this deflection (see Fig. 4). The size of the vector caused by the friction and, moreover, the slightest rotational movement of the wind are, however, proportional to the friction ratio between the moving body (the air) and its support (the earth's surface). This ratio varies greatly depending on the environment over which the wind moves: over the sea, it is much less than on land or on forest-covered surfaces in particular. Nowhere is it zero, however, even for two superimposed layers of air moving in the same direction but at different velocities.

9.2.6 Centrifugal Force

In addition to the two elements described above which deflect the gradient wind from its theoretical path (over a body which does not revolve around itself), a third force must be considered in numerical wind calculations. This force is exerted parallel to the gradient wind and occurs because the differences in pressure causing this gradient are not always rectilinear in their arrangement.

The lines of equal pressure (isobars) that can be drawn around activity centers – whether high or low pressure – describe curves which can be regularly assimilated into a succession of sections of a circle. The gradient wind is deflected by the Coriolis force so that it becomes parallel to the isobars. Since the isobars are rarely rectilinear, an additional force complements the two others and either reduces or accelerates the movement. Centrifugal force is the term for this third force.

In a low-pressure zone (see Fig. 5a), the air mass is driven toward the center by the pressure gradient and displaced to the right by the Coriolis force, diminished by the effects of the friction[1]. The wind resulting from this displacement of air is

---

1 In this illustration, we will limit ourselves to the northern hemisphere, although our example is also valid for the southern hemisphere, provided left and right are reversed

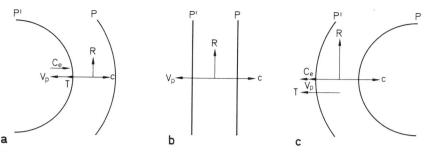

Fig. 5. Centrifugal force

therefore activated by a movement of rotation if these isobars are in the form of concentric circles. However, this situation produces a lateral force which tends to impart a tangential path to the current, i.e., drive it away from the low-pressure center. The action of this force is therefore parallel, but opposite to that caused by the pressure gradient; it consequently decreases the propagation speed of the current. It follows that at equal latitudes and gradients, those winds undergoing cyclonic inflection are slower than the gradient force would have it, the gradient force being deflected by Coriolis force and both being reduced by friction.

If the isobars are rectilinar and parallel, the current flows perpendicular to the gradient with no change in its force (see Fig. 5b).

We also have a curving of the isobars for high pressure, but this time, the centrifugal force acts in the same direction as the gradient, thereby complementing it. This consequently increases the velocity of the wind (see Fig. 5c).

9.3 The Formation of Fronts

In the equatorial zone and the subtropical zones which hem it in on both sides, air movements are caused primarily by pressure gradients resulting from differences in insolation and the consequential uneven warming of the land. Gradually moving further away from the equator toward the poles, the situation becomes more complicated. The greatest disturbances, however, are to be found in the temperate zone.

In this zone, there are two principal movements: warm air flows from the subtropical zones toward the pole, while cold air moves from the polar caps toward the equator (Figs. 3 and 6A). These air masses, the temperature, water vapor content and density differ greatly due to their origin, flow parallel to one another, but in the opposite direction.

Since the warm and moist air is lighter, and the cold and dry air is heavier, billows are formed along the separation line (see Fig. 6B and C). The greater the contrast in density between the air masses present (i.e., the greater their difference), the more violent will be the resulting turbulence. Generally, in the temperate zones, the air masses present mix little by little, and the transition zones (the fronts) are very elongated. All along such fronts one finds rising air, which causes cloud

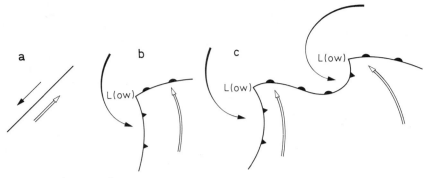

Fig. 6. Development of a wave

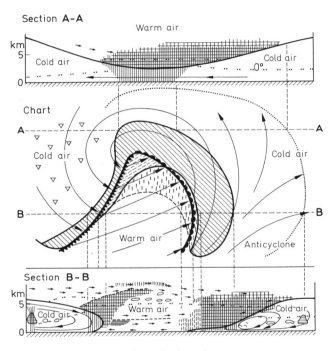

Fig. 7. Projection and sections of a frontal system

cover and precipitation or, on the contrary, air moving down, which causes clearings.

We shall not deal in detail with the development of a complex frontal system, and we shall limit ourselves to giving a general diagram of it in Fig. 7.

In this diagram, the warm air which advances behind a cold air mass rises and cools following the dry adiabatic process until saturation is reached. From then on, as a result of the condensation phenomena which occur, it follows the moist adiabatic process, and this continues up to a very high altitude.

In the case of the cold front, the cold and dry air which follows the warm mass will slide under it and lift it. This rising motion is subject to the same influences (dry adiabatic, saturation, moist adiabatic) as previously, but the latter are limited to a much narrower belt, so that the phenomena resulting from it (vertical development of cloud cover, precipitation, wind gusts, etc.) are infinitely more violent, and may even generate storms.

The resulting low-pressure oscillations and turbulence, although initially due to air flows originating from the great pressure belts, disturb the general flow. They give rise to real winds whose direction and force is no longer a direct function of the sunshine received. Furthermore, an effect of those billows is the production of cloud cover, which reduces direct sunshine and, therefore, affects the formation and persistence of the initial flows.

As a result of this, the study of winds in the temperate zones becomes complicated and, in contrast to what appears at the edges of the equatorial and subtropical zones, the winds flowing from them follow very different directions in the middle latitudes.

9.4 High-Altitude Air Flows

We have seen in Sect. 9.2.5 that the friction between an air mass and the ground slowed down the motion and altered the direction of the air flow. As one rises, the movement of an air mass sliding over a lower layer is less affected by friction phenomena, because their coefficient is low. Consequently, the resulting force (b in Fig. 4) also decreases. The wind originating from an isolated pressure gradient, therefore, is more violent at the same latitude. Furthermore, its direction becomes ever closer to the perpendicular to such pressure gradient.

We see in Fig. 1 that hydrometeors, i.e., clouds and precipitation of all kinds, stop their vertical development at the tropopause. In the standard atmosphere, the tropopause is represented by the transition between the troposphere – the layer lying immediately above the ground and characterized by a linear decrease in temperature as the altitude rises – and the stratosphere, i.e., a region of almost constant temperature and, furthermore, almost devoid of moisture.

This transition zone constitutes a practically insurmountable obstacle not only for hydrometeors, but also for the pressure variations due to the turbulence described in Sect. 9.3. Above, i.e., in the stratosphere, the winds are practically laminar and only follow the gradient laws. This is due to the difference, between the equator and the poles, in the mean temperature of the lower layer (troposphere), which, itself, is subject to the laws of heating determined by the angle of incidence of sunshine on the ground.

In the equatorial zone, the air is warm and its specific gravity is relatively low. Consequently, the height of the air column necessary to exert a given pressure is much greater than at one of the poles, where generally the air is cold and dry, and therefore dense. Thus, at the same altitude, the pressures found in the stratosphere are higher in the equatorial zone than above the polar zones. There follows a pressure gradient which causes a general westerly air flow moving eastward as a result of the Coriolis force. Since the air density is low at high altitude and especially in the stratosphere, the friction exerted by a layer against another is negligible or nearly so. The general air flow which is the result of the pressure gradient and Coriolis force is therefore almost parallel to the equator. Therefore, the exchanges from the equatorial zone toward the polar zones, or vice versa, are of little importance and are generally limited to the troposphere.

9.5 Convection

We have seen in Sect. 9.2.1 the importance, in the lower layers of the atmosphere, of the energy exchanges between solid bodies and the air. As a result of such energy exchange, the air which is in immediate contact with plants and solid bodies is heated or cooled depending on the amount of radiation received. There necessarily follow some variations in the density of that air and, consequently, puffs of warm air which rise in the surrounding air which has remained relatively cool. These "hot air drops" (a kind of balloon without an air bag) are immediately replaced by oblique inflows originating from the areas located directly around and above. Consequently, above cultivated areas and above forests, alternate upward and downward movements take place both by day and by night. They are small-scale movements, which facilitate the diffusion of water vapor, but also the dispersion of pollen, spores, etc. (in this respect, see Chap. 11).

If the thermal gradient between the ground and the air mass lying above it is considerable, these vertical exchanges may, in many cases, assume considerable proportions. We have seen in Sect. 9.2.1 that vertical movements of the same nature give rise to the low-pressure belt which surrounds the equatorial zone. However, even in the temperate or polar climates, a heating of the ground following either a strong general sunshine, or a localized absorption of radiation by a sandy beach, surfacing rocks or, in other terms, by any solid body with appropriate intrinsic properties, may lead to the formation of a hot air drop. The latter, rising, will be capable of triggering the condensation process, which will accelerate its ascending motion, and therefore the formation of clouds with a massive vertical development (cumulus, comulonimbus), and this will continue until the storm phenomena appear.

While the wind can be easily measured in the horizontal plane, the direct measurement of ascending or descending vertical air flows is much more difficult, because it is not performed in a primary direction at a given place. Furthermore, convection does not generate only vertical ascending or descending currents, but more often oblique currents, so that their measurement becomes rather difficult. Therefore, special instruments have been built for such measurements (see Chap. 12.2.5).

Literature

Baumann, H.: Witterungslehre für die Landwirtschaft, S. 139. Berlin, Hamburg: Verlag Paul Parey 1961

Defant, A., Defant, F.: Physikalische Dynamik der Atmosphäre, S. 527. Frankfurt/Main: Akademische Verlagsgesellschaft 1967

Devuyst, P.: Comprendre, interpréter, appliquer la Météorologie, pp. 164. Bruxelles: A. de Vischer 1972

Eimern, J. van: Wetter- und Klimakunde für Landwirtschaft, Garten- und Weinbau, S. 239. Stuttgart: Eugen Ulmer 1971

Emsalem, R.: Climatologie générale. Tome 1: Fondements des équilibres atmosphériques, p. 215. Alger. 1970

Ficker, H.: Wetter und Wetterentwicklung. Verständliche Wissenschaft 15, S. 140. Berlin: Springer 1952

Kuhn, W.: Numerische Wettervorhersage, S. 10. Neue Zürcher Zeitung 159 (1971)

Landsberg, H.E.: Weather and health. An introduction to biometeorology. Anchor Books, p. 148. Garden City, N.Y.: Doubleday & Co. 1969

Liliequist, G.H.: Allgemeine Meteorologie, S. 368. Braunschweig: Vieweg 1974

Miller, A., Thopson, J.C.: Elements of meteorology. 402 pp. Colombus (Ohio): Charles E. Mervill. Publ. Comp., A Bell & Howell Co. 1970

Möller, F.: Einführung in die Meteorologie. I. Physik der Atmosphäre, S. 222. Mannheim: B. I. Hochschultaschenbücher 276 (1973)

Möller, F.: Einführung in die Meteorologie. II. Physik der Atmosphäre, S. 223. Mannheim: B. I. Hochschultaschenbücher 288 (1973)

10 The Wind

B. PRIMAULT

10.1 General

In the preceding chapter, we have shown how the differentiated heating of the ground and the unequal action of sunshine affect the displacements of the air masses on our globe. In our demonstration, however, we have assumed a priori that the earth presented a uniform surface, both with respect to its rugosity and with respect to its color (i.e., its heat absorption properties and their nature, or, in other terms, the possibility of diffusion of the heat absorbed on the surface toward the interior).

This assumption, however, is challenged by the real world. Actually, whoever examines a large-scale map or, even better, a globe, immediately realizes that these theoretical conditions are not fulfilled. Thus, the area of our globe covered by seas is greater than that occupied by the land masses. Furthermore, the latter are distributed differently between the southern hemisphere and the northern hemisphere. Finally, on the continents, there are mountain ranges, sometimes considerable ones.

These first three facts imply that the heating due to sunshine, at the same time of the year, at the same latitude and irrespective of cloud cover, is very different depending on whether it occurs on the ocean, on a continent near the sea, or on a high plateau. Actually, the radiation which reaches us from the sun is in part intercepted by the atmosphere, so that the lower one moves toward sea level, the greater is the portion of that radiation which does not reach the earth, and this even in clear weather.

In addition to the differences caused purely by their geographic position, the land masses are covered by very diverse vegetation. Let us only think of the tropical forests and of the great steppes which cover the subtropical regions which immediately adjoin them. The result is that the heating of the ground and of the vegetation, beginning from the lower layers of the atmosphere, will be very uneven from one place to the next, even over small areas. Furthermore, the variations in rugosity which are the result of local differences are capable of considerably modifying the force and the direction of the great air flows which we have previously described.

Since the pressure decreases with altitude, the temperature follows this general pattern and also decreases (see Chap. 9.1.3). However, as a result of the heating of the slopes under the effect of the sun, or of their cooling by night, the air does not always have the thermal gradient which would apply in the standard atmosphere. Consequently, air exchanges will be found between adjoining regions or between successive altitude levels. Thus, for example, in the case of a stationary anticyclone covering a mountainous region, cold air is trapped in the valleys, giving rise to inversions.

For all these reasons, a great number of local winds are generated each day, especially as long as gradient winds do not disturb the local or regional features.

Furthermore, the great mountain ranges or the seacoast have a disruptive effect upon the normal flow of the great gradient air flows. The result will be winds which have their own features and which are well known by the local populations.

The purpose of the present chapter is to show to what extent regional or local conditions produce particular winds which are not determined by the formation or displacement of low-pressure zones or anticyclones on a continental scale. It will also deal with the modifications to which the great intercontinental air flows are subjected as a result of orographic factors.

10.2 Local Winds

10.2.1 The Cold Air Flow

As a result of the difference in specific gravity between a dense fluid and those which surround it, the former tends to flow by gravity toward the lower levels. This, for example, is the phenomenon which drives the rain water from the mountains toward the sea through brooks, streams, and rivers.

In the absence of large-scale opposing forces, i.e., in the absence of pressure gradients, or, in other terms, in calm weather, a sort of sedimentation takes place in the air: the cold masses accumulate in the vicinity of the ground. This movement is frequently noticed in winter in the anticyclones of the temperate zone. If the lateral flow of the cold air is slowed down, or prevented (for example, in a closed valley), a temperature inversion may then be noticed (the air is warmer at altitude than at ground level).

Furthermore, a corollary of the radiation phenomena is that all objects, solid or gaseous (the ground, vegetation, buildings, even more than the air itself), cool down without compensation. The temperature of the air which lies on cold solid bodies decreases by contact. Certainly, the air also cools, but the heat loss is much less marked. As a result of this, a difference in temperature is established between the air which lies above the ground and that which composes the higher layers; this involves a difference in density between the two masses (Fig. 1). The cold, and therefore heavier, film initially follows all the contours of the terrain. On a slope, no matter how slight, there is an imbalance as soon as the difference in density between the two air masses is produced. This imbalance causes a movement of the cold mass toward the bottom of the slope, and this air is immediately replaced by the slightly warmer mass lying above it, which, in turn, becomes cooler in contact with the ground and the vegetation.

The movement of the cold air is all the more rapid as the slope is steeper and the local thermal gradient the more marked. Generally, it is very slow, and this air mass behaves like a viscous fluid, such as, for example, molasses. The cold air flowing along the slopes accumulates above any obstacle and stagnates there until a wind gust or energy transfer conditions help reheat it again. When we refer to obstacles here, we do not only think of continuous barriers, such as a railroad or road embankment, an uninterrupted line of buildings, or the edge of a forest. A cluster of trees, a hedge of limited thickness, a reed screen, or even, under certain

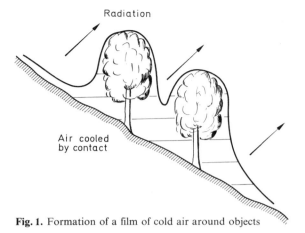

Radiation

Air cooled
by contact

Fig. 1. Formation of a film of cold air around objects

Cold air
flowing along
the slope

Cold air lake

Areas
exposed
to frost

Fig. 2. Cold air lake and danger of frost related thereto

conditions a steel wire screen, may contain a moving cold air mass, just as a fork is capable of catching molasses.

In many cases, an increase in the local danger of frost must be attributed to these phenomena of slowing down of cold air. What actually takes place is the formation of cold air lakes of greater or lesser size, in which the plants, radiating heat energy and thereby cooling, are not reheated by warmer air originating from the upper layers (see Fig. 2).

10.2.2 Land and Sea Breezes

As in the preceding cases, differences in density between neighboring air masses cause regular winds on the seacoast. With low or nonexistent pressure gradients, such winds blow practically every day with the same intensity, in the same direction and at the same time.

By day, the total radiation is absorbed by the coastal land, which is heated. By contact, the temperature of the air lying above it rises as a function of that of the ground, and its density decreases. Then it rises like a bubble and the low pressure caused by this ascending movement sucks in colder air from the surrounding regions, more particularly the colder air which lies above the sea or a lake (see

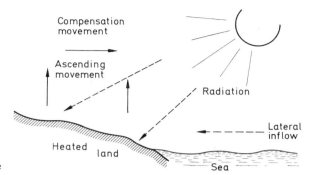

Fig. 3. Formation of the sea breeze

Fig. 3). Actually, the water is heated less rapidly than the ground, and therefore remains comparatively cooler than the coast.

Therefore, by day and near the ground, there is an inflow of air from the sea toward the land. This movement is balanced by an opposite flow (from the land toward the sea) at low altitude (at an altitude of 200–500 m).

By night, similar phenomena occur, but in the opposite direction: the land cools rapidly by radiation and the air which covers it follows that movement, becomes heavier, and flows toward the lower areas, the sea or the lake. Furthermore, the water cools off less rapidly, so that the air lying above it is comparatively warmer. Then denser cold air slides, so to speak, between the water and the mild air, in which convection phenomena take place. As during the day, an opposite balancing flow is then established at low altitude. Therefore, by night and at ground level, there is a flow of air from the land toward the sea. These alternate flows take place around all the large water areas bordering on a relatively flat coastline. If, instead, the coast is steep, these movements are accentuated by the phenomena that we shall discuss later.

10.2.3 Valley Winds and Mountain Winds

Similar phenomena of greater or lesser importance also take place in the interior of the continents whenever differences in air density occur. This is particularly the case between the mountain massifs and the plains.

As a result of the increase in atmospheric turbulence in inverse ratio to the altitude, and especially as a result of the fact that the valleys and lower areas are generally covered with vegetation, one notices that by day the mountains are heated more markedly than the plain. An imbalance in density between the air masses then appears, in the same manner as in preceding case of land/sea opposition. A regular wind then takes place by day, a wind which blows from the lowlands in the direction of the mountains.

By night, and as a result of the same radiation phenomena, an opposite wind blows from the mountains toward the plain. Therefore, one notices in the valleys of the great mountain massifs a regular alternation of winds blowing from the plain toward the mountains by day, and from the mountains toward the plain by night.

10.2.4 Additional Notes

Even though the winds mentioned in this chapter may sometimes, because of their regularity and especially their force, be assimilated to prevailing winds, one should not lose sight of their local or, at most, regional origin. Therefore, they should be clearly distinguished from the prevailing winds of a continental nature and due to pressure gradients originating outside the region.

When determining wind patterns, whether one uses data recorded by instruments or starts from the deformation in the crowns of trees (habitus), these local winds often mask the prevailing winds of a continental nature. Actually, by their regularity, they appear more frequently in the diagrams and their effect on the vegetation may assume a predominant importance.

Furthermore, if in estimating the direction of the most frequent winds one only considers the deformations suffered by the vegetation, it should not be forgotten that the latter are more or less marked depending on the intensity of the wind or the time of day and the period of the year during which it blows. Thus, a violent wind which re-occurs regularly may bend the branches or the tops of trees, irrespective of the time at which it blows, and this as a result of a repeated mechanical pressure.

Much weaker winds – even sometimes almost imperceptible to man – may have the same effects on the shape of the crowns of trees, provided that they blow at the time when the terminal buds are developing, i.e., when the yearly new small branches are lengthening. Thus, in the temperate regions, local winds of low intensity, but blowing regularly in the spring and during the night, cause the same deformations in the crowns of trees as much stronger winds, but blowing by day in all seasons, even with less regularity.

By only considering the shape of the crowns of trees, one can be led to draw totally erroneous conclusions concerning local wind patterns. Consequently, those observations must be examined in the light of the physiological properties of the plants (growth conditions) and of the possibilities of mechanical deformations beyond the periods of growth. These considerations assume their full importance in the study of the possibility of gas exchanges between one region and another or of the dilution of large- or small-scale industrial pollution.

Finally, the prevailing or local winds may be so violent that they prevent any development on the windward side of the trees. Furthermore, a defensive process on the part of the plant is often observed, which increases the thickness of the bark on that side. If the wind is loaded with dust, it will have an abrasive effect on the trunk, the branches, and even on the leaves.

10.3 Some Particular Great Air Flows

10.3.1 General

The winds of a local or regional nature that we have just described (see Sect. 10.2) all show the same feature: a daily alternation. They are the result of differences in the heating of the ground or of the water between day and night.

Generally speaking, their intensity is greater by day than by night as a result of the fact that the thermal regulating mechanism constituted by either the sea or lake water, or by the plain and its vegetation, attenuates the differences in air density due to the emission of heat radiation during the night. By day, on the contrary, the heating of the ground – in the mountains or along the coast – may fully exert all its effects. Consequently, the density gradients are more marked by day than by night.

There are, however, other movements of air masses whose mechanisms may be assimilated to those that we have just described. Such mechanisms, though, are no longer the result of rapid (day/night) fluctuations, but of much slower variations in air density, which depend, this time, on the seasons. Wherever a land/water opposition manifests itself, one discovers the formation of regular air flows which leads to a repetition of similar, or even identical, phenomena. In certain cases, they could be assimilated to the fluctuations in the pressure belts described above (see Chap. 9.2.3).

However, the opposition between warm air and cold air may also manifest itself in a different way. We have seen in Chap. 9.3 that the differences in density between warm and moist air and cold air cause low-pressure areas and frontal systems. In this case, the differences in density are not very marked at the origin. Consequently, the resulting movements are relatively slow. They are only gradually amplified under the effect of a sort of "chain reaction". If, on the other hand, two air masses of very different densities are suddenly placed in the presence of each other, they inevitably cause a very violent turbulence. However, the sudden inflow of cold and dry air into a tropical zone has a seasonal nature. For this reason, the tropical cyclones (typhoons or other tornadoes) are not encountered in all seasons with the same intensity. This is well known by the populations of the region visited by such atmospheric phenomena, who, for generations, have been suffering their ravages and, in many cases, have been able to take protective measures in time.

10.3.2 The Monsoons

We have examined above (see Sect. 10.1.2) certain mechanisms which cause regular winds along the shores of great bodies of water, whether they be lakes of the sea. Those winds follow a night/day pattern, i.e., a complete inflow and outflow oscillation covers a 24-h period, or one consecutive day and night.

One can easily imagine air movements of similar origin, no longer based on the alternation between day and night, but on that of the seasons (summer/winter). Then, there will be an annual pattern, with a land/sea phase covering several weeks, and another sea/land phase, also lasting several weeks. The two phases will then be separated by periods during which other phenomena will be of predominant importance. In order to produce such mechanisms, it is necessary that large continental masses adjoin vast seas. These regular winds which re-occur each year in the same period are called monsooons. They are particularly well known in India, but also in most continents, such as, for example, in Europe.

In Asia, notably in India, the monsoon wind is caused by the opposition between the continent of Asia, and more particularly the great plains of Siberia, and the Indian Ocean. Since the two regions are separated by a very high mountain range, the Himalayas, there occurs the formation of flows affected by obstacles: the original laminar flow is transformed into a semi-turbulent flow which climbs the mountain ranges and then descends again on the other side after having been subjected to the same transformations as described in Sect. 10.4.1. The summer monsoon, which blows from the ocean toward the land, brings abundant precipitation to India and parts of Indochina. The peak season is staggered from July in India to September in Indochina. On the other side of the Himalayas, for example in Tibet, that same monsoon is similar to a foehn wind.

The winter monsoon causes dry weather to the south of the Himalayas, but since the air masses which compose it originate from the plains of Siberia, its temperature, although relatively high as a result of the descending movement, is low in comparison with the masses stagnating over the Indian continent. Actually, because of its arctic and continental origin, this air only contains little condensable water when it is lifted by the mountains. Consequently, only little condensation heat can be found in it.

For the same reasons, the precipitation which, theoretically, should be abundant in winter in central Siberia and in Tibet actually does not occur there. There also, just as to the south of the Himalayas, the peak precipitation period takes place in the summer. It is, however, much less marked.

The existence of frequent and abundant precipitation at the beginning of the summer (May and June) in western and central Europe is due to a similar mechanism. The fact that it does not last longer and does not intensify until July and August is due to the advance in a northerly direction of the subtropical anticyclonic belt.

10.3.3 The Evolution of Tropical Cyclones

In Chap. 9.3, we have seen how two air masses, of different densities and flowing beside each other, give rise to billows, and then to turbulence, the intensity and the extent of which depend precisely on the difference in density. Below the tropics, one of the density factors is of preponderant importance: humidity. As a result of the channeling of cold air flows by the mountain ranges, such as, for example, the Rockies, or by the marine currents, one often notices the unseasonable arrival of drops of cold and dry air into the subtropical subsidence zone. These cold air masses then plunge toward the ocean, where they meet warm and very moist air, and, therefore, of much lower density. There follows immediately a rotary movement of very great intensity, but of relatively small dimensions. All the energy enclosed by the difference in density between the two air masses is consequently gathered in a relatively restricted space. Furthermore, the intense rotary movement generates at altitude a centrifugal effect, which rejects the air toward the outside of the cone formed by the cyclone itself, an inverted cone the axis and the top of which are represented by the "eye of the typhoon". The pressure then decreases enormously in the center of the system, which increases the rotary movement even more.

As a result of the ascending currents and of the expansion of the air that they cause, the atmospheric humidity condenses and condensation heat is released, which increases the upward movement of the entire mass even more. Inside the cyclone, therefore, suction forces and electric phenomena (lightning and thunderstorms) develop, often accompanied by very violent hailstorms. The cyclone thus formed may remain active for several days. The trajectory that it follows is determined by the neighboring anticyclonic cells or low-pressure areas. So long as the principal cyclonic formation is supplied with moist air, its rotary motion is not slowed down, and this happens even less because this development generally occurs over seas whose water is warm (the center of the Pacific, Caribbean Sea, etc.). On the other hand, when the cyclone reaches regions where the sea water is colder (Labrador or Kamchatka Currents, for example), or land masses, where, therefore, the water vapor content in the atmosphere is reduced, the supply of the system is slowed down and the turbulent movement is slowed down until it gradually dies. These mechanisms and these cyclonic cells are those which cause the most violent winds, and, therefore, the greatest damage, both at sea and on land.

The regions where such cyclones are generated are generally known and the trajectories of typhoons can be calculated with sufficient accuracy to make reliable forecasts possible.

10.4 Orographic Modifications of the General Air Flows

10.4.1 Peculiarities of the Effects of Obstacles

The local or regional winds mentioned in Sect. 10.2 only take place under calm conditions, i.e., within stable anticyclones. When the formation of a pressure gradient takes place on a continental scale, these local air flows are attenuated, or even annulled, or, on the other hand, accentuated by the gradient wind.

Over the plains, the gradient wind has an essentially laminar nature, and, while its force is decreased at ground level, this is mainly due to the friction caused by the ground, by buildings, and by the vegetation. As soon as it faces a mountain range, its features are thereby altered. Rising phenomena are produced, which modify the conditions of the air mass as a result of compression and decompression phenomena. Only the air layers flowing relatively close to ground level undergo a deformation. At high altitude, on the other hand, the laminar wind remains approximately unchanged (Fig. 4).

In addition to these compression phenomena affecting the mountain wind and decompression phenomena under its wind, the vertical component which is generated as a result of the general rise of the air mass due to the obstacle itself brings the air mass to higher altitudes. Then its temperature decreases, following the dry adiabatic laws (see Chap. 9.1.4). Consequently, the relative humidity of that air increases and may reach saturation. From then on, condensation phenomena appear: cloud formation and release of condensation heat. The lowering in temperature resulting from decompression is thereby diminished: then it follows the moist adiabatic process (see Chap. 9.1.7). This phenomenon is often accompanied by the formation of clouds, from which precipitation is released. There follows a decrease in the amount of water in suspension in the air mass, in the form of either water vapor or liquid droplets, and an accumulation of latent heat inside the air mass. The rising mass is thus subjected to different direct actions, some tending to increase pressure (dynamic compression as a result of the gathering together of the air streams), the others to decrease it (static decompression due to

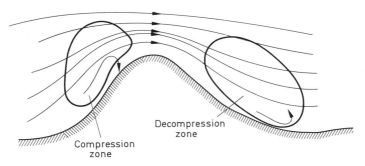

Fig. 4. Air flows altered by passing over a mountain

the increase in altitude). Since the latter are the more important ones, the final result is a release of heat by condensation. Therefore, the air mass which reaches the top of the mountain no longer has the same characteristics as that which covered the plain before its ascent. It is then a new air mass, with quite different characteristics (latent heat, absolute humidity), which is going to plunge into the valleys under the effect of the mountain wind.

Thus, there occur again compression phenomena, due to the difference in altitude, and decompression phenomena, following the staggering of the air streams, of which the majority again assume a laminar nature. But, since the air mass has released a great amount of its humidity while climbing the mountain, the corresponding condensation heat (approximately $590 \, \mathrm{cal \, g^{-1}}$ or $2.5 \cdot 10^6 \, \mathrm{J \, kg^{-1}}$ of water fallen in the wind) is enclosed in the air mass which flows in the direction of the plain. That heat is felt there by its producing much higher temperatures (foehn effect).

If the moving air mass is of polar or continental origin, the final increase in temperature is nevertheless rather small, at the same altitude, after passing the obstacle. Thus, a northerly wind blowing from Scandinavia, or a northeasterly blowing from Russia, which crosses the Alps, gives the inhabitants of Venetia or of Friuli an impression of glacial cold: it is the bora. In India, the winter monsoon which plunges down from the high peaks of the Himalayas is cold, in spite of its catabatic nature, even more marked than that of the bora, or even of the foehn. We could multiply these examples, because in both hemispheres and practically in all five continents there are known to be particular gradient winds, which descend from the mountain ranges after climbing over them.

There are, however, other gradient winds whose trajectory is modified by mountains; in this case, however, in the horizontal plane. Then the characteristics of the air mass are not modified by this. Let us mention the typical example of the mistral, which, in France, blows along the Rhone valley under the impulse of a low pressure located above the Gulf of Genoa.

10.4.2 Effects of the Coastline

We have seen in Chap. 9.2.5, while explaining the physical principles governing the winds, the important effect of the roughness index of the ground both on the

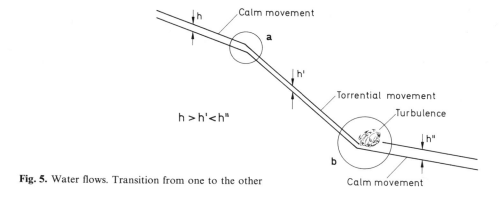

Fig. 5. Water flows. Transition from one to the other

velocity of propagation and on the direction of air flows. Thus, an air flow controlled by large-scale pressure gradients, which passes from the sea to the land or vice versa, will have its velocity of propagation and its direction considerably altered by the change in the roughness index. The same will apply inland, if this air flow passes from a bare region to a region covered with vegetation. Repercussions at altitude will necessarily follow, because, if the velocity at ground level decreases, the moving air mass itself does not diminish in any way. What happens here is similar to what happens in a canal full of water in which the slope, initially steep, suddenly becomes flatter. The water passes from a torrential to a calm movement (see Fig. 5). This transition is accompanied by violent turbulence and the thickness of the liquid flow (volume and section size being equal) considerably increases over a very short distance.

Consequently, just as in the case of a liquid, vertical movements will have to compensate for the variations in the velocity of propagation at ground level.

Any vertical movement of the air, however, is accompanied by decompression, and even condensation processes, once saturation is reached. Therefore, there will often be an increase in cloud cover, or in precipitation, along a coast receiving the winds coming from the open sea, and thereby moist air.

On the other hand, if winds, even when loaded with moisture and accompanied by massive cloud cover, pass from the land to the sea, the cloud cover will decrease as a result of the subsiding movements of the air. The latter are due to the fact that wind velocity increases in proportion to the decrease in the roughness index. If we return to the image of the canal full of water, we can assimilate such a change to a transition from a calm movement to a torrential movement (Fig. 5). The result of it is a sudden decrease in the height of the liquid, but no notable turbulence zone.

However, the velocity of the wind is not the only thing which is modified by the passage from the sea to the land, or vice versa: its direction also varies, again as a result of the variation in the roughness index (in this respect, see Chap. 9.2.5).

If, in a plain, the vegetation suddenly changes over rather considerable areas (for example, passage from a prairie or cultivated fields to a forest), similar processes are activated. Actually, since the roughness index of the mass of vegetation varies between the cultivated fields and the forest, differences are produced in the velocity and direction of the wind, accompanied by rising and subsiding air currents. However, here the changes in roughness are less marked than in the preceding case (land/sea), so that the rising and descending currents, as well as the condensation or evaporation phenomena are also less visible, or even negligible or almost nonexistent.

However, the roughness index can be modified not only by the ground cover, but also, or even more, by orographic changes. Thus, a great air flow passing from a plain to a hilly region will have its structure very considerably modified. Here, however, it is not a case on the same scale, and the modifications to the air flow will be felt, in the second case, up to much more considerable altitudes. Actually, there will be a juxtaposition of two phenomena: the first being the transformation of a laminar movement into a turbulent movement, the second, the change in roughness which takes place on a larger scale.

Both the passage from the sea to the land and that from one type of ground cover to another, as we have seen, involve modifications to the structure of the wind. While this structure can be generally regarded as laminar – except in the layers which are in immediate contact with the liquid or solid surface – this laminar structure is altered by the changes in direction and velocity which result from changes in the roughness index of the ground. Consequently, in the transition area, turbulence is produced, which remains limited to a relatively narrow zone, following very accurately the index change curve.

A turbulence which appears in the wind stream, with its rising and falling, is the cause of considerable, but very localized, fluctuations in velocity: the presence of gusts.

If the turbulence due either to a change in roughness or to the presence of an obstacle is accompanied by phenomena of instability, it may give rise to powerful whirlwind movements, although limited to restricted areas. It is then a case – on a much more reduced scale, of course – of phenomena similar to the formation of tropical cyclones. This explains why certain countries rather far away from the subtropical seas – the regions where typhoons are generally formed – sometimes experience small miniature tornadoes. Although the latter do not have the devastating effects of the great tropical cyclones, they nevertheless leave great damage in nature, mainly in the forests, and among man-made buildings. Passing over lakes or deserts, they lift their water or sand, producing opaque columns called waterspouts.

Literature

Bouët, M.: La pénétration de l'air froid dans une vallée alpine. Arch. Meteorol. Geophys. Bioklimatol., Série A 15, 46–49 (1966)

Hentschel, G., Leitdreiter, W.: Die Häufigkeit von Inversionen im bodennahen Luftraum (15–76 m über Grund) in Abhängigkeit von Jahreszeit, Tageszeit und Windrichtung. Ang. Meteorol. 3, 353–362 (1960)

Johnson, O.: An examination of the vertical wind profil in the lowest layers of the atmosphere. J. Meteorol. 16, 144–148 (1959)

Koch, H.G.: Der Wind als Standortsfaktor im Klimamosaik des Mittelgebirges. Arch. Forstwes. 9, 901–942 (1960)

Mäder, F.: Untersuchungen über die Windverhältnisse in Bodennähe bei verschiedenen Wetterlagen. Veröffentlichungen der Schweizerischen Meteorologischen Zentralanstalt 9, S. 42 (1968)

Panofsky, H. A., Towsend, A. A.: Change of terrain roughness and the wind profile. Quart. J. Roy. Meteorol. Soc. *90*, 147–155 (1964)

Schram, K.: Die Windverhältnisse in der bodennahen Luftschicht an einem Hang von etwa 25° Neigung. Veröffentlichungen der Schweizerischen Meteorologischen Zentralanstalt 10, S. 13 (1968)

Scultetus, H. R.: Bewindung eines Geländes und vertikaler Tamperaturgradient. Meteorol. Rdsch. *12*, 1–10 (1959)

Vaupel, A.: Advektivfrost und Strahlungsfrost. Mitteil. Dtsch. Wetterdienst. *3*, 31 (1959)

Wagner, N. K.: An analysis of some over-water wind profile measurements. Transactions, Am. Geophys. Union *39*, 845–852 (1958)

Yeshino, M. M.: Some local characteristics of the winds as revealed by wind-shaped trees in the Rhone Valley in Switzerland. Erdkunde. Arch. wissensch. Geogr. *XVIII*, 28–29 (1964)

11 Transportation of Particles by Air Masses

B. PRIMAULT

11.1 Origin of Pollutants

The air which covers the surface of the earth is never pure. It always contains in suspension foreign particles, whether they be solid, liquid, or even gaseous. Even the air of the polar zones, which is the clearest known above ground, is not free of them. The nature and origins of such impurities are quite varied.

11.1.1 Natural Pollutants of Mineral or Vegetable Origin

When the wind blows over the sea, it raises waves on it and, if its velocity is sufficient, it causes spray on their crests. The water droplets which are thus removed from the liquid mass and driven into the atmosphere largely evaporate. The salt which they originally contained crystallizes into very fine cubes, which then float in the sea wind. Thus, it is not surprising to find traces of sea salt in greater or lesser amounts in the vicinity of the coast and even quite farther inland.

However, the wind does not act only on the sea: it also raises clouds of sand from the deserts and steppes. These fine mineral particles are often lifted to a very high altitude, depending on the meteorological conditions. We have seen in the previous chapter the mechanism whereby waterspouts and certain cyclones are formed above desert regions. Such rotatory phenomena affect the air masses up to very high altitude. Therefore, it is not surprising that fine particles of sand, just as the salt from the sea, reach regions located several thousand meters above ground level.

However, the wind does not raise only mineral particles, i.e., fine rock particles on the ground or the salt from the sea. It also lifts considerable amounts of organic particles originating from the decomposition of vegetables, whether they be dead leaves or arable land. Well known in this respect are the dust clouds in the prairies of the United States, dust clouds caused by wind erosion, and which ravage the land after the elimination of the grass cover due to inconsiderate farming practices. The pad formed by the intertwined roots of the perennial herbaceous plants of the great prairies is finely cut by the plowshare, and the roots of the cereals (especially wheat and rye) which replace them cannot substitute for it. It follows that the soil, very mobile by nature, is left bare between the harvest (ploughing of fallow ground) and the germination of the following crop. Since generally spring cereals are involved, the soil thus left bare is exposed to the effects of the wind during the fall, the winter, and part of the spring. A period without rain is then sufficient for the first few centimeters to assume a dusty texture under the action of the sun. Thus it is calculated that, on the average, the great plains (or wheat and corn fields) of the United States lose each year from 1 to 2 cm in thickness, which represents millions of tons of good fertile soils pushed toward the seas. Similar phenomena may occur elsewhere also, as man destroys the natural grass cover. Furthermore, it has been found that the soil deteriorates very rapidly.

However, the solid particles of natural origin which are found in suspension in the air do not originate only from the earth itself. Our atmosphere actually is constantly bombarded by an impressive number of meteorites. The matter which composes them is brought to incandescence as a result of friction with the atmosphere, so that it liquefies, or even gasifies. A certain amount of those meteorites – or, better, of the matter which composes them – falls to the ground by gravity in the form of more or less massive blocks, or – and this is what is of particular interest to us here – of dust. The latter then remains in suspension in the atmosphere for a prolonged period of time. Its duration is inversely proportional to the diameter of the particles. Let us also note that part of that dust of meteoric origin is radioactive.

The wind and the meteorites are not the only sources supplying the air with solid pollutants of natural origin: the vegetation provides us with an impressive number of them. First of all, this is the case of the clouds of dense smoke, of very great extension in area and altitude, caused by forest fires. That smoke is composed of myriads of solid particles, which remain in suspension in the atmosphere because they are very fine. Susceptible of being carried very far, they constitute a not inconsiderable source of pollution, even in the polar regions, devoid of vegetation, or in the tropical regions, where the frequent rain prevents forest fires.

In the short run, the most massive source of atmospheric pollutants certainly is the eruption of volcanoes. Following the telluric movements and thermal phenomena which accompany them, at the time of volcanic eruptions one always notices the ejection of smoke and very fine dust. Thus it happened that, at the time of the Krakatoa eruption in the Sunda Islands in 1883, the cloud of smoke that it ejected could be followed all around the earth, which it circled several times before becoming sufficiently diluted to be no longer visible to the human eye.

The solid particles are not, however, the only natural pollutants in our atmosphere. There are also liquid pollutants, of which water certainly is the most important. The drops in suspension in the atmosphere may be very fine and form haze, or larger and give rise to clouds or fog.

In addition to the solid and liquid components in suspension in the atmosphere, our air is contaminated by pollutants of a gaseous nature. Among those which are more important to us, if we do not consider water vapor, because we regard it as one of the natural components of the atmosphere, we find the carbon dioxide produced by the respiration of higher animals and plants, and the methane gas originating from the decomposition of natural organic compounds, be they animal or vegetable, under the action of anaerobic organisms. These two gases are found everywhere in our atmosphere in greater or lesser concentrations; therefore, they may be regarded as atmospheric pollutants of natural origin, just as the solid or liquid particles that we mentioned above.

11.1.2 Pollutants Due to Human Activities

From the most remote antiquity, the activities of man have been an ever-growing source of atmospheric pollutants. As such activities developed, and especially after the immense expansion of industry, the number of sources and, above all, the volume of atmospheric pollutants emitted have increased to an extent that is difficult to imagine. At the beginning, the rare fires of cave dwellers could easily remain unnoticed. Later, with the development of agriculture and the technique of clearing the land by fire, the smoke has assumed ever-increasing proportions, and this not only since the neolithic period, but during the entire Middle Ages, and even to our time, when vast areas of dry grass are burned each year.

The great conquests of the Middle Ages, if not of the Roman age which preceded them, also produced not inconsiderable sources of pollutants, by the burning of castles, of boroughs, or even of entire cities. These repeated fires certainly increased the amount of solid pollutants, and also the carbon dioxide content in our atmosphere.

It is, however, especially with the industrial revolution of the end of the 19th century, the appearance and expansion of railroads (steam traction), and then – and above all – of the automobile, that pollution has increased to an extent that renders it a public danger. The factories first, the locomotives, the automobiles later, have contributed to such a fouling of our air that traces of it are even found on the great inlandsis of Greenland or at the South Pole, where they form real deposists. The latter have an undeniable influence upon the ablation of glaciers in the polar regions, as well as in the higher regions of the land masses.

The pollutants originating from industry and transportation are not only of a solid nature; they are also liquid – to a relatively minor extent, it is true – and, for the most part, gaseous. These gases are often harmful to man, to animals, and to vegetation. The CO_2 content is not the only one that is rising very rapidly. Some gaseous fractions, harmful, such as H_2S, as well as corrosive, such as SO_2, are also assuming alarming proportions.

11.1.3 Nature of the Sources of Pollutants

In discussing sources of pollution, it is desirable, before examining the possibilities of transfer of pollutants, to know the spatial distribution of their origins. If one imagines a volcano, a factory chimney, a locomotive stack, the exhaust pipe of a car or of a truck, such a source of pollution can be assimilated to a single point. The distribution in the atmosphere of the solid or gaseous particles – i.e., their diffusion – will be a function not only of the air flows in the environment, but also of the temperature gradient existing between the gaseous mixture which carries the pollutant – because it is rarely released in its pure state – and the surrounding air. In this case, a plume originating from a single point will be formed, whose shape, size, and displacement will be a function of both the environmental conditions and the characteristics of the effluent itself (Fig. 1).

In the case of wind erosion, of the formation of sea salt crystals, of the clouds of smoke produced by forest fires, or even of the emissions originating from entire industrial regions, one is no longer dealing with a single-point origin of the solid, liquid, or gaseous atmospheric pollutant. That origin then covers wide areas and, for calculation purposes, it could no longer be assimilated to a single point. The cloud of pollutant released from it may possibly have a shape similar to one or another of those shown in Fig. 1, but its dilution and transfer will considerably differ from it each time.

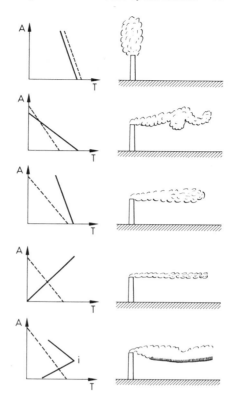

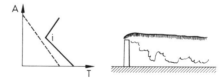

Fig. 1. Some types of plumes of smoke having a single-point source. *A*, altitude; *T*, temperature; *i*, inversion. The *full lines* show the change in temperature as the altitude changes, the *dotted lines* the change in humidity

11.1.4 Transportation of Pollutants

As we have already made clear above, one dispersed into the atmosphere in the immediate vicinity of their source, pollutants do not remain stationary. On the contrary, they are carried by the air flows, both at ground level and at altitude, toward areas often very far from the place of their release.

The velocity and direction of these displacements depend on the pressure gradient prevailing all along their trajectory. However, since pressure gradients at a given level only make it possible to calculate a horizontal laminar wind, it is also necessary to know the vertical gradient, because it governs the ascending or descending movement of the particles. In that movement, it is necessary to consider two elements, which are added up algebraically:

The movement of the air mass itself, sometimes affected by considerable amounts of heat accompanying the release of pollutant, and the sedimentation rate of the particles.

The latter is a function of the size, of the shape and of the weight of each component. In a descending movement, the natural sedimentation movement of the particle, due to the force of gravity of the earth, is accelerated by the motion of the air. In an ascending current, on the other hand, this movement is slowed down, interrupted, or even reversed, depending on the relative velocity of the air in comparison with the natural sedimentation rate of the particles.

In the case of single-point sources of pollution, or of large-area sources, but accompanied by a release of heat, the initial rising of the cloud of particles is due first of all to the temperature gradient existing between the source and the surrounding places.

11.2 Formation of Liquid Pollutants

11.2.1 Water

The best known of these transfer phenomena, both vertical and horizontal, is constituted by the drops of water, i.e., the clouds. This is a typical case of pollution without a point source and with a low, or nonexistent temperature gradient.

Let us study, for example, the processes involved on a fine summer day: the ground is heated unevenly under the action of the sunshine and the air lying above it, in turn, is heated by contact with it. Warm spaces and cooler spaces are then produced, and such unevenness generates temperature gradients. The density of the most heated parts is decreased and they begin to rise like huge balloons. This rising motion takes place initially following the principle of the dry adiabatic process (see Chap. 9.1.5), which means that it is relatively slow, and the temperature gradients remain low. As a result of cooling under the effect of the expansion phenomenon, however, the relative humidity increases. The rising movement gradually accelerates; but this acceleration is relatively small. It continues until the air is saturated. From that moment and as a result of the condensation, the processes change. The temperature change is no longer governed by the dry adiabatic process, but by the moist adiabatic process, and considerable amounts of heat are released. Indeed, during this rising movement, a great part of the moisture contained in the moving mass is diffused toward the surrounding air masses, following the principle described below (see Sect. 11.6).

As a result of these rising movements, localized low-pressure areas appear at ground level. To compensate for this, there follow inflows of air, laterally initially, and later downward. In the latter, opposite processes are involved: moist adiabatic process, evaporation (therefore more rapid cooling of the air mass) and, finally, dry adiabatic process.

To a man who observes these phenomena from the ground, the rising and descending movement are invisible as long as they develop following the principles of the dry adiabatic process. As soon as condensation occurs, and therefore the moist adiabatic process is involved, the phenomena are directly visible, because the drops of water in suspension in the atmosphere (clouds) intercept the visual beams.

Then, in calm weather, it is possible to follow the evolution of the rising clouds, the formation of cumulus, or even cumulonimbus clouds, or, instead, to see the disappearance of a cloud mass caught in a subsiding movement (for example, in the evening, when sunshine no longer heats the ground and, therefore, the rising movements stop and an equilibrium is established).

What we have described is of an essentially vertical nature so long as no gradient wind or horizontal movement is involved. On the other hand, when there is a gradient wind, the rising or descending movements are altered and become oblique; the clouds do not remain in the same place, but are displaced laterally. Since it often happens that these vertical movements of the air are small (for example, in the case of the arrival of a warm front, or of the formation of low stratus clouds), the horizontal air flow prevails and the clouds are driven by the wind from place to place, which to a great extent makes weather forecasting possible.

11.2.2 Other Liquids

What we have just said concerned the behavior of water in the atmosphere as a result of the vertical movements and horizontal displacements accompanied by changes in the temperature of the air mass, and secondarily in the physical state of the water.

The same phenomena are repeated irrespective of the liquid present. Then, however, considerable differences appear, which essentially are due to the condensation heat, which varies from one liquid to the next. The higher the boiling point of the liquid under consideration, the greater is the number of calories released per unit of mass.

It is not at all necessary, however, as in the case of water, that the boiling point be reached for evaporation or condensation phenomena to occur. Then we would never have evaporation on the surface of the ground, since the temperature nowhere reaches or exceeds 100 °C at sea level.

The condensation and evaporation phenomena take place at any temperature and are the result of the equilibrium between vapor pressures and saturation pressures. As soon as a difference occurs between the partial pressure exerted by the vapor and the saturation pressure (valid at that temperature and at that pressure), evaporation phenomena appear. On the other hand, as soon as the amount of vapor in suspension in an air mass exceeds the upper limit, i.e., saturation is reached (taking into account the pressure and temperature of the environment), we have condensation phenomena.

In the case of atmospheric pollution by liquids other than water, the droplets which are formed as a result of condensation generally are too few in number to give rise to a compact cloud, visible to the eye.

11.3 Gaseous Pollutants

It often happens that the gases mixed with the atmosphere do not reach the saturation point, in spite of powerful convective movements. Then they are carried, like the air itself, from place to place, and, if they are harmful, may cause considerable damage, although they cannot be detected by the naked eye. Man perceives them only if they are corrosive for the nasal system (for example, SO_2) or have olfactory properties (for example, H_2S). But even if their low concentration renders them undetectable, their prolonged action may undermine the health of entire populations or destroy all vegetable life over vast regions.

11.4 Solid Pollutants

Gases and liquids are not the only pollutants found in the atmosphere. We have seen above how nature each year releases great amounts of solid particles into the atmosphere, and these particles are carried by the wind and by the convective movements, and finally are gradually deposited as sediment. The greatest amounts of mineral products are probably due to volcanic eruptions. At the time of such telluric phenomena, tons of dust are projected to a very high altitude, carried by the wind over long distances, recovered, lifted by convective movements, or pushed to the ground by the natural katabatic movements. Nevertheless, even after the clouds of volcanic ash are sufficiently diluted in the atmosphere to be no longer visible to the naked eye, large amounts of dust are still traveling in the air.

The second source of atmospheric pollutants is constituted by the strong winds (simoon, khamsin, etc.) which drive the air in the desert and lift the sand, carrying it to very high altitudes, whence it is taken up by horizontal air flows and carried far from its place of origin. Thus it happens that, in the Alps, one finds reddish layers of desert sand in the snowfields: furthermore, "red snow" is a phenomenon known since the most remote antiquity.

Forest fires also are a source of solid pollution in the atmosphere, and those pollutants remain for weeks sufficiently compact to be visible. At this point, we recall the "blue sun" observed in Europe in 1958, caused by the passing of smoke clouds in front of that star, generated by forest fires which had revaged Canada and the United States.

At the present time, and as a result of the ever-increasing advance of industrialization, the sources of solid pollutants have greatly increased. While the first mentioned practically cause no damage, in view of their physiologically neutral nature, the new agents that industry releases into the atmosphere have irritant properties on the mucous membranes, and are corrosive or directly harmful to plants and living beings. We are at this point thinking more specifically of the lead released by road traffic. In spite of these particular properties of the new pollutants, their dissemination is governed by the same principles as in the case of the natural and generally neutral pollutants.

Until now, we have only discussed mineral solid particles. There are others, of which, in agriculture, some play a beneficial role, others a harmful role, but which are of vegetable origin: they are spores and pollen.

Many plants could not be fecundated, i.e., they could not reproduce, without a certain air pollution due to the grains of pollen disseminated into the air by the male flowers and received by the pistils of the female flowers. This pollution generally shows a difference from mineral pollution, by the fact that the solid grains in suspension in the atmosphere are much larger. Furthermore, in order to facilitate their transportation by the vertical or horizontal movements of the atmosphere, they are often equipped with either ailerons or air sacs, which decrease their total specific gravity, which makes them more maneuverable. Thus, in the areas of the dispersion belt of pollen, one finds at regular intervals (generally every four years) yellowish deposits on lakes and streams. Such deposits are due to the sedimentation of the pollen spread into the atmosphere by millions, billions, and trillions of individual plants. Nature has deployed considerable forces to ensure the fecundation by the wind which is a feature of certain plants, of which cereals constitute an important part.

However, the grains of pollen are not the only pollutants of vegetable origin: mushrooms also spread in this manner. Thus, their spores play the same role of solid organic pollutants as the grains of pollen and their diffusion is subject to the same rules.

11.5 Aerosols

The air pollutants that we have discussed so far were of a respectable size, whether they were liquid or solid (over $1\,\mu$ in diameter). Therefore, they could be easily collected by filters, so that the air could be purified of them. Their microscopic examination and the analysis of their specific properties were thereby facilated.

There are, however, much finer agents of air pollution, whose origin may be mineral, industrial, or natural biological. Those pollution agents could no longer be collected by filters, because of their small size (under $1\,\mu$ in diameter). Their sedimentation rate is extremely low, as a result of a less favorable ratio between their volume and their surface, and of electrophysical phenomena which increase their lifting properties. They are the aerosols.

Aerosols remain in suspension in the atmosphere for a very long time, even without the assistance of vertical air flows. They are the constituents of dry haze. Thus, they can attain sufficient concentration to be visible to the naked eye, i.e., they can intercept radiation in the visible sector of the spectrum. Their origin is most varied, and it is approximately the same as those mentioned with reference to solid pollutants, namely: volcanic eruptions, sandstorms, forest fires, industry, etc.

In agriculture, however, there is a source of air pollutants of this size which is worthy of particular attention: viruses.

Many diseases of plants and especially of animals, both wild and domestic, are due to extremely small pathogenic agents called viruses. While they are generally disseminated by carriers larger than they are (insects, trucks, game, etc.) to which they are attached, nevertheless, in certain cases, they also appear in the atmosphere in the form of isolated protein molecules. They are then exposed to various natural destructive agents, such as, above all, dryness, and the ultraviolet rays originating from the sun. Their transportation by the wind, however, can play a considerable role in the case of epizootics, such an important role that it may render questionable most of the prophylactic measures mandated by governments. This natural transportation by the wind is the reason why the provisions presently in force will probably have to be completely revised.

11.6 Principles of Diffusion

The principal gases which affect the development of vegetation are water vapor and carbon dioxide. The principles of diffusion concerning perfect gases are therefore valid for both in the same manner.

The atmosphere is a relatively homogeneous medium, without visible and especially without impermeable partitions. Consequently, the exchanges among the different parts of this complex may take place freely. Preferably, they will follow the gradient laws. We have seen above, when we were discussing the formation of winds (see Chap. 9), that nature seeks an equilibrium among atmospheric pressures. This equilibrium is then attained by a displacement of air masses, the consequence of which is the formation of flows of wind. The situation is not different when, in two neighboring air masses, the partial pressure exerted by one of the components of the air mixture differs between the two masses. If on one side a moist air mass is present, and on the other side a drier mass, the pressure exerted by the water vapor contained in the former is relatively high, as compared with that exerted by the same water vapor – and it alone – enclosed in the latter. Nature will attempt to reestablish the equilibrium, and the passage of a certain amount of water vapor from one point (humid) to another (dry) will take place. As in the case of the wind, which displaces air masses from place to place, one notices a staggering of gaseous masses, selected by the partial pressure that they exert. This tendency to reestablish the equilibrium is called diffusion.

Since, between the two masses, there is neither a change in total pressure, nor, generally, in temperature, this displacement of the water vapor mass from one point to another is imperceptible to the eye. For this reason, in this case, one does not refer to wind, but to diffusion. The principles which govern such diffusion, however, theoretically are the same as those which govern the formation of wind. Therefore, diffusion requires that there be a partial pressure gradient, and its rapidity is a function of such a gradient.

The gradient is established as soon as, in one part of the atmosphere, water vapor is introduced or part of it is removed. The greater the difference in vapor pressure, the more considerable will diffusion be.

What we have stated concerning water vapor is also applicable, and in a similar manner, to carbon dioxide. Actually, gases behave in a similar manner irrespective of their chemical composition.

11.7 Expansion and Dilution of Plumes of Smoke

As stated above (see Sect. 11.2), a great part of air pollution, whether it be liquid, solid, or gaseous, or even by aerosols, is due to industry. Therefore, just as in the case of that produced by volcanoes, its source, compared to a world-wide scale, is a single point. It forms a plume which has its origin either along the top of the factory, at the level of the windows, or only at the top of a stack. The plume then expands, depending on meteorological conditions, and while increasing in size its concentration decreases.

Apart from the difference in the geometric size of the source, there is no notable difference between the dilution of that plume and that of the pollution due to natural sources (evaporation from seas, lakes and streams, forest fires, etc.). The same principles of dissemination by the wind, or taking into account convective air movements or the principles of diffusion, are applicable in this case.

Furthermore, in order to calculate the fallout originating from industrial smoke, there are formulas which take into account, on the one hand, the gradients, whether temperature or pressure gradients, and, on the other hand, the size of the ejected particles. Those formulas consider the amount of particles projected into the atmosphere, the amplitude, the initial velocity, and the temperature of the stream of pollutant.

Literature

Alther, E. W.: Chemisch-biologische Untersuchungen zur Fluorose des Rindes, S. 126. Diss. Landwirtschaftliche Hochschule Hohenheim (1961)

Berge, H.: Phytotoxische Immissionen, S. 100. Berlin, Hamburg: Paul Parey 1963

Cole, G.-A.: Vegetation survey methods in air pollution studies. Agronomy Journal 50, 553–555 (1958)

Dickson, R. R.: Meteorological factors affecting particulate air pollution of a city. Bull. Am. Meteorol. Soc. 42, 556–560 (1961)

Diem, M.: Staubniederschlagsmessungen vor und bei Betrieb eines Dampfkraftwerkes. Meteorol. Rdsch. 10, 145–151 (1957)

Flemming, G.: Meteorologische Überlegungen zum forstlichen Rauchschadengebiet am Erzgebirgskamm. Wissensch. Z. Tech. Univ. Dresden 13, 1531–1538 (1965)

Fortmann, H.: Luftverunreinigungen und deren Wirkung auf Pflanzen und Tiere. Z. Aerosolforsch. u. Therapie 7, 61–73 (1958)

Gorham, E.: Atmospheric pollution by hydrochloric acid. Quart. J. Roy. Meteorol. Soc. 84, 274–277 (1956)

Harrington, J. B., Metzger, K.: Ragweed pollen density. Am. J. Bot. 50, 532–539 (1963)

Havlicek, V.: L'influence des émananations d'origine industrielle et d'éléments météorologiques sur la chute de poussière. Acta Univ. Agr. Brno 4, 331–345 (1965)

Just, J.: Messtechnische Probleme bei der Überwachung von Immissionen. Angew. Meteorol. 4, 281–290 (1963)

Lebbe, J.: Résultats récents obtenus dans l'étude de la pollution atmosphérique par les gaz d'échappement des véhicules automobiles de l'atmosphère de Paris. Pollution atmosphérique 7, 316–325 (1965)

Leuschner, R. M.: Luftpollen-Untersuchungen in der Schweiz – Methoden und Ergebnisse: eine kurze Übersicht. Verh. Schw. Naturforsch. Ges. 130–134 (1972)

Lodge, R. W., McLean, A., Johnston, A.: Stock-poisoning plants of Western Canada. Canada Dep. Agr. Publication 1361, 34 (1968)

Lomas, J., Gat, Z.: The effect of windborne salt on citrus production near the sea in israel. Agr. Meteor. 4, 415–425 (1967)

Mukammal, E. I., Brandst, C. S., Neuwirth, R., Pack, D. H., Swinbank, W. C.: Air polluants, meteorology, and plant injury, p. 73. OMM. Note technique 96 (1968)

Primault, B.: La propagation d'une épizootie de fièvre aphteuse dépend-elle des conditions météorologiques? Schw. Arch. Tierheilk. 116, 7–19 (1974)

Steinhauser, F.: Messungen der Luftverschmutzung in Wien. Archiv für Meteorologie, Geophys. Bioklimatol. 10, 200–209 (1960)

Vadot, L., Belle, P., Inard, A.: Etude de la diffusion des fumées et détermination de la hauteur d'une cheminée. Pollution atmosphérique 43, 3–14 (1969)

Valko, P.: Vereinfachtes Auswerteverfahren für die Schüeppsche Methode zur Bestimmung der atmosphärischen Trübung. Arch. Meteorol. Geophys. Bioklimatol. Série B 11, 75–107 (1961)

12 Wind Measurement

B. PRIMAULT

12.1 General Remarks

In Chaps. 9 and 10, we have explained to what extent air masses are displaced as a function of various particular features, either orographic, or related to the pressure gradient. Those more or less considerable displacements of air masses are commonly called winds. The wind may have beneficial effects (supply of carbon dioxide, fecundation of certain crops) or harmful effects (damage, rainstorms, dryness, air transportation of parasites or of viruses).

In any case, the wind is of very great significance in the determination of a climate, and this has been recognized since the most remote antiquity. Hippocrates already instructed his disciples settled in a new city to observe its wind system before beginning to practice their art.

Probably, this basic principle is responsible for the most ancient monument known in our time and devoted to the observation of this phenomenon: the Tower of the Winds, built in Athens in the first century before our era.

At all times, the winds and their destructive or psychological effects have impressed man, who often deified them. Let us recall the other Aeolus in the Odyssey, or the number of poets who have praised the light breezes of spring.

At the present time, thanks to the progress of science, it is possible to obtain protection from the former in most cases, but the latter remain, and modern man feels their effects just as primitive man did.

Therefore, it is important thoroughly to know this meteorological parameter, if one wishes to adapt an agricultural concern to the natural conditions of a given place.

12.1.1 Units

In physics, the CGS system is generally used to define dimensions. In order to describe a movement, therefore, the "centimeter-second" system is used. This unit, however, is only little used in practice. As a general rule, it is preferred to use instead the "meter-second" (i.e., one-hundredfold), or the "kilometer-hour", which is very close to it by definition, but represents a certain multiple of the former (see Table 1). This latter unit is mainly the result of the speeds measured in traffic on the ground. At sea and in the air, instead, the "knot"[1] is most often used as the unit of velocity. Therefore, it is not surprising that this unit is also used to measure the wind, a meteorological element which has immediate repercussions on the rate of movement of vessels and especially of aircraft.

---

1 1 knot = 15.43 m in 0.5 min = 1 nautical mile an hour = a 1-min arc of the equatorial sexagesimal degree (1.855 m) an hour

Table 1. Comparison of units

cm s^{-1}	m s^{-1}	km h^{-1}	knots
1	0.01	0.036	0.01943
100	1	3.6	1.943
27.78	0.2778	1	0.5396
51.48	0.5148	1.8533	1

Table 2. The Beaufort scale

Force	Definition	Conversion scale (mean of the force category)		
		m s^{-1}	km h^{-1}	knots
0	*Calm or almost calm.* Smoke rises almost vertically. The surface of calm water is as smooth as a mirror	0.5	1	1
1	*Light air.* Its direction is shown by the smoke. Flags are almost still. The surface of the water shows occasional ripples	1.0	3	2
2	*Light breeze.* It is felt on the face and moves the leaves on the trees. Flags are half-flying. The surface of the water shows many ripples (wavelets)	2.6	9	5
3	*Gentle breeze.* It constantly shakes leaves and small twigs. Flags are three-fourths flying. The surface of the water shows small waves	4.6	17	9
4	*Moderate breeze.* Dust and light debris billow. Small branches and bare twigs are shaken. Flags are fully extended. On the water, the waves rise and, on large lakes, show occasional whitecaps	6.7	24	13
5	*Fresh breeze.* Bushed, leaves, and large bare branches begin to oscillate. Flags snap. The waves form white caps in places, more noticeably on large lakes than on small ones	9.3	33	18
6	*Strong breeze.* Large branches oscillate. Telephone wires ring. The wind blows around buildings and obstacles. The waves often foam and form some spindrift when breaking; on small lakes, the white-crested waves are more rare	12.4	45	24
7	*Moderate gale.* The trees bend. Walking into the wind is difficult. On large lakes, the waves break into foam and in places the crests form spindrifts in lines parallel to the wind	15.5	56	30
8	*Fresh gale.* Small branches and twigs break. Walking is difficult. The spindrifts become more numerous and denser; they are less pronounced on small lakes	19.0	68	37
9	*Strong gale.* Large bare branches break. Buildings are damaged (tiles and chimney tops are torn away). The lakes are striated with spindrift; visibility on the water is slightly reduced in places	22.4	81	44
10	*Whole gale.* Whole trees are broken or uprooted. Serious damage to buildings. The surface of large lakes whitens as a result of widespread spindrift; visibility is reduced	26.8	96	52
11	*Storm.* Much destruction. This intensity is very rare inland	30.9	111	60
12	*Hurricane.* General devastation	> 33.0	> 115	> 63

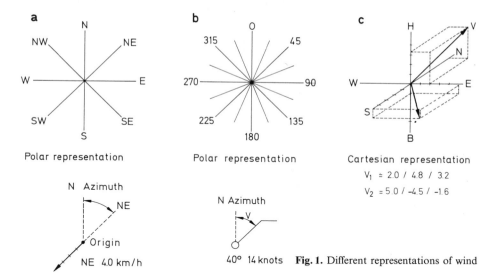

Fig. 1. Different representations of wind

When ships still used to move under sail, a British admiral had conceived a wind unit directly applicable to determine the amount of sail that it was still possible to carry sailing in the open sea with the wind abeam. That scale, therefore, is not linear, and its use ashore presents certain difficulties. Nevertheless, for a long time many meteorological services have been using this scale to define wind velocity. The name of the admiral referred to above was Beaufort and the scale invented by him bears his name (Table 2).

12.1.2 Data in Polar Coordinates

The most common method of describing a wind is to define it by two magnitudes: the direction from which it blows and its velocity. The most commonly used system of coordinates, therefore, is the polar system. The wind is then assimilated to a vector and defined by an azimuth – the one from where it blows – and a distance covered in the time unit: its velocity. The wind is thus represented in the form of an arrow, the origin of which is located at the measuring point; its axis follows the direction of the symmetric prolongation of the azimuth, and its length is proportional to its velocity (Fig. 1a).

In synoptic meteorology, the international conventions prescribe a different presentation, in which the azimuth is indicated by a short staff and the force by a conventional number of barbs (under the WMO rules, a full barb is equivalent to 10 knots, 0.5 barb to 5 knots). The velocity, although reported in units, is rounded off to the nearest 5 kn (see Fig. 1b). To define the azimuth, its direction can be given either following the cardinal points and the intermediate points (Fig. 1a), or in degrees following the nonagesimal scale (Fig. 1b).

The direct representation of a wind expressed in Beaufort degrees is difficult, in view of the nonlinear nature of that scale. Consequently, to express it graphically, the Beaufort degrees are transformed into one of the other three units, and this new velocity is the one recorded on the diagram (see Table 2).

12.1.3 Data in Cartesian Coordinates

The data expressed in polar coordinates, as described above, are only suitable to a bidimensional representation of the wind, i.e., on one plane. As soon as one wishes to introduce the third dimension, i.e., vertical convection, the system of polar coordinates is no longer suitable for a simple expression of the wind vector. In order to do so, it would be necessary to introduce into the expression two angles: the azimuth, as above, and the angle between the vector and the horizontal plane. Actually, a graphic representation of such an expression is practically impossible, and its calculation requires the application of spherical trigonometry.

For these reasons, it is preferred to use a different expression of the wind vector when the latter must be considered outside the plane, i.e., in space: the Cartesian coordinates (axes XYZ). Such an expression of the wind vector, however, requires the construction of special instruments, which we shall examine later (see Sect. 2.2.5). The wind then is no longer expressed by an angle and a velocity, but by three linear magnitudes. The latter are algebraic magnitudes, and therefore may have negative values. They are represented according to Monge's principle, by projecting the recorded values along three axes perpendicular to one another (see Fig. 1c). The axes are most often oriented along the directions North–South (X), West–East (Y), and an upward vertical (Z).

12.1.4 Wind Representation on a Climatological Scale

For practical purposes, it is desirable to be able to represent the wind that has been blowing over a long period of time at a given place. For that purpose, two general ideas can be followed: the first consists in determining a mean wind, the second in proceeding to a frequency analysis.

Mean Wind. The calculation of a mean – and, therefore, theoretically stable – wind in polar coordinates requires the prior tracing of a polygon in which the wind vectors are added up graphically. The length of the resultant of such polygon is then divided by the number of observations, while its azimuth remains unchanged (Fig. 2). Such a method of proceeding requires long work and therefore is seldom used. With records expressed in Cartesian coordinates, it is possible to determine mean winds very easily. It is sufficient to express the arithmetic mean of the data for axes X, Y, and Z, respectively, in order to obtain the mean wind. A mean wind, however, has a practical meaning only if it expresses the totality of the air flow. This requires measurements taken at very short intervals, which is seldom done.

Frequency Analysis. Since the measurement of the wind is taken regularly at relatively long intervals (every 3 h, for example, in the case of a synoptic station), a mean wind could not fulfill its intended function. Furthermore, in nature, it is very rare for a wind always to blow from the same direction. Then a mean wind could not represent the combined or successive effects of the wind. In order to do so, it is preferable to obtain a representation of the frequency and not of the mean.

Distribution of Directional Frequencies. Operating on the basis of records taken either in polar coordinates, or in Cartesian coordinates, the number of observations made in different directions is itemized. For that purpose, the horizon is divided into equal sectors (compass card with 8, 16, 32, 36, etc., points). Each observation, the azimuth of which is contained within one of the sectors, is

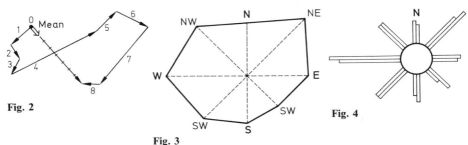

Fig. 2

Fig. 3

Fig. 4

Fig. 2. Mean wind polygon

Fig. 3. Directional distribution in Zurich

Fig. 4. Frequency distribution of strong winds in Zurich. (To the *left*, in the morning, to the *right*, at noon)

counted as an additional unit, irrespective of velocity. In order to indicate the distribution of the wind, it is sufficient to record in each direction the number of observations counted. The cases of very light wind or of total absence of wind will be counted separately and indicated on the margin (number of "calms"). Thus one obtains, around the observation point, a polygon which shows the frequency of the winds observed in each direction around the horizon (Fig. 3).

Distribution of Intensity Frequencies. The count described above gives an exact idea of the distribution of the directions from which the wind blows, but it provides no information at all concerning the force of those winds. The force of the wind, however, is the basis of the beneficial or harmful effects of this meteorological element. Therefore, a new manner of representing the wind will be sought: showing its intensity or its velocity. One will proceed in an identical manner, but noting for each group of azimuths or sector of the horizon the corresponding force. This force may be expressed by a mean value or by frequency analysis. For each principal direction, therefore, either the mean value or a histrogram will be determined. The graphic representation of these calculations will then give us the force distribution (Fig. 4).

12.2 Mechanical Measuring Instruments

12.2.1 Wind Vane (Direction), Fig. 5

If one does not consider smoke plumes, the most commonly used instrument, and the one used since the most ancient times to determine the direction of the wind is the weathercock. A vane, the shape of which can vary in infinite ways, acts upon a needle. The assembly revolves around a vertical axis. As a result of the mechanical action of the wind upon the vane, the point of the needle is always turned in the direction from which the wind is blowing. In the entire Christian world, the weathercock placed on the top of church towers has been used as a wind vane.

Such an instrument has certain drawbacks. The most unpleasant is the fact that the indicated direction is a delayed equilibrium point around the air flow. Actually, in nature, it is rare to deal with a perfectly laminar wind. Most often, the

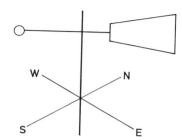

Fig. 5. Simple weather vane

wind vector changes its direction very rapidly over time. The successive gusts all blow from a direction slightly different from the preceding and from the following ones. Therefore, the wind vane will have a tendency to jump from one cardinal point to another depending on the different gusts to which it is subjected, without, however, instantaneously adjusting to them. Furthermore, the greater its weight, the greater its mechanical inertia as well. Thus, the wind vane is subjected to a more or less long oscillating movement around its axis at each wind change, even if the wind is laminar for a certain time. Therefore, there will be a tendency to build small and light devices, which will more easily follow the fluctuations of the wind.

12.2.2 Swinging Plate (Instantaneous Force), Fig. 6

The mechanical force applied by the wind against a solid force is not limited to the vertical plane. A wind vane whose axis of rotation is placed horizontally will find a balance corresponding to the force of the wind acting upon it. Therefore, the two components of the wind can be measured by placing two weather vanes with axes perpendicular to each other, but joined together on a vertical axis. Actually, the pivoting plate only indicates the effective force of the wind if it is permanently placed perpendicular to the air flow.

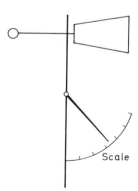

Fig. 6. Swinging plate

12.2.3 Windmill with Cups (Distance Covered), Fig. 7

In order to eliminate the requirement of keeping the wind force detector perpendicular to the air flow, it has been thought that one could place in the plane of the wind (generally, the horizontal plane) a small windmill with cups, the axis of

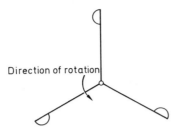

Fig. 7. Windmill with cups

rotation of which is vertical. The windmill may have three or more arms. Its angular velocity then is the function of the distance covered by the wind in the time unit, irrespective of its direction. The direction of its rotation is always the same. It is generally used to integrate wind velocity over rather long periodes (at least 10 min). As in the case of the weather vane, its reaction time (the time necessary to reach the equilibrium speed of rotation) depends on the moving mass. In order to measure gusts, small and light devices will be used. The same will apply to the measurement of light winds.

12.2.4 The Simple Propeller (Integrated Velocity), Fig. 8

In order to attenuate the movements of a swinging plate, or, more simply, to integrate the distance covered by the wind while maintaining the plane of the detector perpendicular to the air flow, it is possible to use a propeller with a horizontal axis mounted on a weather vane. The rotation speed of the propeller will be a function of the velocity of the air flow. In this case, the inertia or delay in its reaction will be felt in the positioning weather vane as well as in the propeller itself.

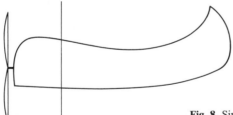

Fig. 8. Simple propeller

12.2.5 Free Propellers, Generally Tridimensional (Integrated Velocity and Direction), Fig. 9

In order to eliminate the requirement of keeping the plane of the propeller perpendicular to the air flow and directly to obtain wind data in Cartesian coordinates, it has been thought that one could place three propellers along the three principal axes X, Y, and Z mentioned above (Sect. 2.1.3). In contrast with the windmill, the propeller has the property of revolving in both direction, depending on the direction from which it is activated. Therefore, from the direction in which the propeller is turning, we obtain a distinction between the positive and negative values required by the Cartesian recording system.

Furthermore, the speed of rotation is proportional, on the one hand, to the velocity of the wind, and, on the other hand, to the sine of the angle formed by the direction of the flow with the plane of the propeller. Thus, an air flow blowing perpendicularly to the plane of the propeller will cause it to turn in the positive or negative direction depending on whether it attacks it from the front or from the rear, but at a speed directly proportional to its own ($\sin 90° = 1$). If, instead, the air flow is parallel to the plane of the propeller, the latter will not turn ($\sin 0° = 0$). By mounting on the system revolution meters making it possible to differentiate between directions of rotation, one directly obtains the wind data in Cartesian coordinates and in digital magnitudes.

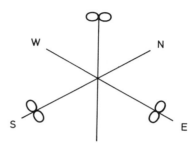

Fig. 9. Tridimensional propellers

12.3 Static Instruments

12.3.1 Weathercock Without Vane (Direction), Fig. 10

We have stated above, under Sect. 12.2.1, that a classic weathercock, i.e., one which is oriented into the wind by the mechanical action upon its vane, had a tendency to oscillate around its axis of rotation. This pendulum-like movement takes place at each wind shift, and therefore constantly or at least very often during the time unit. In order to remedy this defect, a weathercock without vane has been invented, which is based on the following principle:

A vertical cylinder with openings on both sides is exposed to the wind. The openings are connected to compression chambers (one for each of them), separated from each other by an oscillating blade. The blade is mounted on a stress gauge which activates an amplifier, and then a servomotor connected to the cylinder mentioned above. If the plane of the openings is perpendicular to the wind, no difference in static pressure between the compression chambers will be detected. Then the blade will remain in its vertical position, and the servomotor turned off. As soon as the wind attacks the cylinder mentioned above along the perpendicular to a plane other than that of the openings, a difference in static pressure between them will be detected. The blade which separates the compression chambers will automatically assume an oblique position and the stress gauge will be activated in one or the other direction. It will then start the servomotor through the amplifier, until the difference in pressure is eliminated. The cylinder, rigidly connected to the motor, will revolve around its axis. The new position of the axis of the openings corresponds to a plane perpendicular to the axis of the external air flow. Better

Master cylinder

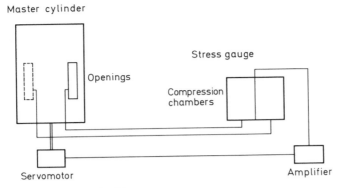

Fig. 10. Weathercock without vane

than a weather vane, such an instrument makes it possible to follow the fluctuations in the wind, even during a very turbulent movement, because its reaction is slower. Furthermore, by the control of the pressure chambers, there will be no pendulum-like movement while seeking the equilibrium position, and therefore in the recording. This constitutes an undeniable advantage in case of sudden changes in the direction of an air flow. As in the case of the weather vane, this instrument only shows movements taking place in one plane (usually horizontal).

12.3.2 Pitot Tube (Instantaneous and Integrated Velocity), Fig. 11

The Pitot tube controls two pressure-recording devices: one always exposed in front of the air flow and showing the dynamic pressure of the air; the other, parallel to the air flow and showing the static pressure. The difference between these pressures is the stopping pressure. By knowing the specific mass of the air, the velocity of the flow can be deduced. In practice, this is done by connecting the two pressure intakes to a gauge. From the resulting difference in levels, the wind velocity is obtained. The instrument is very sensitive and makes it possible to measure the instantaneous velocity of the air flow rather easily, even during gusts. In order to provide accurate recordings, however, the Pitot tube must be kept in the plane of the wind.

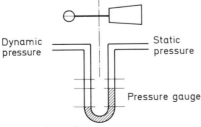

Fig. 11. Pitot tube

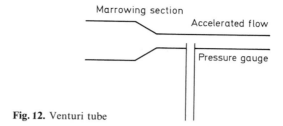

Fig. 12. Venturi tube

12.3.3 Venturi Tube (Instantaneous Velocity), Fig. 12

Its basic principle is the evaluation of the difference in pressure existing on either side of a narrow throat inserted in a tube placed in the path of the air flow. As in the preceding case, the measuring instrument is a differential anemometer and the principle applied here is that of the acceleration of the flow by a narrowing of the section of the stream. Like the preceding one, this instrument makes it possible to measure the instantaneous velocity, and must always be placed parallel to the air flow.

12.4 Physical Instruments

12.4.1 Hot Wire Anemometer (Velocity), Fig. 13

The cooling or heating of a body by contact with the air depends, on the one hand, on the temperature gradient existing between that body and the surrounding air, and, on the other hand, on the rate of displacement of the latter, or, in other words, on the wind. This property of bodies is used to measure that parameter. For that purpose, a vertically placed heated wire is exposed to the action of the wind. Its equilibrium temperature is measured. A second wire, identical to the first, but sheltered from the wind, makes it possible to measure the temperature of the surrounding air. Thus, one obtains a temperature gradient between the control wire (the temperature that any identical wire would have in the absence of wind) and that exposed to the air movement. The heat loss is proportional to the wind velocity. The vertical mounting of the measuring wire frees the instrument from its dependence upon the direction of the wind. Like other instruments, this makes possible an excellent measurement of the instantaneous velocity, but works satisfactorily only in one plane.

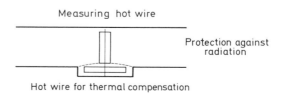

Fig. 13. Hot wire anemometer

$G_{1,2,3,4,}$ = Sound wave generators

$R_{1,2,3,4,}$ = Receivers

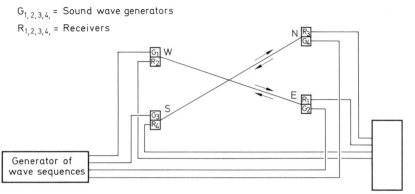

Fig. 14. Acoustic anemometer

12.4.2 Acoustic Anemometer (Velocity, Possibly Direction), Fig. 14

The rate of propagation of sound waves is accelerated or slowed down by the displacement of the air depending on whether the latter is oriented in the same direction as the sound wave or against it. If the air flow is oblique with reference to the propagation of the wave, the latter will be thereby distorted as a function of the sine of the angle formed between the two directions.

The detector is composed of two perpendicular axes. On each of those axes are mounted on opposite sides a transmitter and a receiver of sound waves (generally, different frequencies are used). Then sequences of sound waves are transmitted alternately, and the time required by each of them to reach the receiver is measured. The time difference, calculated initially on each axis, and then between the two, indicates the alterations due to the air flow. As in the case of the free propellers (see Sect. 2.2.5), the changes can be positive or negative.

Since the rate of propagation of a sound wave in the air is also a function of the temperature, it will be desirable to measure it. Since the distance between two pairs of transmitters-receivers is known, the sequences of sound waves can be used directly for that purpose.

It is possible to build tridimensional measuring instruments using this technique. For that purpose, it is sufficient to add a vertical axis to the instrument described above.

12.5 Data Transmission and Recording

12.5.1 Mechanical Instruments

Any force exerted on a body can be transmitted by a mechanical process to a recording system based either on a drum or on an endless belt. This, for example, may be the case of the position of a weather vane or of the pressure difference in a Pitot tube. Even the rotation of a windmill may, by the insertion of a reduction gear and of an eccentric, be transformed into a linear motion which can be graphically recorded. The number of mechanical systems used to record the wind, either in a polar system (direction and velocity), or in a Cartesian system (two or three perpendicular directions), is practically unlimited.

12.5.2 Electric Instruments

If the distance between the detector and the recording point is considerable (generally, over 8 m), electric signals will preferably be used to transmit the information.

For all the instruments indicating an angle (generally, a weather vane), a potentiometer will simply be placed on the measurement axis, and the desired signal will be a variable resistance.

A potentiometer can also be operated by a piston measuring the pressure difference in the Pitot tube or in the Venturi tube. Then, one will have a system making it possible to transmit a velocity measurement.

In the case of a windmill or of a simple propeller (turning only in one direction), of which one wishes to know the speed of rotation, it will be advisable to mount a generator or a magneto on the axis of rotation. The intensity of the current supplied will then be a function of the speed of rotation of the windmill or of the propeller. If it is desired to integrate the wind velocity over a rather long period (measurement of the distance covered) – irrespective of whether it is done by a windmill or by a propeller – it will be preferable to mount a switch on its axis. The switch will produce an impulse at each revolution. In the case of tri-dimensional propellers, in addition to the number of revolutions, it is necessary to know the direction of rotation. It is then essential to add an electric system based on successive impulses staggered in the same sector of rotation and making it possible to determine by electronic means a criterion to discriminate between directions of rotation. The impulses will thereby receive a distinctive signal.

With respect to wind measurements by hot wire or by acoustic anemometer, the signal produced by the detector is ipso facto an electric signal. Depending on the nature of the signal (resistance, difference in potential, voltage), it may be re-translated into units that can be used by a digital voltmeter. Such an instrument also offers the advantage of making possible the transscription of the signal on a carrier which can be directly used by a computer. In the case of an impulse signal, a digital detector will display the information either visually or on a carrier usable by electronic computer.

Literature

Anderson, L.J.: Hot-wire anemometer for laboratory and field use. Bull. Meteorol. Soc. *40*, 49–52 (1959)

Bergeiro, J.M.: Topicos meteorologicos de applicacion practica. Serie de Cartillas de Divulgacion meteorologica. Fasciculo E. Montevideo *16* (1949)

Bergen, J.D.: An inexpensive heated thermistor anemometer. Agricultural Meteorology 8, 395–405 (1971)

Bradley, E.F.: A small, sensitive anemometer system for agricultural meteorology. Agricultural Meteorology 6, 185–193 (1969)

Businger, J.A., Miyake, M., Inoue, E., Mitsuka, Y., Hanafusa, T.: Sonic anemometer comparaison and measurements in the atmospheric surface layer. Jap. J. Meteorol. Soc. Jap. *XXXXVII*, 1–12 (1969)

Ersking, S.S.: A friction-free potentiometer for determining wind directions by the use of reed contacts. Agr. Meteorol. *9*, 105–108 (1971)

Gloyne, R.W., MacSween, N., Allen, M.J.: The assessment of exposure by the rate of disintegration of a standardised textile flag. Joint Shelter Research Committee. Ministry of Agriculture Fisheries and Food. U.K. *39* (1969)

Hallaire, M., Perrin, de Brichambaut, C., Goillot, C.: Techniques d'étude des facteurs physiques de la biosphère. Institut national de la recherche agronomique. Paris. I.N.R.A. *70-4*, 543 (1970)

Platt, R.B., Griffiths, J.: Environmental measurement and interpretation Rheinhold Books in the biological sciences, p. 235. New York: Rheinhold Publ. Corp. 1964

Rutter, N.: Tattering of flags at different sites in relation to wind and weather. Agr. Meteorol. *5*, 163–181 (1968)

Sternzat, M.S.: Instruments and methods for meteorological observations, p. 424. Israel Program for Scientific Translations 1966

13 Future Development of Measuring Techniques and Research in Agrometeorology

B. PRIMAULT

13.1 Introduction

As far as one goes back in the history of peoples, the chronicles that they have left us about their development, their sorrows and their rejoicing, mention years of famine or of abundance. Until the last century, such phenomena remained the lot of limited areas, and the people living in distant places did not know about it and, therefore, were not affected by it.

With the colonial conquests, but especially with the development of maritime trade and the establishment of vast railroad networks, it is possible, if not easy, to transport large amounts of merchandise over very long distances. Therefore, it would be possible to imagine, on a continental scale, or even on a world-wide scale, a balancing between overproducing regions and the regions with insufficient production, and thereby the gradual reduction of the risk of famine.

With the introduction of the telegraph, and later of the telephone, and finally of radio and television, and the use of satellites for communications, a case of famine or of overproduction is very rapidly known all over the world. Consequently, peoples living at the antipodes may share the distress visited on a population hit by famine.

One might think that with the modern development of ships of great tonnage (displacement of over 100,000 brt), or of long-range transport aircraft with large cargo compartments, famine should no longer exist in any part of our planet.

The eradication of a disease, whether it be caused by bacteria or viruses, is easier than the elimination of famine, even if surplus production were equivalent to the deficits and this applied to the entire planet.

Actually, many considerations militate against a world-wide system of harvest estimates, and thereby of famine prevention. On the one hand, the peoples most subject to poverty do not have the physical resources to obtain the foodstuff that they need, and, on the other hand, the major producing countries have a tendency to make use of this situation, some for political, others for commercial purposes. Furthermore, such a balancing system would require the establishment, at several points in the world, of massive reserves of foodstuff, and the volume of present production is not sufficient for doing so, in addition to fulfilling immediate requirements. Finally, the lack of education that is found in vast territories and certain religious practices prevent the elimination of the present demographic explosion, which, by itself, causes those requirements to grow at such a rate that the efforts undertaken to increase production are immediately reduced to nothing.

Nevertheless, a sensible view of the situation, the transfer of scientific knowledge, and a well-understood modernization of farming methods would certainly make it possible to reduce the growing disparity between total production and the needs of all mankind.

To that effect, everybody's efforts must tend to the adaptation of crops to the *actual climate*, and agricultural research must suggest to us new varieties which are more productive and better adapted to the different climates of this planet.

Thus, our efforts must aim in two directions: the improvement of measuring techniques and planned research. Only under these conditions will it be possible to establish sufficient stocks to avert future famines.

Furthermore, it must be admitted a priori that the dietary habits of some peoples are not necessarily the best for others, and it must be relearned, in a number of regions, how to cultivate the old native crops.

13.2 Measuring Techniques

13.2.1 Old Data

In most countries, there are available meteorological data covering more or less long periods.

The adaptation of farming to local conditions (soil *and* climate) requires climatological observation which can only rely on very long series of observations.

Up to the present time, the utilization of these data has followed the pattern of classic climatology, which means extracting from them mean values valid for the entire year, the seasons, or the months. This rigid partition of time, however, does not correspond at all to the actual cycles of vegetable life. The months have a varying number of days and even the year does not have a fixed number of days, but, every four years, and also every one hundred years, we have a modification to this arbitrary cycle. Consequently, calculations made on such a rigid basis would not be able to reflect the conditions to which plants or animals are exposed, as they are placed in an environmental context in which meteorological conditions are constantly changing.

Consequently, wherever long series of observations are available, their content should be re-compiled, for the purpose of much better adapting their presentation to the requirements of agriculture.

First of all, it would be necessary to leave aside the mean values, because the more the value of a factor varies from year to year or at a certain time during the year, the more the mean value loses its significance. Actually, of what use is it to us to know that the mean precipitation in the month of February at Tillabery in Niger is 1 mm, if that 1 mm is the result of a single 31-mm precipitation which fell in a 30-year period?

A first step in this direction has been taken by the WMO, which has published the probabilities of precipitation, in five categories (CLINO), for a very large number of stations.

In agricultural meteorology, however, precipitation alone is not sufficient to define living conditions. It is also necessary to know other parameters, such as temperature, duration of sunshine, etc. In one and all cases, we need those values not in the form of mean values, but in the form of frequency analyses or tables.

Consequently, all the documentation available to us will have to be reprocessed, and, for that purpose, the use of high-speed computers is the most rational method. In each country or in each region, it is necessary to establish data banks containing *all* available meteorological information.

This will require an immense effort to check the reliability of the recorded data, and then their organization in the form of data banks, in order to have all available information at hand at any time.

When we are referring here to all available information, it should not be limited to the classic figures resulting from systematic observations of exact phenomena. Often chronicles are a not inconsiderable source of information. These chronicles may refer to a definite phenomenon of a meteorological nature or to visible or economically relevant consequences of a definite meteorological phenomenon (movements of glaciers, revenue collection, etc.). It is not necessary for the phenomenon to have taken place where it has been recorded, because it may be the result of facts that occurred in regions very remote from where it has been observed. As an example, let us mention the height of the Nile flood. The ancient Egyptians prayed their gods to grant them sixteen cubits, which for them meant an abundant harvest. But the number of cubits reached by the Nile flood depended on the precipitation in Abyssinia. Consequently, it would be possible to conduct a calculation of the frequency of probability of high or very low precipitation on the high plateaux of Abyssinia by using the data available to us concerning the height of the Nile flood in ancient times.

13.2.2 Observation Networks

In many countries, the density of existing observation networks is insufficient to give an accurate picture of the prevailing climatic conditions.

Certainly, it is possible, on the bais of existing series of data, to perform extrapolations in order to obtain acceptable data on a mesoclimatic scale. The unreliability of such extrapolation, however, rapidly increases with distance, especially in regions of uneven terrain. Consequently, it is of the greatest urgency to establish more numerous observation stations, in order to compare the data recorded there with the long series that are available to us. The establishment of new stations, and especially the extent and diversification of the instruments allocated to them must match the variability of the phenomena to be observed.

While temperatures are subject to a normal distribution pattern (Gauss's bell curve), the same does not apply to the other meteorological factors. The latter cannot be easily extrapolated by simple statistical operations. Direct observation is still the best method to obtain values for comparison, and, consequently, to be able to extend the validity of existing data series to wider areas.

In many cases, observers are available on the spot, and it will be sufficient to train them, because the observation itself does not require great knowledge. Thus, it will be possible greatly to reduce the cost of setting up networks.

In the less populated regions or in those of difficult access, only automated stations can bring a solution to the existing problems. Furthermore, such stations offer the advantage of being capable of operating just as well by day as by night, which means that, without worrying about local conditions (for example, availability of observers), it is possible to increase the volume of information.

Such information will have to be centralized, checked, and processed. In order to do so, mail or telegraph communications lines must be planned to reach the observation stations. If the station is automated, it will be possible to transmit the information immediately. Otherwise, it will be transmitted after a more or less long delay, which may range from a few minutes to several months.

When it is planned to complete or to modernize a network, it would be desirable to design at the same time a rapid data-processing system, which means the use of very powerful computers in order to reduce the sources of errors. Furthermore, it will be possible immediately to enter the information into a data bank as soon as it has been checked, because *checking remains essential.*

13.2.3 Parameters to be Measured

When it is planned to establish any part of an observation network, there is the temptation to increase the number of parameters to be observed, in order to make use of the observer and of the communications network as "rationally" as possible. Actually, experience has shown that a limited number of parameters in many cases was to be preferred to a profusion of them. For this reason, it would be desirable to be able to limit those parameters to the principal ones, which generally are precipitation, temperature – in the air as well as in the ground – humidity, and radiation. If it is feasible, further classic factors can be added to them, such as wind, cloud cover, soil moisture, etc.

In certain areas, precipitation is the variable factor par excellence. In that case, the number of stations observing precipitation alone should be greater than that of the stations where the other meteorological parameters are recorded.

In establishing such an observation network, one should never forget its intended purpose, which is to increase the production of foodstuff, or to increase the production of vegetable products usable by man (textiles for clothing, wood for construction or heating, etc.). Since in all these cases nature shows us to what extent the vegetation reacts to variations in meteorological factors, phenological observation should not be omitted at any place. It would even be desirable to have a phenological observation network larger than the climatological network itself, with respect to the number of observation stations.

Together with the phenological observations, which, by their very nature, are essentially qualitative, it is desirable regularly to proceed to quantitative observations of natural or cultivated plants (lengthening of the stems, number of flowers per sprout, dry matter, etc.).

13.2.4 Local Adaptation

Such a program requires massive capital investments to be usefully implemented. Such funds are often unavailable. Thus, the programs must be adapted to the resources and, therefore, reduced.

The politician (government official or member of parliament) most often is not in a position to decide in which direction such reduction must be performed: whether it is in the number of observation stations, or in their equipment, or even in the measurement program. In our opinion, it is preferable to have a smaller number of well-equipped stations – quality and not number of instruments – which are well supervised, rather than a considerable number of isolated posts where the observer is left to his own devices and, consequently, will soon grow tired of an activity of which he can see neither the specific purpose nor the reason.

13.3 Research

13.3.1 Need for Research

In consulting the literature, it is astonishing to see the number of studies which have been undertaken and successfully completed in the field of agricultural meteorology. Thus, one would be tempted to believe that at the present time research is less necessary than observation.

Certainly, it is not possible to carry out research in agricultural meteorology if, at the same time, accurate and voluminous meteorological information is not available. However, most, not to say all, of the research carried out is not transferable, i.e., if it was possible to obtain quantitative results in a certain country by a certain method or a given model, that method or model cannot be used successfully in any other place, unless soil and climate conditions are very similar to those prevailing in the first place.

It is absolutely wrong to believe that it is possible to use a good model for calculating agricultural production or for struggling against parasites which has been successfully tested in one country, by simply applying to the production or antiparasite campaign in another region. Such a method of operation can only lead to failure. A repetition of such failures will raise doubts as to the soundness of agricultural meteorology, or, more generally, of applied meteorology in its broadest meaning. Consequently, before applying any model, it will be necessary to check its soundness.

13.3.2 Planned Research

Any plant or any animal living in the open is affected by meteorological factors.

Their influence may be related to the development of crops (rapidity and quantity) or to their quality, but also involves the evolution of the populations of predators. In any planned research, therefore, it will be necessary, as a first priority, clearly to define its intended purpose, i.e., to know whether that research is intended to calculate the amounts produced, the quality of the products, or the date on which they will become available. In all cases, the two principal factors – soil and climate – both play their roles. In the case of quantity and quality, farming operations, including fertilizing, often predominate over the meteorological effects, while in the case of the time of availability the meteorological influence is always predominant.

Quantity. In the examination of the volume of production, in addition to the meteorological factors, it will be necessary to take into account the nature and texture of the soil, as well as the fertilizer spread over it – amount and nutritional value. The plant variety, or even a strain within the same variety, may considerably alter the calculations and, therefore, the harvest forecasts. The latter generally will be of a negative nature, i.e., by means of appropriate meteorological considerations, it will be possible to predict the deficits in production due to a given meteorological accident occurring at a specified time during the development of the plant. Such methods of estimating make it possible to detect in advance the areas in which scarcities will develop. On the other hand, it is very difficult, or even impossible, to estimate by this method where surpluses will occur and their extent. Therefore, its general application to a world-wide system of yearly distribution of total food production is highly questionable. If, on the other hand, reserves wisely distributed over the world are available, it will be possible to draw from them according to the needs thus determined in advance. Later, the surpluses will be used to replenish the reserves thus depleted.

Quality. Depending on farming practices and on the meteorological conditions prevailing during the growth cycle, the quality of the products varies greatly. That quality may involve their texture, their color, their nutritional value, and even the possibility of preserving them.

In planned research, an attempt will be made to determine at what time the plants are most susceptible to accidents caused by given meteorological parameters. Then the producers and the market will be informed accordingly, so that both may take adequate measure in case of occurrence of the suspected accidents.

Timing. The more a crop is perishable, the more important it is to know the time of its ripening. The transportation system, the market, the processing plants, as well as the consumer, must be prepared to absorb the production, so that nothing will be lost.

Consequently, part of the research in agricultural meterorology must be directed to clarify the problems of growth and ripening.

Such research is not only useful for the movement of perishable crops, but is also used to prepare production probability charts, especially in countries with highly variable climate, such as mountain countries.

Special attention will therefore be paid to the study and observation of the ripening rate of farm products as a function of weather changes. Actually, farming methods and fertilizing affect those phenomena less than the variations in meteorological conditions.

Predators. Scientific research in agricultural meteorology must not be limited to production alone. It must take into account the biological factors which limit that production, namely predators, whether they be insects or fungi.

These two kinds of living creatures, just as the host plants or animals, are subject to the meteorological conditions of the time, and, in the longer run, to climate fluctuations. Consequently, their individual development and the evolution of the populations are governed by meteorological conditions.

Therefore, also in this case, it is necessary to carry out planned research, for the purpose of knowing to what extent predators are likely entirely to destroy or at least to reduce the harvests.

The struggle against these predators, whether it be chemical or integrated, can only be carried out satisfactorily by knowing the weather pattern and its impact on the predator to be contained.

Furthermore, this is not a question of destroying a predator by means of an appropriate treatment, but rather of maintaining its population within economically tolerable limits. Inconsiderate treatment always leads to the selection of resistant populations, which multiply and require more and more violent interventions, which may become harmful to man himself.

In most cases, the procedure will be similar, if not identical. It will be possible to follow the same patterns and to use the same basic data.

Furthermore, this research must be used first of all to estimate, on the basis of climatological data, the risks of predator and epiphytic infestation, respectively. It will then be necessary retrospectively to calculate the time of the onset of the disease and its extent (number), and to compare it with the phenological stage reached by the host, in order to estimate whether there is a risk of unacceptable economic damage or not. If there is such a risk, the required number of interventions will be obtained. Since the cost of each of them is known, their economic impact will be deduced from it.

An annual application of the results of this research will make it possible to inform the operators in the field in time of the need for treatment (whatever the treatment may be: chemical, biological, or integrated). By proceeding knowledgeably, the interventions will be rendered more effective.

Fungi. The effects of meteorological factors on the development of fungi have been known for a long time. Much remains to be done, however, to be able to transfer the present knowledge from region to region.

The crucial moments with respect to the effect of the weather on the appearance and propagation of an epiphytic infestation are the formation and the hatching of the spores. Often the entire harvest depends on those two periods of time, which may be very brief (only a few hours in certain cases).

Insects. Just as in the case of fungi, insects in their development are subject to the effects of weather patterns. In this particular case, however, the evolution of each individual, and therefore of the entire population, is not so dependent on particular conditions. It is more a question of rate of growth than of an absolute prerequisite for development. Therefore, the meteorological thresholds are less imperative. Nevertheless, defining them should not be neglected for this reason.

13.3.3 Models

When one studies the literature devoted to the description of models developed to simulate the growth of plants or the volume of production, generally one is astonished at the complexity of the formulas that are proposed. The same applies to the models concerning the development of insects and fungi. It might be deduced from this that such complexity is necessary, because nature never reacts in a simple manner. In many cases, however, the impact of certain parameters is negligible, in economic terms.

Furthermore, these models often involve meteorological parameters which are not observed by the ordinary networks. This is the result of the fact that those models are prepared by especially well-equipped research stations.

If the model uses one or more parameters or time intervals not included in the general program of climatological stations, it has no chance of practical application.

The purpose of using the models, however, is essentially practical. Consequently, the simpler the model, the more limited the number of parameters appearing in the formula, and the greater its chances of being applicable and applied in practice.

Often there is an excessive emphasis on the search for accuracy. Actually, in agriculture, the necessary level of accuracy is much lower than in the exact sciences, and particularly in physics. Consequently, a variability of $\pm 5\%$ in the results is already highly significant in agriculture. Therefore, it is not necessary to obtain less than $1\%_{00}$ significance to have a valid result. The possibility of variations both in quantity and in ripening date is proportionally much greater in nature.

The reason for the complexity of the proposed models is to be attributed, on the one hand, to the complexity of the installations available at the research stations, and, on the other hand, to the presence of a high-performance computer. While the computer is a very welcome working tool in research in agricultural meteorology, it may easily act as a trap for the researcher, by enabling him to diversify his model almost infinitely to obtain ever-smaller increases in the accuracy of the results. However, as we have seen, such accuracy generally is not necessary in practice, and in many cases it is even a hindrance.

Summarizing:

The simpler the models, the more practical they will be.

In any case, a model should never involve a parameter that is not measured everywhere, or an unusual interval between observations.

13.3.4 Adaptation

There is often the temptation to believe that a model which has been successfully tested in one place can be applied without further work in another. Generally, such a line of reasoning is not correct, because climatic and pedological conditions often are very different. Therefore, it is very dangerous to attempt to transfer models directly from place to place, irrespective of the nature of the subject of the simulated development (fungus, vegetable, or animal).

It might be believed that such transfer may be carried out better if the model has been prepared in the laboratory rather than in the field, because there the various actions of meteorological or pedological factors are better defined, better known, and better controlled.

Nature acts not only by its continuous action, but mainly by the ever-changing fluctuations in environmental conditions. Thus, a certain growth is caused less by the mean temperature of one day than by the temperature variation between day and night. The variation stimulates the plant, and the stimulation is necessary to enable the plant fully to take advantage of the mean temperature observed. Thus, if the temperature does not vary during the day–night cycle, the result obtained in a controlled-climate chamber will not correspond at all to the result obtained under apparently similar conditions, but in real life in the field. For this reason, the models prepared on the bases of data collected in the field have a greater practical value than the model resulting from laboratory observations.

Furthermore, all models, no matter how sophisticated, and even if they are based on a great number of physical axioms, always contain an empirical part. That empirical part greatly reduces the possibility of applying them to other areas. Thus, an evaporation model prepared in a humid region of the world cannot be used without adaptation in a dry region.

Consequently, whenever one wants to use a model taken from the literature, it will be necessary first of all to know where the model was developed and how. Then it is essential to check whether growth, soil, and climate conditions in the new region are similar to those in the original region. Thus, any model will first have to be tested and adapted, if necessary, before being used in practice in new areas.

Briefly, the simple transfer of a model from one place to another can only lead to a failure.

13.3.5 Adaptation of Farming Methods

Any revolution, whether it takes place in the political, industrial, craft, or agricultural field, only leads to the destruction of the existing structures, without making any improvement in them. Only after a painful reconstruction work on the ruins is a new equilibrium attained.

Consequently, an evolution is always preferable to a revolution. Agriculture is no exception to this rule.

The expected and desirable improvement in production will not be obtained by simply importing methods successfully tested elsewhere. Each new method must first be patiently tested, improved, and adapted before being extended to large areas. The sudden introduction of new farming methods, the importation of great numbers of new machines – even ultramodern ones – of new species – whether they be animal or vegetable – or of massive amounts of fertilizer, without being preceded by a thorough study, will not lead to the desired results. Actually, it is extremely rare to obtain good results by simply transferring methods, machinery, or species, even if they have been successfully tested.

In many developing countries in the world, research in agricultural meteorology must focus on this adaptation of techniques from elsewhere. The adaptation of such techniques must take into account not transitory meteorological conditions (the term transitory may extend to several consecutive years), but the real very long-range variability in climatological conditions. Certain burning failures of industrial farming in Africa are a formal proof of this fact.

Let us not wait until our life is threatened before analyzing the cause of these failures. They are not only due to political or commercial causes; very often they are the result of bad scientific analysis.

13.4 Conclusion

From time immemorial, man has been attempting to predict his future. Even now, there are still legions of fortune-tellers and authors of horoscopes. We do not in any way wish to join their cohort.

106 Physical and Meteorological Principles of Agrometeorology

We believe, however, that, in the more or less near future, the application of the methods of modern agricultural meteorology is the best way to obtain an increase in crops and to provide the peoples of our world with a means to fight poverty. Famine is a scourge that can be checked, or even eliminated; but to do so it is necessary to be able to rely on two essential factors: man's intelligence further to explore the mysteries of life and to make use of the knowledge thus acquired; and the will on the part of governments to act *together* for the welfare of all mankind.

Revolution has never been a solution to national or international problems: in agriculture, perhaps even less than in other fields.

Part III   Applied Agrometeorology

1 Agricultural Climatology

J. SEEMANN

1.1 General

Agricultural climatology is a branch of science that concerns itself with the influence of climate on the cultural conditions of agricultural plants, animal husbandry, the occurrence of detrimental influences (both biological and weather-conditioned) and especially on agricultural operating methods in general. The results of appropriate investigations represent the collective term "planning information". The method is almost exclusively a statistical one. Originally, one used almost exclusively the so-called climate values of general "climatic statistics", e.g., daily average figures, decade averages, monthly averages, or the deviations of the individual climatic elements (shorter-period averages or totals), from averages from extended periods. It is common in climatic statistics to use the annual divisions derived from the general calendar for the time periods. It was found, with increasing experience in the field, that an application of the usual climatic statistics offered only very limited application possibilities. In the comparison of climatic elements over shorter time periods, with average values collected over many years, it is usual in general climatology to indicate a positive deviation from the long-term average, in the case of precipitation as "too wet", and in the case of temperature, as "too warm", while in the case of negative deviations one speaks, correspondingly, of "too dry" or "too cold". Application of such concepts are not appropriate to agricultural climatology, since they cannot, in this way, sufficiently characterize the actual situation. It can easily be seen from the Tables 1–3 in the chapter Water requirements of plants that precipitation cannot simply be evaluated on the basis of comparisons with long-term average values, if for no other reason than the soil conditions alone. Similar considerations apply to the use of temperature averages or corresponding temperature sums. Temperature sums or averages of various locations or terrain areas can be composed of very different individual values, each of which has differing biological effects, especially if they extend over longer periods of time. This author (Seemann, 1951) has been able to show in an investigation on the starting time of fruit-tree blossoming in its dependence on temperature totals, that, for example in Germany, locations around 50° latitude "require" a far higher temperature sum for the blossoming of fruit trees, then, perhaps, in areas about 54° latitude. Initially, this appears to be unreasonable. Upon closer examination of the corresponding temperature sums, it is shown that these very sums are composed of different individual values. If one wants to consider general meteorological or climatic elements as ecological factors, one will learn to evaluate critically direct transfers of simple climatic statistics, and find a more suitable form for describing climatic conditions for agricultural climatological methods. Data on the frequency of the occurrence of certain meteorological phenomena will generally play an important role. For example, frequent dry periods, the frequency of Spring- or Fall-frosts, duration of the frost-free period, slow warming in Spring, duration of the vegetation period, and similar occurrences, make the climate, for example, a limiting factor in agricultural operating methods of certain regions. Hashemi (1974) has attempted to establish agricultural climatological zones in Iran, according to similar considerations.

In order to answer certain questions, it will be necessary to form combinations from various climatic elements with the aid of models, such as, for example, in the question: "how frequently is it possible in individual areas, to harvest grain with a combine with 20, 18, 16, or 14% moisture content" (Heger, 1973). For this, it is necessary first to establish a model for the determination of the drying processes in respect of the applicable course of the weather. In similar manner, Pfau (1972), has evaluated agricultural climatological questions in connection with optimal hay harvests in open country or with artificial drying of hay.

A portion of the problems has already been indicated with these indications. One other important aid that is of special importance to agricultural climatology, phenology, must still be pointed out. Phenology plays an important role in two ways in agricultural climatology. On the one hand, data on phenological occurrences indirectly contain certain weather conditions, on the other hand, phenology offers the possibility of separating individual plant development phases in biologically useful manner. From this results, for the purposes of agricultural climatology, a "natural" calendar in contrast to the "astronomical" one. Before discussing the application of phenology in agricultural climatology, it is proper, initially, to discuss phenology more specifically.

1.2 Phenology

In respect to phenology, it is necessary generally to differentiate between plant phenology and animal phenology. Plant phenology represents the science of developmental occurrences in plants. The same applies to animal phenology. While plant-phenological observations are being carried out in many countries, animal-phenological observations are available only very rarely. For this reason, as well as because of the fact that special importance is attached to plant phenology in agricultural climatology, we will here concern ourselves only with plant phenology. Corresponding with common usage, this will in the following be simply designated as "phenology".

In respect to network observations in phenology, three groups of plants are differentiated. Wild-growing plants (trees, bushes, and herbage), agricultural plants, and fruit-woods. Which of these plants is ultimately selected for observation in various areas, will differ, depending on climatic regions; the plants in the tropics will be different from those in moderate climates. It is however to be recommended that plants be included that are distributed as widely as possible, so that one will be able to compare the development of these plants over large areas. This applies, for example, to wheat. Schnelle and Volkert (1974) have established an observational network with their International Phenological Gardens that extends from 41° to 63° latitude and from 0° West to 27° East longitude. Here the observations are limited to woods that all originate from the same original plants.

In the selection of plants for phenological observations, plants should be included, if possible, in which several development conditions can be simply observed in order to be able to determine also the duration of the time periods between individual development phases. In the case of grain these are, for example, the following phases: seeding, germination, leafing, shooting, spiking, blooming, juice ripeness, and yellow ripeness. Such sections are especially useful for "calendar" limitations in various agricultural meteorological problems. Even the knowledge of the duration of vegetative and generative phases that are to be determined in this manner is of considerable importance. However, the duration periods between developmental stages of two different plants can also be utilized for certain purposes, e.g., the period of time between the flowering of apple trees and the blooming of rye. Schnelle, in his book *Plant Phenology* (1955), has thoroughly reported on the methods for phenological observations and their evaluation.

In the evaluation of phenological observations, one can proceed approximately according to the following points:

1) Comparison of developmental phases of the same plant in the same location, from year to year.

2) Comparison of similar phenological occurrences in one and the same plant in various locations (geographical distribution).

3) Duration periods between various starting phases in the same location, e.g., number of days from the germination of a type of grain to blooming, and between blooming and yellow ripeness in one year in comparison to other years.

4) Comparison of the duration periods between phenological occurrences during one year at various locations (regional comparisons).

5) Absolutely earliest and latest, as well as most frequent start of certain phenological occurrences from observation series over many years.

Map presentations are generally used for regional comparisons. By means of lines connecting equal phase starts or equal periods of duration, so-called isophanes, those areas are delineated on the maps under consideration of the orography, in which a certain growth phase occurred during a certain time period. This time-delineation usually takes place at intervals of 10 days. It is common in phenology not to indicate the starting date of a certain phase with the calendar date, but rather to express this date by the number of the day from the beginning of the year.

The regional differentiation in respect to the beginning of certain developmental phases of plants is determined to a high degree by the geographical latitude and the elevation. In Germany, the average term for apple blossom at 54° latitude is approximately 30 days later than at 48° latitude. Such a delay from South to North can, in France, be determined as being an average of 3.2 days per 100 km. Based upon 10 years of observations of various woods, Angot (1882–1892) has calculated an elevation-conditioned delay of the bloom in France of 3.9–4.3 days per 100 m elevation difference. Minio (1932) determined an average of 4.4 days for the Dolomites and Fiori (1905) 3.0–3.6 days for the Appenines per 100 m. In the German central mountains, a delay of the blossoming of apples of up to 5 days per 100 m takes place.

The differentiation of data for the developmental phases in the same location are determined exclusively, from year to year, by varying weather conditions. Based upon a delayed beginning of spring-warming, for example, the beginning of the bloom of fruit-woods can occur as much as 20 days later in contrast to years with early warming. It is understandable that regionally conditioned differences (latitude, elevation) can be expanded even more by weather conditions, if one considers that the course of the weather can, even on a regional basis, still be very different.

1.3 Phenological Climatology

As has already been mentioned, phenological developmental phases, or the duration between two phases always reflects the complex course of wheather conditions or – considered over longer periods of time – that of the climate in the location of the observed plants. The temperature course will primarily affect the temporally differing occurrence of a developmental phase and, to a certain extent, so will the precipitations. The presentation in the form of a map of the beginning of a developmental phase therefore results in a picture of the regionally differing course of temperature throughout a certain developmental period of the plants.

Figure 1 presents the average beginning of the bloom of apple trees in the Federal Republic of Germany. This phase, established on the average over many years, occurs in the geographical latitudes during the time before April 20 until after May 20. The most early regions are on one hand in the South-West and, on the other in the West (in those parts of the country to the northwest of Bonn). Apple blossoming begins latest in the mountainous areas and in the North. Such a phenological map, therefore, presents the areas with differing early spring-warming and expresses the differences in the regional course of the effective temperature. Schnelle (1965) has published corresponding phenological maps for all of Europe. The period between Spring and Fall is completely determined by means of the phenological phases of the average beginning of planting of summer grains, apple blossoming, winter wheat spiking, winter wheat harvest and winter wheat planting. By means of a map of the average time periods between summer wheat planting and winter wheat planting by Schnelle (1970), the great differences in the duration of the vegetative period can be recognized. The agriculturally useful period, therefore, lies between less than 100 days in the Northern USSR and in Scandinavia to more than 260 days along the Mediterranean coast and in the Iberian and French Atlantic Coastal Regions. The map, therefore, provides knowledge concerning the average periods of agriculturally useful times, something that would be only very difficult to obtain on the basis of climatic values.

Naturally, shorter vegetative periods can also be determined by means of phenological phases. Thus, Seemann (1949), determined the average duration of the vegetative period for intermediate crops after (the harvest of) winter rye in Northwest Germany. The beginning is established as the time 5 days after the harvest of winter rye. Since, for the determination of the vegetative period, no useful phenological phases exist, the climatological criterium was used: the end of the occurrence of daily temperature averages of 10 °C. Such combinations of

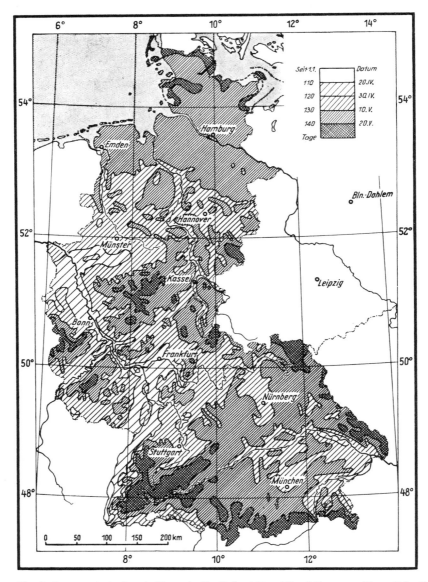

Fig. 1. Average start of apple-bloom in the Federal Republic of Germany. (From Schnelle, 1953)

phenological and climatological values are often applied in agricultural climatology. If one wishes, for example, to determine the frequency of frost damage during fruit-wood blossoming for a certain territorial area, then an appropriate combination of flowering dates with temperatures below 0 °C is quite logical. Seemann (1950) has carried out such determinations for several areas in Germany.

An entire series of other phenological-climatological combinations has been presented by Freitag (1965) in the publication *Studies on the Phenological*

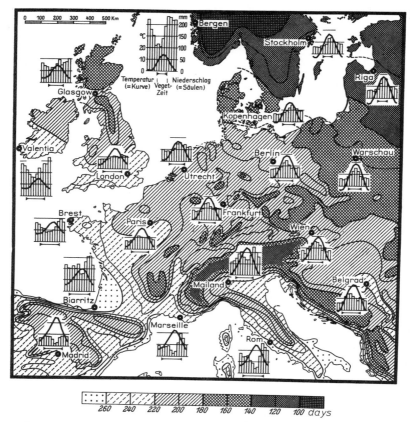

Fig. 2. Climate and vegetation period in Europe. Number of days between seeding of summer grains and winter wheat

Agricultural Meteorology of Europe. Further examples for the synthesis of phenological and climatological observation results were presented by Schnelle (1961) in his publication *Agricultural-Phenological Annual Course in German and European Agricultural Regions.* It is from this publication that Fig. 2 was obtained. This map presents, simultaneously, the duration of the total vegetative periods in Europe and the climatological characterization of various locations. The total vegetative period was calculated from the difference between the sowing of summer grains in Spring and the sowing of winter wheat in Fall. In addition, the annual course of air temperature and the monthly amounts of precipitation are also entered for a number of locations. The duration of the vegetation period is, again, entered for each station especially by a horizontal line below the precipitation columns, so that the precipitations and the temperature conditions can be compared for the vegetation periods in the varying locations. Of the stations mentioned there, Bergen in Norway and Rome (Italy) show, for example, among others, the greatest differences. Bergen is characterized by a short vegetative period, low temperatures, and high precipitation, while Rome, on the other hand, stands out by a long vegetative period, high temperatures, and small amounts of precipitation (especially in summer).

As a conclusion for these discussions, a cartographic evaluation by Schnelle (1953) is to be pointed out as an example for a phenological – agro-economical combination. This map points to its base of the beginning of the wheat harvest in monthly steps. In practice, wheat is being harvested somewhere in the world during every month. This phase migrates with a certain rhythm, initially, during the months of February to September northward from the equator, and then from September to February southward from the equator. In addition, points indicate the distribution of the growing areas of wheat. Finally, the map also contains information on amounts of wheat that reached world trade during the years 1948–1950 and that were transported overseas.

It was intended to show by means of these examples what possibilities exist in handling agricultural climatology by means of phenology. At the present time, there is not yet any agricultural climatological treatment for entire countries or for continents. For this reason, what has been written here concerning agricultural climatology can only serve as a suggestion toward corresponding work.

Literature

Angot, A.: Study on the progress of vegetative phenomena and the bird migration in France. Ann. Bur. Cent. Meteorol. (1882–1892) (French)

Fiori, A.: Phenological observations in respect to actual altitude at Veldarno. Nuovo. G. Bot. Ital. *12* (1905) (Italian)

Freitag, E.: Studies on the phenological agricultural climatology of Europe. Ber. D.W.D. *98* (1965)

Hashemi, F.: Agroclimatic zoning of Iran. Iran. Meteorol. Org. DCC *630*, 2515 (1974)

Heger, K.: Estimates of the weather risk on grain harvesting with the combine under utilization of an agro-meteorological model. Ber. Landwirtsch. *51* (1973)

Minio, M.: Quelques valeurs du retard des floraisons dû à l'altitude triées des séries d'observations italiennes. Acta phänologica *1* (1932)

Pfau, R.: Available field working days for various methods of feed-harvesting. Ms-print No. 38 of the Curatorium for Technology and Construction in Agriculture, Frankfurt/Main 1972

Schnelle, F.: Contributions to the phenology of Germany. Ber. D.W.D. *1* (1953)

Schnelle, F.: Phenological world map.: beginning of the wheat harvest. Geogr. Pocketbook (1953)

Schnelle, F.: Plant Phenology. Leipzig: Akad. Verlagsanstalt Geest and Portig 1955

Schnelle, F.: Agro-phenological annual course of the german and European agricultural regions. German Geographic Meeting. Wiesbaden: Fr. Steiner 1961

Schnelle, F.: Contribution to the phenology Europe. I. Ber. D.W.D. *101* (1965)

Schnelle, F.: Contribution to the phenology Europe. II. Ber. D.W.D. *118* (1970)

Schnelle, F., Volkert, E.: International phenological gardens in Europe: the basic network for international phenological observations. In: Phenology and seasonality modeling. Lieth, H. (ed.) Ecological studies, Vol. XIII. Berlin, Heidelberg, New York: Springer 1974

Seemann, J.: The climatic-phenological conditions for intermediate crops after winter rye in North-West Germany. Hamburg 1949

Seemann, J.: Late Frost danger in NW-Germany under consideration of fruit-growing. Z. Baumschulen (1950)

Seemann, J.: Are values of general climatic statistics useful for agro-meteorological investigations? Arch. Wiss. Ges. Land- Forstwirtsch. (1951)

2 Agrometeorological Statistics and Models

J. SEEMANN

2.1 General

Great importance has been assigned in past decades to statistical investigative methods in agricultural meteorology. The desire to be able to understand the relationships between weather and biological developments (plant development, yield, incidence of pests), has made the application of statistics necessary to start with. In this respect, work with statistical problems was, and still is today, a special working branch in agricultural meteorology. The fact that the success of such efforts did not always meet the expectations completely may well have been inherent in the complexities of the biological processes and the magnitudes of the influencing values that were not satified by the statistical methods that have been applied. This does not mean, however, that great importance is no longer to be assigned to statistics in agricultural meteorology today. On the contrary: in order to maintain and increase agricultural yields, one requires today especially detailed knowledge of the processes that influence the formation of yields, especially of the meteorological influences. Added to this are increasing knowledge of the decisive physiological and physical factors for plant material production, which, in combination with the advances in calculation technology, now offer the possibility of usefully applying statistical methods by way of the biophysical model.

Statistics have been applied for a long time in all branches of science, technology, and economy. The (available) statistical methods become ever more versatile. It is obvious that not all statistical methods or even the basics of statistics can be treated within the framework of this chapter. In this, the reader must be referred to the appropriate literature. Basics on statistics in general are available, for example, from the publications of Fischer (1948) and Mood (1950). One can gain orientation on variance analysis from Ahrens (1967) and Scheffé (1959). Basic on biological and agricultural statistics are present in the books of the following authors: Finney (1952), Mudra (1958), Snedecor (1948), and Weber (1967). In all of these voluminous publications, there are numerous references to work concerning specific statistical problems of other authors. In this context, it is possible only to discuss in a very general manner, the statistical methods that are applicable to agrometeorological purposes. For statistical work in agrometeorology, the following investigative methods are of importance: variance analysis, regression analysis, correlation calculation, the physicobiological model, and the simulation model.

2.2 Variance Analysis

Variance analysis can be applied to every type of experimental result that was obtained by quantitative measurements. It was originally developed for the evaluation of field experiments by Fisher (1948). In more recent time, variance analysis has found application in many other areas. Today, it represents one of the general statistical methods. Often several factors contribute to the origin of differences between statistical measurement values. The yields per ha of various grains can be caused, for example, by differences in the varieties themselves, by the weather conditions, fertilization, etc. In such a case, a variance analysis must first be carried out, before the examination of the average value differences – e.g., the average variety yields – is performed. Variance analysis, in this case, is a preparatory step to the examination of a hypothesis on the difference between a series of average values. Variance analysis primarily serves the dependable determination of the experimental error. Experimental error is that portion of the total variability that cannot be further analyzed, that is composed only of incidental influences.

These brief indications must suffice here. As concerns the execution of variance analysis, reference is made to the appropriate literature (especially Ahrens, 1967; Scheffé, 1959).

2.3 Regression Analysis

In the execution of experiments, one determines measured values (x), that are in a certain relationship to another variable (y). If these values are entered as points in a coordinate system with coordinates X and y, then a line can be established without difficulty, if the y-values increase essentially proportionally to the rising x-values, from which the points deviate only slightly. This line is called a regression curve. If this increase of the y-line is not strictly proportional, or if varyingly great y-values are assigned to an x-value, then a "point cloud" develops from the entering of the obtained measuring points. An example for this is given in Fig. 1. In this case, it concerns investigations of the problem as to how the quality of wines is influenced by varyingly great energy inputs from solar radiation during the vegetation period. In the ordinate, the wine weights (in ° Oechsle) of grapes are entered and in the abscissa the correspondingly determined intensity of solar radiation. The term Oechsle degree indicates the data of the values of the specific weight of the new wine that are to the right of the comma, e.g., 75° Oechsle = 1.075 specific weight. The disuniform measuring accuracies for the y-values can be caused from measuring accuracy or from the fact, that the effect of a (certain) factor is present in all determined growth results (perhaps of a meteorological element), but that the investigation results are superimposed by other influences (different soil conditions, a different location, care, etc.). In this case one will not be as successful in visually placing a regression curve into the point "cloud" in such a manner that it shows the average dependency of the y-values on the x-values. In order then to determine a regression curve, one will require a regression analysis. Accordingly, the regression analysis is always to be applied when the dependence of two variants upon each other is to be tested, in which one of the variables is an independent variable, whose values are determined prior to the experiment.

A regression analysis is subject to the following conditions:

a) For each value of the independent variable, the dependent accidental variable y must be normally distributed.

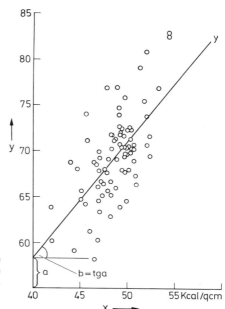

Fig. 1. Regression curve for the relation between the new wine weight (in °Oechsle) of grapes and the energy take-up from solar radiation during the vegetation period $(Y = 1.3x + 6.7)$

b) The average value of the y-values in the basic total must be a function of $x (y = f(x))$.

c) The form of the function for the regression analysis must be known.

d) The variance of y must be constant or proportional to a known function of x.

The position of the regression curve in the "point-cloud" is determined by the normal "regression equation":

$$y = a + bx,$$

where a indicates the point of intersection of the regression curve with the Y-axis, and b the angle which the curve forms with the X-axis. b is calculated according to the following formula:

$$b = \frac{\sum\limits_{i=1}^{n} x_i \cdot y_i - \bar{y} \cdot \sum\limits_{i=1}^{n} x_i}{\sum\limits_{i=1}^{n} x_i^2 - \bar{x} \cdot \sum\limits_{i=1}^{n} x_i} .$$

b is called the regression coefficient. It is the tangent of the angle that the regression curve forms with the X-axis. The regression coefficient indicates by how much, on the average, y increases when x is increased by one unit.

The equation for the regression curve is:

$$Y = \bar{y}_i + b(x - \bar{x}_i).$$

If one introduces in sequence corresponding values $(x_1, x_2, ..., x_n)$ in place of x, in our example the values for the energy-intake, the appropriate y-values can

Table 1. Regression table

x_i	y_i	$x_i \cdot y_i$	x_i^2
x_1	y_1	$x_1 \cdot y_1$	x_1^2
x_2	y_2	$x_2 \cdot y_2$	x_2^2
$\vdots$	$\vdots$	$\vdots$	
x_n	y_n	$x_n \cdot y_n$	x_n^2
$\sum\limits_{i=1}^{n} x_i$	$\sum\limits_{i=1}^{n} y_i$	$\sum\limits_{i=1}^{n} x_i \cdot y_i$	$\sum\limits_{i=1}^{n} x_i^2$
$\bar{x}$	$\bar{y}$		

therefore be calculated. In doing this, one can draw the regression curve into the "point-cloud".

One establishes a table in which the corresponding values are entered as an aid in the calculation.

For the practical calculation process, under x_i, all values $(x_1, x_2, ..., x_n)$ of the one (independent) variable would be entered, corresponding to the example in Fig. 1, the energy intake from solar radiation. The second column would contain the y-values that were determined from the investigation $(y_1, y_2, ..., y_n)$, these would be the determined new-wine weights of the grapes from vineyards with appropriately different energy-intakes during the vegetation period (X-values). Into the third column are entered the products of the x and y values, and in the fourth column the calculated squares of the x-values. The appropriate sum is formed from the values of each column, as well as the arithmetic averages for the x- and y-values of the first two columns.

In addition, one determines the scattering of the points about the regression curve (S) according to the formula:

$$S^2 = \frac{1}{n-2} \sum (y_i - Y_i)^2.$$

Here, n is the number of the measured values. The numerator represents the sum of the squares of the deviations of the observed y_i-values from the corresponding values of the regression curve (Y).

2.4 Correlation Calculation

While regression analysis concerns itself with the type of the correlation, inasmuch as it exists between coincidental variables, correlation analysis determines the degree of the correlations. This degree is measured with the correlation coefficient. The smaller the scattering of the observed value pairs around the regression curve, i.e., the more points fall onto the curve, the more precisely it is determined. The correlation coefficient is considered a measure for the dependence of two observation series. It was introduced into statistics by Bravais. Correlation

Table 2. Correlation table

x_i	y_i	$\bar{x}_i - x$	$\bar{y} - y_i$
x_1	y_1	Δx_1	Δy_1
x_2	y_2	Δx_2	Δy_2
$\vdots$	$\vdots$	$\vdots$	$\vdots$
x_n	y_n	Δx_n	Δy_n
$\sum\limits_{i=1}^{n} x_i$	$\sum\limits_{i=1}^{n} y_i$	$\sum\limits_{i=1}^{n}(\bar{x} - x_i)$	$\sum\limits_{i=1}^{n}(\bar{y} - y_i)$
$\bar{x}$	$\bar{y}$		

calculation has, for decades, and is in part still today used for the investigation of the dependence of plant development or of the yields on weather conditions. It seems to be appropriate, therefore, to discuss at least the important parts of the determination of the correlation coefficient.

The correlation coefficient is determined according to the following formula:

$$r = \frac{\sum (\bar{x}_i - x)(\bar{y}_i - y)}{\sqrt{\sum(\bar{x}_i - x)^2 \sum(\bar{y}_i - y)^2}}.$$

One establishes a "correlation table" (Table 2) for the execution of the calculation. If one wants to determine, for example, the relationships between temperature or precipitation during certain growth periods and the yield of grain, one notes, for example, in the table in column x_i, in the form of x-values $(x_1, x_2, ..., x_n)$, the determined yields of the individual harvests (perhaps of different years or of different areas), and, in similar manner, in column y_i the corresponding Y-values of the temperature or the precipitation during specific time periods (e.g., months, decades, the vegetative or the generative phase etc.). The arithmetic average is formed from the values of each column. One introduces the differences between the average values and the measured values $(\bar{x}_i - x_1, \bar{x}_i - x_2,$ etc. as well as $\bar{y}_i - y_1,$ $\bar{y}_i - y_2,$ etc.). In doing this, one has compiled the necessary values for the calculation of the correlation coefficient.

The correlation coefficient becomes obviously $= 1$ or $= -1$, if both value series $(x_i$ and $y_i)$ are in a strictly linear correlation. It assumes the value of 0 when no correlation at all exists. The closer r approaches the value 1 or -1, the closer is the correlation between the two value series. If one obtains, for example, a correlation coefficient of 0.98 for the relation between the amount of precipitation during a specific part of the vegetative period and the yield, this relationship is very close and one states that these precipitations had a "positive" effect upon the yield. According to present-day knowledge, correlation calculations can only find limited applicability in agrometeorology. Other methods are required for the investigation of the complicated relationships between the meteorological growth factors and the physical processes in the plant. For this, the physicobiological model promises better success.

2.5 The Physicobiological Model

Under a physicobiological model, we understand a conceptual determination of a situation that is connected to elements that are related to each other, so that it can be described in a physicomathematical manner. A relatively simple consideration can be established in the analysis of the formation of yield. On the one hand, we have the supply of nutrients and the meteorological environmental factors as input values and, on the other hand, the yield as the output factor. The problem now is to determine the dependence of the output value upon the input values.

The first model that takes in all growth factors was established by Mitscherlich. In this, a simple differential equation is used. Mitscherlich makes the differential quotient, formed from yield and the amount of nutrients offered, proportional to the difference between maximum and actual yield. The formulation of this differential equation can count as the standard example for the introduction of models into agricultural research. Von Boguslawski and Schneider (1964) have developed a considerably improved modification on the basis of Mitscherlich's model which also permits consideration of yield depressions.

As an example for a model that has been developed especially in agro-meteorology, Baier's (1973) *Crop-Weather Analysis Model* is to be discussed somewhat more thoroughly. It is the primary task of this model to analyze the daily influence of up to three selected meteorological variables on the intermediate growth condition or the final yield of agricultural plants. In this connection consideration is also given to the differing influence on various developmental phases (e.g., vegetative or generative phase). The effect of each of the three variables on growth forms either a positive or negative straight line or a curve. The quantitative reaction interaction during the course of the life of annual plants is a function of "biometeorological periods" as defined by Robertson (1968). The reaction characteristics are unknown and also not predetermined. Instead, it is assumed that reaction interactions take place in step-wise distribution throughout the entire growth period, and that the daily added-growth coefficients in each variable can be fitted to a polynomial of the 4th degree as a function of the "biometeorological period". The decision on the functional relationship, as it is used in the model, is based on the assumption that plant development is basically dependent upon three meteorological parameters, solar energy, air temperature, and soil-moisture (or evapotranspiration). The three variables modify each other. They must be available on each individual day in order to produce a positive or negative effect for the final yield. The "crop-weather analysis model" is formed on the basic equation:

$$Y = \sum_{t=0}^{m} V_1 \cdot V_2 \cdot V_3 \,,$$

where Y is an independent variable, that corresponds to the finally determined harvest or to a growth component over a specific period of time. Y is, for example, the seasonally dependent grain yield in kg/ha.

$\sum_{t=0}^{m}$ is the sum of the daily V-values of the biometeorological period $t = 0$ to $t = m$.

V_1, V_2, V_3 are functions of the selected independent variable of the meteorological parameters.

Each V-function has the general form

$$V = (u_1 t + u_2 t^2 + u_3 t^3 + u_4 t^4) + (u_5 t + u_6 t^2 + u_7 t^3$$
$$+ u_8 t^4) x + (u_9 t + u_{10} t^2 + u_{11} t^3 + u_{12} t^4) x^2,$$

where U_1, U_2, ..., U_{12} are calculated coefficients. Each V is determined in a repeated regression analysis. X in V_1, V_2, or V_3 represents one of the variables that is selected for the analysis in a separate calculative process.

A further physicobiological model that has been used in the German Federal Republic since 1968 for practical consultation purposes, the *Phytophthora-negative-prognosis*, should be mentioned. This model was developed by Schrödter and Ullrich (1966, 1967). The basis of this model is the understanding of the ecology of the fungus *Phytophthora infestans*, primarily the understanding of van der Zaag (1956) and van der Plank (1963). Schrödter and Ullrich base their model on the concept that all meteorological constellations that are given within a climatic area during the time between the formation of an initial focus and the beginning of epidemic spreading of the fungus have an effect on the development of the fungus in some manner, and thus determine the length of that period. The manner in which the meteorological constellations can influence the fungus and its spread from the primary infection focus is presented in the form of an evaluation function for the meteorological parameter combination. Hourly values of temperature, air humidity, and precipitation are used as parameters. An evaluation number is determined for the potential influence of the meteorological conditions that are given within a week, which can be approximately defined as the Logit-value (see van der Plank, 1963) of that infestation that would have occurred if the development of the infestation were dependent only upon this factor.

The evaluation function has been established by means of mathematical-statistical evaluations of observation results of several test areas. It is based on the interconnecting regression equation under consideration of the biology of the fungus.

The evaluation function itself is represented by the following equation

$$Y = c_1 y(Kp) + c_2 y(Sp) + c_3 y(M) + c_4 y(U).$$

Y is the evaluation number in logits of 10^{-2}, $y(Kp)$ indicates the portion that results from the influence of the temperature on spore germination and infection under consideration of the duration of high humidity or duration of wetting. $y(Sp)$ is that portion that is derived from the influence of the temperature on sporangium production under consideration of the duration of high humidity. $y(M)$ is that portion that results from the influence of the temperature of mycelial growth of the fungus. $y(U)$ is a correction value that contains the inhibition of spreading by intermediate dry periods. c_1, c_2, c_3, and c_4 are constants.

The calculation can be considerably simplified by an appropriately compiled table of the meteorological values. The preparation of the meteorological values consists in the processing of the hourly obtained values of temperature, relative humidity, and precipitation, according to prescribed viewpoints, into weekly frequency distributions. The sum of valence times (X) frequency, multiplied with 14 different "evaluation numbers" (multiplication factors), produces the so-called weekly evaluation number. The summation of these evaluation numbers from week to week for the period that is still free of epidemic, beginning with the week of the sprouting of the potatoes, produces a corresponding sum. Since it was found from experimental investigations that the observed beginning of the epidemic never occurred before the evaluation number sum of 150 was exceeded, as a rule only above a sum-value of 270, these limit values are especially characteristic. These limit values are used in the manner for a plant-protection warning service, so that it can be stated that no chemical control measures are required in potato fields against *Phytophthora infestans* prior to reaching the evaluation number 150 (therefore negative prognosis!). When the number 150 is reached, a preventive, and at 270 a directed, control effort is recommended. The great value of this prognosis activity is in the elimination of unnecessary prophylactic control measures.

2.6 The Simulation Model

The physicobiological model for the *Phytophthora*-negative prognosis introduced an innovation in comparison to the "crop-weather analysis Model", in that known ecological factors are applied. Certain biological reactions to the influence of meteorological elements are known from experimental investigations, while the crop-weather analysis model either does not consider reaction characteristics or these are not known. It should be clear today that a solution of a problem in which the parameters can be interpreted physically or physiologically, and whose values can be checked by means of direct determinations, is more convincing than one with statistical formulas, in which unknown laws with unexplainable constants lead to correct results. With the presently still wide-spread application of regression calculations, a certain relationship between weather and yield can be made obvious, but this relationship is not necessarily applicable in other climatic regions. In the development of simulation models, on the other hand, one attempts, to as great an extent as possible, to consider known biological reactions to the influences of the weather complex, which have general validity and must, therefore, be applicable in all climatic regions. In this the mutual dependences of the influence values must be included as much as possible.

In order to clarify what has been said here, Fig. 2 represents a simple simulation model. Two systems influence the closed plant system from the outside, the atmospheric system and the soil system. The atmospheric system is the weather with its meteorological elements as ecological factors. The soil system is the reservoir for nutrients and water. Only the most important life processes are presented for the plant-system. In order to be able to make statements concerning the effects of the weather on the growth processes in the plant, or even on the end product of these processes, the harvest yield, a series of sub-models is required. Initially, the atmospheric system has an effect not only on the plant, but also on the soil system. The soil moisture is influenced by precipitation and evaporation, which, again, indirectly means an influence on the nutrient uptake. The first submodel must offer a possibility for the determination of soil moisture from the corresponding weather conditions. The effect, for example, of varyingly high soil

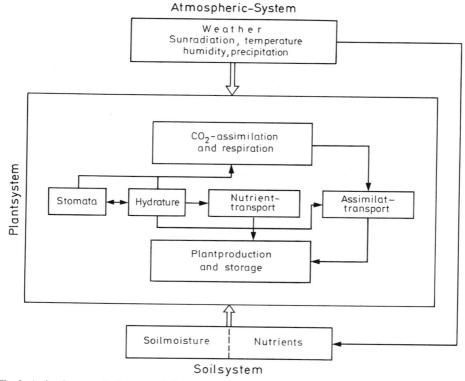

Fig. 2. A simple example for a simulation model

moisture on their development is known for many agricultural plants from numerous field experiments. This, then, is an evaluation basis for the soil system, for nutrient uptake, as well as for the growth possibilities for the plants in general. As the next model, we shall consider the effect of weather on carbondioxide assimilation and respiration. Here, it is a matter of determining the assimilate increase on a daily basis. However CO_2-assimilation is not only determined directly by light intensity and temperature as exogenic factors, but also by the degree of openness of the stomata and the hydration of the plant, which moreover influence each other in a mutual relationship. Openings of the stomata and hydration are, again, influenced from the outside by the heat conversion on the leaves and the humidity or transpiration. These effects would be the contents of a further model. Finally, the transport of nutrients and assimilation products is required for the production of plant-mass. These are influenced by the water supply and the hydration of the plant. The sum of all of these processes – and only the most important have been mentioned here – finally results in a picture of the interdependence between weather and plant development, whereby the goal is to determine these processes quantitatively.

The reader will surely recognize from this brief description how complicated the structure of such a model must be, in order to be able to determine, in this manner, the influence of the weather on the yield, especially as all preconditions for this are certainly not yet given. Nonetheless, a series of authors have published promising investigate results in connection with this problem. A quite complete presentation of a simulation model for plant production was produced by van Keulen and Louwerse (1974). Visser (1968) has published considerations for a yield model for suitable and detrimental ecological factors. Added to this are proposals for sub-models. Cartier (1968) describes a model for the photosynthesis of the leaf, Unger et al. (1968) report on a model for the water household of agricultural plants and Raschke (1968) on a model for the description of the energy- and gas-exchange of a corn leaf.

It will surely still require time for sufficiently good results, e.g., of the effect of weather on harvest yields to be obtained by means of simulation models. Until then, one will have to work with models similar to the one that was used by Schrödter and Ullrich (1967) for the *Phytophthora*-negative-prognosis, i.e., with physicobiological models, in which the biological "content" must take in an ever greater volume. The great value that can, even today, be assigned to the presently available simulation models is initially in the logical arrangement of the biological processes and their relationships to each other, that is contained in such a model, as a basis for further research on the problem of weather and plant development.

Literature

Ahrens, H.: Variance analysis. Berlin 1967

Baier, W.: Crop-weather analysis model: review and model development. J. Appl. Meteorol. 9 (1973)

Boguslawski, E.v., Schneider, B.: The application of the parameters in Mitscherling's yield curve to the evaluation of field experiments. Z. Acker Pflanzenbau 2 (1968)

Cartier, Ph.: Biophysical model of photosynthetic exchange in leaf surfaces. (French) Stud. Biophys. Biophys. Pflanz. Syst. 11 (1968)

Finney, D.J.: Statistical method in biological assay. London 1952

Fisher, R.A.: Statistical methods for research workers. London 1948

Keulen, H. van, Louwerse, W.: Simulation models for plant production – agrometeorology of the wheat crop. WMO-symposium, Braunschweig, BRD, 1973. WMO No. 396, Offenbach/Main 1974

Mood, A.M.: Introduction to the theory of statistics. New York 1950

Mudra, A.: Statistical methods for agricultural experiments. Berlin 1958

Plank, J.E. van der: Plant diseases: epidemics and control. New York, London 1963

Raschke, K.: A model for the description of the energy- and gas-exchange of a corn leaf. Stud. Biophys. Biophys. Pflanz. Syst. 11 (1968)

Robertson, G.W.: A biometeorological time scale for a cerial crop involving day- and night-temperatures and photoperiods. J. Biometeorol. 12 (1968)

Scheffé, H.: Analysis of variance. New York: Wiley 1959

Schrödter, H., Ullrich, J.: A mathematical-statistical solution of the problem of the prognosis of epidemics with the aid of meteorological parameters, as represented on the example of the potato leaf rot (phytophthora infestans). Agric. Meteorol. 4 (1967)

Snedecor, G.W.: Statistical methods applied to experiments in agriculture and biology. Iowa: Ames 1948

Ullrich, J., Schrödter, H.: The problem of predicting the occurrence of potato leaf rot (phyophthora infestans) and the possibility of its solution by a "negative prognosis". Nachrichtenbl. Dtsch. Pflanzenschutzdienst 3 (1966)

Unger, K., Claus, St., Grau, I.: Models on the water household of agricultural plants. Stud. Biophys. Biophys. Pflanz. Syst. 11 (1968)

Visser, W.C.: A yield model for suitable and detrimental environmental factors. Stud. Biol. Biophys. Pflanz. Syst. 11 (1968)

Weber, E.: Basics of biological statistics. Stuttgart: Fischer 1967

Zaag, D.E. an der: Wintering and epidemiology of *Phytophthora infestans*, together with some new control possibilities. (Dutch.) Tijdschr. Plantenziekten 62 (1956)

3 Agrotopoclimatology

J. SEEMANN

The topoclimate has an intermediate position between the macroclimate and microclimate. Macroclimatology is based on a wide network of measurements and does not register the special features resulting from the topographical differentiation of the terrain, whereas the microclimate comprises areas which are far too small. In order to obtain knowledge on the topoclimate of a regional area, methods of measurement other than those used in general climatology are required, above all by much more closely spaced measurement grids. The agrotopoclimatology with which we are concerned here is in turn a special part of general topoclimatology. It is characterized by a specific statement of the problem. It describes the climatic suitability of a tract of land as a location for particular agricultural plants or as living space for farm animals. The importance of topoclimatological investigations for practical work in agrometeorology is indicated by the large number of pertinent publications in this field. Mac Hattie and Schnelle (1974), who give a general review of the most important problems of agrotopoclimatology in a Technical Note of the World Meteorological Organization list more than 1000 titles on this topic in the appended bibliography alone. The entire complex of topoclimatology has been treated by Geiger (1965, 1969) as well as Molga (1962). Van Eimern (1971) has given a coherent description of the specific agricultural aspects of topoclimatology. In addition, the synoptic treatment of topoclimatology by Quitt (1972) will be of special interest.

Practically all meteorological elements will be modified compared to the free atmosphere or even the macroclimate by the varying conformation of the terrain: altitude differences, slopes, valley locations, hollows and flat areas. Temperature is known generally to decrease with altitude. On average, one can reckon with a temperature decrease of about 0.6 °C per 100 m when ascending from the bottom to the top of a valley. However, Schnelle (1972) observed in investigations in highland areas that the extent of the temperature decrease between valleys and peaks depends both on the total topography and the season, and can display greatly differing values. Table 1 gives a good idea as to how large the differences

Table 1. Decrease of the temperature extremes with every 100 m increase in altitude (°C approximate values)

	Winter			Summer		
	On comparison of stations in					
	Narrow valleys	Wide valleys	Open high ground	Narrow valleys	Wide valleys	Open high ground
Maximum	−0.7	−0.7	−1.0	−1.0	−1.2	−1.4
Minimum	−0.8	−0.5	−0.6	−1.2	−0.5	0.4

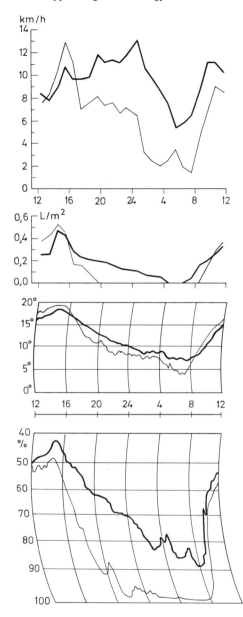

Fig. 1. Daily sequence of wind, evaporation, temperature and relative humidity (graph from above to below) on the surface of a 100 m high tip (⸺) and on flat ground (⸺) in "fine weather" (Seemann, 1970)

can be in the temperature decrease with each 100 m increase in altitude, with differentiation of maximum and minimum temperatures. The seasonal fluctuations in air temperature also show appreciable differences depending on the overall topography, i.e., on the width of the valleys and the adjacent mountains.

As can be seen from Table 1, the decrease of minimal temperatures with altitude is less than that of maximum temperatures, above all on exposed high ground: indeed, the differences are positive in summer. This is a manifestation of

Table 2. Number of buds formed in pea plants in %

	1964 vegetative period		1965 vegetative period		
	10.5.–25.6.	27.5–8.7.	28.5.–10.8.	16.6.–1.9.	10.7.–14.9.
Flat ground	100 %	100 %	100 %	100 %	100 %
High ground (tip)	95 %	69 %	47 %	57 %	69 %

the frequently occurring weather inversions, in which temperature rises at night up to certain altitudes. An example of this will be given in Fig. 1, which shows measurements made in an agrometeorological study on waste tips which were recultivated for agricultural use (Seemann, 1970). The measurements were made on a tip 100 m higher than the surrounding area with a largely flat surface 140 ha in size as well as at a certain distance from the tip in a flat approach. Registration of temperature (third graph from above) shows the curve typical in "fine weather". In the daylight hours, it is about 1.5°–2 °C colder on the tip surface, whereas at night the temperature on the tip is up to 3 °C higher than in the surrounding area below. The example in Fig. 1 also shows the alteration of practically all meteorological elements due to the influence of topography which has already been mentioned. Under the weather conditions when the measurements were made, the wind speed (upper graph) on the tip was at first lower (noon to 5 o'clock in the afternoon) than in the approach. In the inversion situation, the wind speed in the approach falls appreciably, whereas it at first increases further at the measurement site 100 m higher. Evaporation (second graph from above) is less on the surface 100 m higher during the afternoon than in the approach area. However, whereas evaporation ceases almost completely from about 7 o'clock in the evening in the area surrounding the tip, evaporation continues almost throughout the night on the tableland. These differences in evaporation are clearly a consequence of differences in temperature, wind speed and relative humidity (see lower graph). The characteristic feature in the climatic contrasts between low and high ground is the difference in the daily temperature amplitude. At least on the days with inversions at night, the daily amplitude of temperature in the high plateau is lower than on the lower ground. The question now arises as to whether such differences between low and high ground will affect plant development when corresponding weather conditions occur frequently. Together with the meteorological investigations on waste tips mentioned here, experiments in plant cultivation were also carried out. Peas were used as trial plants. It was shown that the meteorological differences of the two locations primarily affected the reproductive phase. In Table 2, the numbers of buds formed on pea plants from five different sowings are listed. Apart from the meteorological conditions, all experimental conditions were identical at both locations.

In the same way, corresponding differences in the dry weight of the pea plants could be detected in these trials. Just as the number of buds was lower on the tip, the dry weight of the plants was also below that on the low ground. The experimental results mentioned here show the practical importance of agrotopoclimatological studies.

Table 3. Total monthly insolation from direct sunlight in kcal/cm^2 in cloud-free weather for slopes of various inclinations and orientations for $\varphi = 50°$. (After Morgan, 1957)

Month	Inclination, slope	N	NE NW	E W	SE SW	S	Difference between north and south slope
December	0° = Flat	2.0	2.0	2.0	2.0	2.0	0.0
	10° = Slope	0.7	1.2	2.0	3.0	3.5	2.8
	20°	0.0	0.6	2.0	4.0	5.0	5.0
	30°	–	0.1	2.0	4.8	6.0	6.0
	90° = Cliff	–	–	1.4	5.7	8.3	8.3
March	0° = Flat	8.8	8.8	8.8	8.8	8.8	0.0
	10° = Slope	7.1	7.9	9.2	10.6	11.2	4.1
	20°	4.9	5.6	8.9	11.8	12.8	7.9
	30°	2.1	4.6	8.7	12.6	14.1	12.0
	90° = Cliff	–	1.2	5.6	8.8	12.2	12.2
June	0° = Flat	18.6	18.6	18.6	18.6	18.6	0.0
	10° = Slope	17.8	18.1	18.7	19.4	19.5	1.7
	20°	16.5	17.0	18.4	19.7	19.6	3.1
	30°	14.2	15.0	17.5	18.8	18.8	4.6
	90° = Cliff	2.0	5.4	8.8	8.0	7.0	5.0
September	0° = Flat	10.8	10.8	10.8	10.8	10.8	0.0
	10° = Slope	9.2	9.6	11.0	12.1	12.9	3.7
	20°	6.5	8.0	10.9	12.2	14.4	7.9
	30°	3.9	6.3	10.6	13.8	15.4	11.5
	90° = Cliff	–	1.8	6.3	9.9	11.2	11.2

Of particular significance for topoclimatology is the different energy input from solar radiation resulting from the dissimilar orientation of the intercepting surface: the re-radiation from these areas in relation to the zenith and the narrowing of the horizon. In the section on (p. 13) and (p. 14), the physical basis has already been referred to. The calculation of the alteration in incident radiation due to topography is also illustrated there. Here it will be shown once more in Table 3 to what a large extent the monthly radiation totals of direct solar radiation deviate on slopes with different inclinations and compass orientations. This is an extract from an extensive calculation carried out by Morgen (1957) according to which in December the south wall, in March the south slope with about 50° inclination, receive the strongest incident radiation, whereas in June it is the southwest or the southeast slope with an inclination of 20°. The differences between the south and north slopes are greatest between the seasons and least in June. There is a certain compensation in practice from diffuse radiation for these in some cases great differences in energy received from direct radiation.

Differences in the radiation enjoyed by the slopes affect above all the temperature of the upper layer of the soil and the layer of air near to the soil. Although an eastern slope has practically the same incident radiation as a west slope, it remains somewhat cooler as a whole than the west slope. Similar observations also apply to southeast and southwest slopes. The energy of the morning sun is partly used to evaporate the dew from the night. On the other hand, the southwest and west slopes receive their greatest incident radiation during the course of the day when the air has already been more strongly warmed.

The effect of the slope climate is recognized in the cultivation and agricultural use, as well as in the phytosociology. The south slopes mostly have an advantage of 5–10 days with regard to drying up in spring. Through the more intensive warming, they also permit an earlier agricultural cultivation, thus enabling a longer vegetative period. However, utilization of favored slope climates is of special importance for the growing of cultivating plants in climatic regions in which the general climatic conditions mostly no longer afford sufficient warmth for favorable development. For example, this is the case for the majority of German quality viticulture, which is therefore mainly practised on slopes.

Topography plays a role not only for a more differentiated warming up during the day, but there is a highly dissimilar local cooling in the layer of air near to the soil at night. A particular agrometeorological problem arises here above all with regard to the generation and distribution of cold air, especially in connection with late and early frosts. The temperature near to the ground is known to fall especially greatly during clear nights when the soil has only stored a little of the heat of the day because of poor heat conduction or unfavorable insolation. Because of the different times of the true sunset for individual spots (slopes, valleys, and hollows), as well as the dissimilar properties of soils and their cover, night cooling begins earlier at some spots on flat as well as on hilly ground. It also progresses with differing rapidity. Areas with differing degrees of cooling are thus formed near to the soil already at sunset. Temperature differences finally induce compensatory airflow. However, on flat ground this is made more difficult by the very great stability of the cold air near to the soil. The cold air so to speak adheres to the soil. The roughness of the surface which is mostly present due to plant growth has an appreciable role in this phenomenon. Wide areas of flat ground are thus often cold on a clear still night because the cold air formed here cannot flow away. On sloping ground, the cold air moves down the slope as soon as it has attained sufficient force. The flow of cold air begins already in the early evening. With only a few meters of head pressure, the cold air is mostly held up by obstacles in the terrain, and then flows over these obstacles. The idea that areas with reduced danger of frost on the lee side of the obstacle would result from planting strips of forest across slopes or valleys is only valid when there is a relatively large gradient on the lee side, so that the cold air below the obstacle can flow away. Investigations by King (1973) have shown that in contrast to earlier ideas, the damming up effect only occurs mainly at the beginning of cold air flow but later flows over the obstacle so that no significant temperature differences can be detected in the weather side and lee of the obstacle by accumulation of the cold air which has formed or which has flowed into the lower valley. Schneider (1972) found that in side valleys the cold air cannot flow away at all when large amounts are already present in the main valley. The fear of great frost danger in the valley areas above the dam when dams are built across valleys is thus not justified in most cases. Already during the night or at the time of the nightly temperature minimum equally low temperatures occur on both sides of the dam by accumulation from the part of the valley below the dam. Nevertheless, the flow of cold air and its collection in valleys or hollows remain an effect of topography which has great significance in agrotopoclimatology because of the resulting intensification of the frost danger. In planning in cultivation areas in which frosts occur at the beginning or at the end of the vegetative period, one of the most important functions of applied agrotopoclimatology is to find these spots with accumulation of cold air and if possible to eliminate them from use for cultivated plants which are susceptible to frost. Schnelle (1965) has reported in details on methods of investigation and measurement for finding spots susceptible to frost.

So far we have mainly pointed out the differentiation of temperature conditions due to topography. There is no doubt that this has the greatest importance for planning of agricultural planting. However, it should not be overlooked here that the particular warming of air near to the soil on the upper soil layers in certain spots is not derived alone from the different energy input from solar radiation, but is also affected by different heat exchange conditions or by wind. Everyone is aware of the fact that spots protected from wind become very much warmer in sunshine than spots with stronger wind. These differences in wind speed are also essentially due to topography. The terrain forms can affect the wind field in two ways. On the one hand, there are local wind systems such as mountain and valley wind, land and sea wind due to terrain. However, these only regularly develop differences in strength under quite special conditions. In the Rhone valley in Valais or in other Alpine valleys, they can become so strong that windbreak plantations are to be recommended. However, every wind is directly influenced by topography, by weatherside and lee effects which are present in any hilly region. With weak winds, the direction of the wind is frequently influenced by valleys, so that quite different incidences of wind direction result compared to the adjacent high ground. A particular phenomenon in country with freely exposed elevations is the large rise in wind speed at the upper edge of a slope when the wind is directly blowing towards the slope. This effect can frequently lead to direct damage to the vegetation. A considerable deformation of the crowns of trees is a typical sign of this.

Literature

Eimern, J. van: The topoclimate and the microclimate. In: Meteorology and climatology for agriculture, horticulture, and viticulture. Stuttgart: E. Ulmer 1971

Geiger, R.: The climate near the ground. Cambridge: Harvard University Press 1965

Geiger, R.: Topolimates. In: General climatology 2. Flohn (ed.). Amsterdam: Elsevier 1969

King, E.: Untersuchungen über kleinräumige Änderungen des Kaltluftflusses und der Frostgefährdung durch Straßenbauten. Bericht des Deutschen Wetterdienstes 130 (1973)

Mac Hattie, L.B., Schnelle, F.: An introduction to agrotopoclimatology – Tech. Note No. 133, Secretariat of the World Met. Organization, Geneva 1974

Molga, M.: Agricultural meteorology, Part 2. Available from the Office of Technical Services, U.S. Department of Commerce (1962)

Morgen, A.: Die Besonnung und ihre Verminderung durch Horizontbegrenzung. Veröffentlichung des Meteorologisch-Hydrologischen Dienstes der DDR (1957)

Schneider, M.: Cold air an road embankments? Not always! Meteorolog. Rundschau 25 (1972)

Schnelle, F.: Frostschutz im Pflanzenbau, Band 2. Die Praxis der Frostschadensverhinderung. München: Bayerischer Landwirtschaftsverlag 1965

Schnelle, F.: Lokalklimatische Studien im Odenwald. Berichte des Deutschen Wetterdienstes 128 (1972)

Seemann, J.: Die agrarmeteorologischen Verhältnisse auf Hochhalden des Rheinischen Braunkohlengebietes. In: Die Landschaftspflege in der Raumordnung. Bonn: Landwirtschaftskammer Rheinland 1970

Quitt, E.: Mesoclimate in the complex of environment. Studia Geographica 26 (1972)

4 Soil Climate

Y. I. CHIRKOV

Soil climate, this perennial cycle of temperature and humidity, as well as of soil air conditions, is caused by the properties of the soil, and is dependent upon a complex system of natural and anthropogenic factors.

The concept of a soil climate was developed in the 19th century by the Russian scientist P. A. Kostychev, who saw soil as a distinctive medium for the manifestation of an atmospheric climate. Subsequently, soil climate became the subject of research in a number of sciences (physics, pedology, agroclimatology, etc.) in many countries, once the tremendous importance of soil climate in agriculture had been established.

Soil climate is the result of atmospheric climate, the properties of the soil, relief, vegetation, and man's productive activity; it is therefore, complex research subject. Soil climate affects the microclimate of the near-earth air layer, as well as soil processes, and also determines, to a considerable extent, the productive capacity of growing plants.

Basically, soil climate is made up of three elements: (1) soil temperature, (2) soil humidity, and (3) soil aeration. Diurnal and annual variations are characteristic for the components of a soil climate, as is the distribution of these variations by territory, depending on atmospheric climate. The composition of the atmosphere is primarily uniform, while the annual composition of soil air varies within considerable limits (CO_2 – from 0.01 to 10–15%, O_2 – from 10 to 20%, etc.), depending on the volume of organic matter in the soil and the intensity of matter decomposition processes.

A temperature and humidity field in the atmosphere is characterized by a considerably lesser degree of variability than that of a soil which is inconsistent with respect to composition and properties, even within the confines of the smaller territories.

Depending on the properties and humidity of the soil, its aeration and thermal characteristics vary significantly.

Under the influence of vegetation, microorganisms, and other natural and anthropogenic factors, soil climate varies in time considerably faster than an atmospheric climate. Finally, soil climate can be controlled by man to a considerably greater extent than the climate of the near-earth air layer. For this reason, reclamative measures aimed at optimizing a soil climate (irrigation, drainage, management, etc.) are quite effective.

4.1 Methods for Studying a Soil Climate

Agroclimatology studies soil climate from a geographical point of view, over a dynamic range defined by the seasons of the year, with allowance for the conditions of the physical-geographical medium which determines its formation.

Experimental data from meteorological stations which conduct observations on soil temperature and humidity, soil freezing depths, and the effect of soil climate conditions on plant development and productivity serve as basic materials for the study of soil climate.

Temperature measurements are conducted several times a day on the soil surface and in the topsoil layer, where temperature variability is especially great (in the USSR, the measurements are conducted every 3 h, 8 times a day), in order to obtain an idea of the diurnal variation in soil temperature. Periodic, maximum, and minimum thermometers are set up on the soil surface. The temperature of the topsoil layer is measured by means of special angle thermometers at depths of 5, 10, 15, and 20 cm.

The temperature of the deeper soil layers is measured by means of removable or electric thermometers (see Chap. 8). Due to the fact that measurement methods are not standardized in the different countries, the depths at which the temperature is measured vary. For example, at stations in the Soviet Union, soil temperature is measured at depths of 20, 40, 80, 120, 240, and 320 cm, in West Germany, at 50 and 100 cm.

Soil humidity is the most important component of a soil climate, usually determined by the thermostat-weight method, although the isotope methods have come into widespread use over the past ten years.

Using the thermostat-weight method, soil humidity is determined as a percentage of the weight of absolutely dry soil, then the amount of productive moisture is computed from the values obtained, using the formula:

$$W = 0.1\,qh(u - k),$$

where W, productive moisture reserves, in mm; q, the volumetric weight of the soil (g/cm^3); h, the thickness of the soil layer (cm); u, soil humidity (as a percentage of absolutely dry soil); k, stable wilting humidity (as a percentage of absolutely dry soil); 0.1, a coefficient for converting a water layer into mm.

The full-scale determination of soil moisture reserves is still very limited in many countries. Systematic observations on the dynamics of soil moisture in principal agricultural crop stands have already been in use for more than 40 years at meteorological stations in the Soviet Union. At present, approximately 2000 stations conduct these observations. The study of the features of soil humidity conditions in different climatic zones makes it possible to base soil reclamation efforts scientifically, and to use a diverse agronomic technology in the cultivation of agricultural crops.

4.2 The Principal Components of a Soil Climate

Soil Temperature. The primary source of the heat entering the soil is sunlight, which is absorbed by the soil surface, converted into thermal energy, and transmitted to the underlying soil layers. At night, the discharge of heat by means of radiation exceeds the influx; the soil surface and the underlying soil layers are cooled.

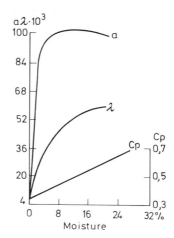

Fig. 1. The dependence of the thermal characteristics of soil on humidity (from A. F. Chudnovskiy). *a* thermal diffusivity, $(\lambda) \rightarrow \delta$ thermal conductivity, *Cp* volumetric heat capacity

The intensity of soil heating and cooling depends on the properties of the soil – heat capacity, thermal conductivity, and thermal diffusivity – as well as on the temperature difference between the soil surface and its underlying layers. The greater the difference, the greater the amount of heat entering (with respect to a negative difference) or leaving the soil.

The heat capacity (volumetric) of dry mineral soils is approximately two times smaller than that of water. Therefore, heat capacity grows with an increase in humidity. An increase in the amount of air in the soil decreases its heat capacity. For this reason, the heat capacity of dry peat soils is less that that of mineral soils, while the heat capacity of humid peat soils is greater.

The thermal conductivity of the soil depends primarily on its porosity, as well as its humidity. Since the thermal conductivity of soil air is approximately 50 times less than that of the mineral particles in the soil, the lower the density of dry soil (the greater the porosity), the lower its thermal conductivity. When water fills the soils pores, its thermal conductivity is increased to an unknown value limit, inasmuch as the thermal conductivity of water is 24 times higher than that of air.

The thermal diffusivity of the soil is characterized by the ratio of the thermal conductivity coefficient to volumetric heat capacity. The thermal diffusivity coefficient is a function of soil humidity and the content of air within the soil. At low soil humidity values, thermal diffusivity grows; however, with an increase in humidity, it is inhibited, since the thermal diffusivity of air (0.16) is higher than that of water (0.0013). Changes in termal diffusivity are the result of a combined change in heat capacity and thermal conductivity. The monogram for determining the thermal characteristics of soil by means of its humidity, developed by the Soviet scientist A. F. Chudnovskiy, may serve as an apt illustration of this interaction (Fig. 1).

Diurnal and Annual Variations in Soil Temperature. A soil temperature change over the course of a 24-h period is called a diurnal variation, representing periodic fluctuations from a maximum to a minimum. The temperature minimum is observed on a soil surface before sunrise, when the radiation balance is negative, and the exchange of heat in the air-soil boundary layer is insignificant. The

Table 1. A time frame for the lateness of maximum and minimum soil temperatures at different depths

Depth (in cm)	Minimum	Maximum	Temperature fluctuation amplitude (in °C)
Nukus (a desert area near the Aral Sea)			
Soil surface	4 h 20 min	12 h 45 min	40.0
5	5 h 30 min	16 h 30 min	15.3
10	6 h 25 min	18 h 10 min	11.1
20	9 h 36 min	20 h 35 min	5.0
Leningrad			
Soil surface	3 h 24 min	13 h 12 min	14.3
20	8 h 06 min	18 h 12 min	2.7
40	12 h 48 min	23 h 42 min	1.0
80	19 h 00 min	7 h 00 min	0.2

temperature maximum is observed for approximately 13 h, then begins its decline, continuing until it reaches the morning minimum. The difference between the maximum and the minimum is called the diurnal variation amplitude.

A number of factors affect the diurnal amplitude value:

1) Time of year. Amplitudes are highest in summer and lowest in winter.

2) Geographic latitude. The amplitude value is related to sun height. Sun height increases from the pole to the equator; therefore, the amplitude is not high in the polar regions, while in tropical deserts, where effective radiation is also high, the amplitude reaches its highest values (50°).

3) Cloudiness. During overcast weather, the amplitude is significantly lower than during clear weather.

4) Soil color. The surface temperature amplitude of dark soils is higher than that of light soils, since the absorption and radiation of dark surfaces is greater than that of light surfaces.

5) Heat capacity and thermal conductivity. The amplitude value is found to be an inverse function of heat capacity and thermal conductivity.

6) Plant and snow covers decrease the amplitude, pretecting the soil from intense heating during the day and from cooling at night.

7) Slope exposure. In the northern hemisphere, southern slopes are more intensely heated than their northern counterparts, while western slopes are somewhat more intensely heated than eastern slopes. In the southern hemisphere, the inverse is true.

The annual variation in soil surface temperature is primarily determined by the influx of sunlight. The maximum mean monthly soil surface temperatures are observed in July, when the influx of heat is the greatest, while the minimum temperatures are observed in January. The amplitude of the annual soil temperature variation increases with geographic latitude. In the equatorial region, it comes to approximately 3° on the average, while in continental polar regions (for example, in Yakutia), it exceeds 70°.

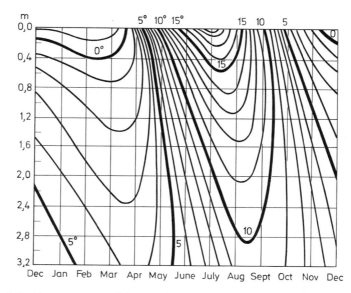

Fig. 2. Soil temperature isopleths (the steppe zone of the Ukrainian Republic USSR)

Diurnal and annual soil surface temperature fluctuations due to thermal conductivity are transmitted to the deeper soil layers. A soil layer, the temperature of which displays diurnal and annual variations, is called an active layer.

The distribution of depth-related soil temperature fluctuations (in the presence of a uniform composition) takes place in accordance with the following Fourier laws:

1) The time frame of the depth-related fluctuations does not vary; that is to say, both on the soil surface and at subsequent depths, the interval between two successive temperature minima or maxima works out to a diurnal variation of 24 h or an annual variation of 12 months.

2) If the depth increases in an arithmetic progression, then the amplitude is decreased in a geometric progression. Consequently, the amplitude deteriorates quickly with an increase in depth. A soil layer, the temperature of which does not vary over the course of a 24-h period, is called a constant diurnal temperature layer. At the middle latitudes, this layer is found at a depth of 70–100 cm. A constant annual temperature layer in the middle latitudes is found at a depth of 15–20 m.

3) The maximum and minimum temperatures set in later at these depths than on the soil surface (Table 1). This delay is directly proportional to depth. According to the data derived from long-term observations, it has been established that the diurnal maxima and minima are delayed an average of 2.5–3.5 h for each 10 cm of depth, while the annual maxima and minima are delayed 20–30 days for each meter of depth.

The study of the depth-related dynamics of soil temperature over the course of a year has great practical significance. These data are used in agriculture and public services, as well as in commercial and highway construction operations. For example, when installing water mains, it is necessary to know the depth to which the soil in a given location freezes. When installing pipes above the freezing depth, a water main freezes. When installing pipes at a considerably greater depth, excavation expenses are increased.

Materials derived from long-term soil temperature observations at different depths may be presented in the form of a special *thermoisopleth graph* (Fig. 2). Three variable values – soil temperature, depth, and time – are linked on this graph. The graph is constructed in the following manner: depth is plotted along the vertical axis (ordinate), while time intervals (usually months) are plotted along the horizontal axis (abscissa). Mean monthly soil temperature values at all depths are entered in the graph field, while points having similar temperatures are connected by means of unbroken lines, which are also called *thermoisopleths*. These thermoisopleths provide a graphic representation of the temperature in the active soil layers of agricultural areas at any depth for each month.

4.3 The Effect of Relief, Vegetation, and Snow Cover on Soil Temperature

Relief has a considerable effect on soil temperature conditions. The influx of sunlight to a soil surface depends on the grade and exposure of the slope. In the northern hemisphere, slopes with a southern component receive more heat, while northern slopes receive less than the horizontally situated flatlands of the region.

Voyeykov ascertained that daytime heating and nighttime cooling values are highest for concave relief shapes (valleys and hollows) and lowest for convex relief shapes (hillcrests), which is explained by the intensity of air interchange, lowest in hollows and greatest on hillcrests.

The exposed soil on slopes is more intensely heated than soil which is covered by vegetation. For example, observations on both southern and northern slopes with a grade of 20°–22° near Leningrad, in both exposed and plant-covered soils, show that the difference in soil temperatures between the northern and southern slopes at a depth of 10 cm reaches 16 °C in the exposed section and only 7 °C in the section covered by vegetation (Table 2).

The temperature in the topsoil layer is higher on a ridge-like surface than on a level surface by an average of 1.0°–1.5 °C.

Plant cover shades a soil surface, absorbing an essential amount of sunlight, but at the same time, protecting the soil from radiation. Nevertheless, the soil under a plant cover is colder on the whole in summer than exposed soil, and warmer in winter. A topsoil layer is 5°–6 °C warmer when lying fallow in summer than the soil under field crops. A forest has a peculiar effect on soil temperature. The mean annual temperature of the soil in a forest is 1 °C colder, at a depth of 1 m, than field soil. In summer, the soil in a forest, at a depth of 20 cm, is 5°–6 °C colder than that in a timber-free plot.

Snow has a heat-insulating effect on soil, since its thermal conductivity is quite low. The freezing depth is decreased, depending on the thickness of the snow layer.

Table 2. Temperature differences of soils on southern and northern slopes at a depth of 10 cm in July (steep slopes 20–22°)

Soil surface	Time of day			
	10	12	14	16
Bare-stripped	8.4	11.8	16.1	15.7
Grass-covered	3.2	4.3	6.2	7.4

The protective property of snow is especially important for young winter crops and fruit trees. Under a snow cover with a thickness of more than 30 cm, young winter crops do not freeze, even during heavy frosts.

4.4 Methods of Influencing Soil Temperature Conditions for Agricultural Purposes

The regulation of soil temperature conditions in different climatic regions is carried out for various reasons. In the north, it is advisable to increase the temperature of the soil, especially in spring, in order to facilitate early sowing and planting, and to create more favorable conditions for the germination and rooting of seedlings, as well as accelerate plant development. Conversely, in the south, an excessive amount of heat can overburden a plant; therefore, it is advisable to employ methods which are intended to decrease the temperature of both the soil surface and the topsoil layer.

The tilling and loosening of soil facilitates a more rapid exchange of heat within it. The loosening of a surface soil layer at a depth of 2–4 cm decreases the temperature of the under soil levels by 1°–3 °C (more in summer than in spring), while the rolling of soil increases its temperature by 1°–2 °C. As an example, data derived from observations in Northern Caucasia (Nalchik) are presented in Table 3.

The regulation of soil temperature is also accomplished by means of mulching; that is to say, by means of a special soil surface covering, consisting of peat, straw, bituminous emulsions, etc.

Mulching reduces soil temperature fluctuations. Depending on the color of the mulch, soil temperature is decreased or increased (dark mulch). For example, the mean monthly maximum temperature of the soil at a depth of 3 cm (July) on a control plot came to 32.0 °C; in an area sprinkled with coal dust, 36.2 °C; and in an area sprinkled with lime, only 25.6 °C. According to data from the Gorkiy Meteorological Observatory, soil humidity was 6–7% higher under a straw mulch during hot drought years, while soil temperature during the daylight hours was 6°–7 °C lower than in an unmulched area.

To increase soil temperature, light soils are darkened with peat mulch and bituminous emulsions. Darkening a soil during clear weather decreases the albedo by as much as 5%, and increases the absorption of radiation by as much as 15%.

Table 3. Effect of individual agricultural technology methods on the maximum (mean monthly) soil temperature at a depth of 3 cm. Nalchik (USSR)

Control plot	April	May	June	July
Control	17.7	18.9	28.0	32.0
Loosened	16.9	18.1	26.7	29.0
Rolled	18.3	20.2	29.4	33.4

Tests in the Kola Branch of the Soviet Academy of Sciences showed that the content of nitrates, mobile phosphorus, and potassium is increased in dark-colored soils, which enhances plant development.

Transparent polyethylene films, used as mulch, facilitate a temperature and humidity increase in the upper soil layer. In the mountainous regions of Soviet Georgia, the temperature at a depth of 10 cm was 8°–10 °C higher during the daylight hours, using a film of this type, than on a control plot. In coldframes (without manure), covered with glass frames, soil temperature is increased 5°–6 °C.

One of the methods for increasing soil temperature is the formation of ridges and rows, which increases the active surface by 20–25% and, consequently, boosts the absorption of radiation from the sun. According to the data derived from tests conducted in the Khibiny Mountains, the soil temperature on ridges was an average of 2°–3 °C higher over the course of the growing season than that of a level field.

Irrigation has a considerable effect on soil temperature. The tests of Chirkov in the steppe zone of the Soviet Union showed that the temperature of a soil surface is 16°–19 °C lower in irrigated areas following water release than in unirrigated areas. At a depth of 10 cm, this difference came to 5°–7 °C, while at a depth of 20 cm, the difference was 2°–3 °C. The drainage of marshy areas increases soil temperature (especially on the surface) during the summer months.

Some soil temperature decreases are also observed in the presence of forest strips, due to the emergence of thicker crop stands in the spaces between the strips.

At present, the previously cited methods for regulating temperature conditions are being more and more widely used in agricultural production, thus taking on great practical significance.

5 Microclimate and Phytoclimate

Y. I. CHIRKOV

Microclimate – the climate of a smaller region within a territory – is caused by the effect of terrain relief, the subjacent surface, and other factors which determine the disparity between air and soil temperature conditions, humidification, and wind intensity. The microclimate of slopes, valleys, hillcrests, etc., are singled out as a function of these relief features. The microclimate of meadows, fields, forest fringes, glades, shore regions lakes, etc., is produced by a disparity in the radiative heating of the subjacent surface. The microclimate of a city is formed as a result of development. A special type of microclimate, called a phytoclimate, is produced amongst the plants.

The features of a microclimate are most notably manifested in the near-soil air layers. In the presence of a marked relief, noticeable differences in soil and air temperatures are produced between areas at a distance of several tens of meters. These differences, which diminish with altitude, exist up to a height of several meters, then level off under the influence of air mixing due to turbulence and wind activity.

Under the moderate climatic conditions associated with clear calm weather, these differences are especially marked, reaching several tens of meters in height. On such days, the soil surface temperature difference between northern and southern slopes in fields with slightly developed vegetation reaches 10°–12 °C. At a height of 20 cm in the near-earth air layer, this difference reaches 3°–5 °C. On clear calm nights, a considerable microclimate disparity is produced, due to both radiative cooling and the escape of the cooled air into low-lying areas. For this reason, the minimum air temperature in the lower reaches of steep slopes, as well as in hollows, is 3°–5 °C lower than that of a level location. During light spring or fall frosts, the microclimate of these relief shapes is most hazardous for agricultural crops, causing a reduction in the length of the frost-free period as compared to a level location. Conversely, a lower intensity of light spring and fall frosts is observed, in conjunction with a longer frost-free period, in the upper reaches of the slopes, on the banks of large rivers, and along the shores of lakes.

The microclimate of slopes depends on the exposure and grade of the slope, conditional upon the variable influx of solar heat to an active surface. For this reason, snow thaws earlier on the southern slopes, the soil is heated and dries more quickly, earlier sowing is possible, and the probability of agricultural crop maturation is increased. The microclimate of southern slopes is dryer than that of plateaus and northern slopes, which is a favorable factor for agricultural crops in regions with excessive humidification. In a steppe region, a favorable microclimate is produced by irrigation.

It is necessary to study daytime and nighttime microclimatic features separately, since the physical processes of the daytime heating and the nighttime cooling of a subjacent surface take place in different manners.

A phytoclimate – the meteorological conditions produced amongst plants (grass stands, treetops, etc.) – is a modified microclimate. A phytoclimate is controlled by the structure of the plant cover; that is to say, by plant height, the density of stand (the number of plants in a unit of area of a field), the leaf surface area of the plants, the manner in which plants are distributed within a stand (a cluster stand, a row stand, etc.), and by the width of the interrow spaces. The phytoclimate of identical crops may vary over a wide range, depending on the type of stand structure used, or may be similar, as in the case of different crops which utilize a similar stand structure. Depending on the species, habitus, and age of a plant community, the density of the stand (plantation), and the sowing (planting) method, illumination intensity, air temperature, soil temperature, air humidity, soil humidity, and wind intensity values differ substantially from the values of these same factors in an open location.

In a developed stand of tall-stalked crops (corn, sugar cane, hemp), the intensity of illumination on the soil surface may be 5–10 times less than on the stand itself. Under dense tapered leaves, the air temperature at noon on a hot day is 4°–5 °C lower, and the soil surface temperature is 15°–25 °C lower, than in an unshaded area, while the relative humidity of the air is 10–20 % higher. Properly to assess the growth conditions of agricultural crops, consideration of the phytoclimate carries substantial significance.

The features of the phytoclimate of individual crops are determined by means of measuring the basic meteorological components (air temperature and humidity, illumination intensity, wind intensity) in a vertical section among plants. Here, measurements are performed both in the interrow spaces and in the rows or clusters themselves. Plant height, leaf surface area, and the number of plants in a 1 m^2 area of the field are gaged simultaneously.

The study of microclimate and phytoclimate in the presence of different relief shapes is accomplished by means of measuring temperature and humidity conditions, as well as wind activity, at several points – on hillcrests, on slopes, in valleys, etc. Here, the effect of environmental conditions is taken into account: the distance from a reservoir, from a forest strip, etc. The results of observations conducted at different points are compared to each other and to a reference location (a meteorological site), characterized by a level area. Minimum and maximum temperatures, both on the soil surface and in the near-earth air layer, air humidity, soil humidity, wind direction, and wind speed are usually determined.

Large-scale maps, characterizing the microclimate of agricultural fields and gardens, are compiled on the basis of microclimatic observations. The compilation of such maps has great practical importance in agriculture, since microclimatic disparities have a substantial effect on both plant development and the formation of a crop harvest.

Improving the microclimate of farming areas is very important with respect to increasing the yield of agricultural crops. In the northern regions, where it is excessively cold and humid, the principal methods of improving a microclimate are to reduce overhumidification of the soil, to give the soil a ridge-like surface (for potatoes, tomatoes, etc.), and to cut off the escape of cold air on slopes. In arid regions, various moisture preservation, irrigation, and timber-thinning methods for the protection of soil are used to improve the microclimate.

Forest strips change the structure and speed of wind, bringing about a wind speed decrease in the interstrip spaces, which, during hot, dry, windy periods, diminishes the disruption of the phytoclimate of field, since the phytoclimate has more humidity than the surrounding air. In winter, forest strips facilitate snow accumulation, as well as the uniform distribution of snow in the interstrip spaces.

Openwork strips with spaces along the entire strip height are considered to be superior in design. An air flow passing through such a strip is greatly weakened and its turbulence is decreased. Beyond the strip, at a distance of three to five strip heights, an area of calm is observed, while a decrease in turbulent exchange in the near-soil air layer is seen at a distance of 15–20 strip heights. This process protects the soil from wind erosion during dust storms. As a result of the decrease in turbulence in the interstrip space, a more favorable microclimate is produced than the one existing in unprotected fields.

The greatest effect of a forest strip is afforded at a distance of approximately 25 strip heights between the strips. That is to say, at a strip height of 20 m, the distance between the strips should come to 500 m.

6 The Climate of Meadows and Pastures

Y. I. Chirkov

Climate and soil conditions determine the propagation of meadow and pasture plant life around the world. Depending on heat and moisture requirements, three groups of grasses may be discerned:

1) Hygrophilous grasses (hygrophytes). This group of grasses grows in locations with well-humidified and over-humidified soils (marshes, flooded meadows, etc.). When the earth dries, these grasses die.

2) Mesohygrophilous grasses (mesophytes) grow in forest and forest-steppe regions, consisting primarily of meadow phytocenoses.

3) Drought-resistant grasses (xerophytes), growing predominantly in the steppes and in semiarid regions. Plants in this group possess a number of protective attributes which help them endure periods of dry air and soil.

Light, heat, and moisture constitute an ecological basis for the growth, development, and yield of all plants, including grasses. To ascertain the agroclimatic characteristics of a territory with respect to conditions for the growth of meadow and pasture grasses, it is necessary to be familiar with the numerical expressions which describe the requirements of these grasses for the abovementioned climatic factors.

In a moderate climate belt, a great number of natural and cultivated grass species start to grow at mean diurnal air temperatures of $+3°$ to $+5°C$, which coincides with the build up of positive total air temperatures in the $25°–45°C$ range. In the steppes, as well as in semiarid regions, certain species of pasture plants start to grow in the presence of mean diurnal air temperatures in the $+9°$ to $+14°C$ range.

Following the commencement of growth, the rate of pasture plant development depends primarily on air temperature, while the accumulation of a biomass depends on the availability of moisture. In the north, in addition to moisture, a close tie is also seen between biomass accumulation and temperature.

In drought regions, an air temperature of more than $35°–40°C$ in unsuitable for the majority of grasses. During biomass accumulation in Kazakhstan, following the disappearance of snow at a positive total temperature of $115°C$, it is possible to start grazing. Grass height in this case reaches 4–6 cm. In European countries, total temperatures are around $80°–120°C$ during biomass accumulation. Grass height reaches 8–10 cm in pastures. In spring, during the resumption of vegetative processes in semiarid regions and deserts, productive moisture reserves in a soil layer 0–20 cm deep come to approximately 20 mm, which guarantees the development of a temporary pasture plant community that subsequently dies out. Moisture reserves of less than 10 mm in a soil layer 0–20 cm deep, with an air temperature of more than 18 °C, averaged over a decade, and an air humidity deficit of 10 mm, are agrometeorological indicators of the commencement of the extinction process for these temporary plants.

Table 1. A comparative evaluation of climate and soil-climate conditions

Agroclimatic zone	Corresponding natural zone	Mean humidity index M	Comparative climatological evaluation of conditions (in points)
Moderately humid	Forested steppe	0.50	100
Moderately dry	Steppe	0.35	70
Dry	Semi-azid	0.22	44
Very dry	Typical desert	0.13	26
Dry foothill	Foothill desert	0.23	46
Very dry southern exposure	Seasonal desert	0.18	36
Moderately dry and wet foothill	Foothill steppe	0.40	80
Wet mountain	Mountain steppe	0.60	120
Very wet	Mountain forest	0.95	190
Excessively wet	High hills	1.66	75

To isolate agroclimatic humidification zones, A. P. Fedoseyev (of the Soviet Union) used the M humidification index, which was computed by means of the formula:

$$M = \frac{Bb + Oc}{\sum d},$$

where Bb = spring reserves of available moisture in a soil layer 1 m deep (in mm); Oc = the amount of precipitation from spring until the time of the maximum grass stand yield (in mm); and $\sum d$ = mean total diurnal air humidity deficits over the same period (in mm).

The agroclimatic zoning of the Kazakhstan territory is carried out in accordance with the value of the M humidification index. Conditionally, assuming that the M index is over 100 points for a moderately humid forest-steppe region, A. P. Fedoseyev made a comparative evaluation of the climate in different regions (Table 1).

It follows from the table that, depending on the magnitude of soil heat and humidity, the productivity of grasses in natural regions rarely changes (in relative units, from 26–28 to 170–190 points). Grass productivity may fluctuate considerably in the same natural region during contrasting weather conditions.

Perennial grasses yield the most plant mass growth at a soil humidity of 80–85% of the total moisture capacity. For this reason, the irrigation of pastures in the summer months, even under forest-steppe and forest conditions, has a considerable effect.

The necessity of irrigating pastures and hay fields in the northwest portion of the Soviet European territory has been confirmed by numerous researchers. According to the calculations of P. Aksomaytis, the cultivated pastures of Latvia require irrigation at an overall rate of 90–210 mm, depending on climatic precipitation conditions.

Similar results have been obtained in many countries. In England, where the irrigation of grasses was practically unknown until 1945, the area of irrigated pastures and hay fields had reached 42,000 ha by 1963. J. North demonstrated that, under the conditions prevalent in England, the water requirements of grasses fluctuated within the limits of 450–500 mm, while the average amount of precipitation over this same period of time (the growing season) came to 230–370 mm. Consequently, according to North, the estimated shortage of moisture, under the climatic conditions native to England, comes to an average of approximately 200 mm for grasses over the course of the growing season.

In Scandinavian countries, the irrigation of pastures and hay fields is carried out chiefly on large agricultural farms at present. A continued increase in the productivity of cultivated grasses in this region linked to the prospects for expanded irrigation.

In Central Asia and Kazakhstan, it is possible to produce five to six grass crops per year through the use of irrigation, while the harvesting of hay reaches 20,000 kg per ha.

7 The Climate of Grain Crops

Y. I. Chirkov

The most widely planted grain crops are wheat, corn, and rice.

Wheat is grown under various climatic conditions, from the tropics to the Arctic Circle. The area of lands in which wheat is planted now exceeds 210 million ha. The diversity of wheat varieties and strains is extremely large. Early-maturing and late-maturing wheat strains of wheat require a varying supply of heat. With respect to moisture, a considerable difference is seen between wheat strains which are drought-resistant and those which are not. Wheats are divided into the following varieties, according to the conditions required for the occurrence of the vernalization stage: the winter type, which is planted in the fall and in which the vernalization process takes place at reduced temperatures; and the spring varieties, which are planted during the vernal period.

The characteristic curve of climatic conditions over a geographical area in which winter wheat is grown can be traced, with allowance for conditions during both the active vegetation and winter periods. With respect to hibernation conditions, three types of region are isolated as the most variable climatically: I – regions with a stable winter dormancy period for wheat (below-zero temperatures during the coldest month); II – regions with an unstable winter dormancy period; and III – regions without a winter dormancy period.

In region I, located primarily in Eastern and Central Europe, the Northern portion of Asia, and the Norther portion of America, the length of the active growing season amounts to approximately 4 months, almost of which comes within the fall growing season.

The northern boundary of this geographical area, as well as the eastern boundary in Europe and the western boundary in Asia, is determined by the length and severity of the hibernation period (a mean monthly air temperature of below $-10\,°C$ in January, and a snow cover residence time of more than 4 months). In the northern reaches of region I, planting is carried out during the second half of August, while the harvesting season starts at the beginning of August. The dormancy period, wherein mean diurnal air temperatures are lower than $+5\,°C$, continues for more than 6 months.

In region II, having an unstable winter dormancy period, the continuous snow cover often melts in winter. Summer is characterized by great humidification differences: a number of areas within the region (Central Asia, Iran, Pakistan) are very dry, while the southeastern portion of Asia is very humid.

Region III, having no winter plant dormancy period, encompasses the subtropics and the mountainous areas of the tropics. In this region, wheat is grown during the cold rainy season. Both winter and spring wheat are grown in this region simultaneously. Here, the planting time for wheat depends, to a considerable extent, on humidification conditions, as well as the onset of the rainy season. Consequently, winter heat lingers among winter crops in the subtropics, and therefore, these crops mature later than spring wheat. Total active air temperatures (the temperature total over a period with mean diurnal temperatures higher than 10 °C) are commonly used as agroclimatic indices for heat assurance characteristics during the growth of winter wheat. The northern heat assurance boundary for winter wheat during summer extends over the isotherm of the temperature total of 1500°–1800 °C for 80 % of the year (depending on the strain). In winter, the recurrence of minimum soil temperatures on the order of − 15° to − 18 °C at the wheat nodule tillering depth is a limiting factor in the hibernation of winter wheat. At these temperature values, the winter killing of the wheat occurs. The soil temperature minimum depends on the air temperature and the snow cover height. In areas with light winter snowfall, at a mean temperature of − 10° to − 12 °C in the coldest month, the probability of winter wheat crop damage is significant. A complicated index developed by Ye. S. Ulanova is used to assess agroclimatic conditions for the formation of a winter wheat harvest in the Soviet Union:

$$K = \frac{W + R}{0.01 \sum t},$$

where W = spring reserves of productive moisture in a soil layer 0–100 cm deep during the resumption of growth; R = total precipitation from the resumption of growth until full maturity; and $\sum t$ = air temperature totals over the same period.

Spring wheat is grown almost exclusively in the Northern Hemisphere. Agroclimatic conditions for the cultivation of spring wheat are characterized by a smaller degree of difference than those for winter wheat. The northern boundary of the geographical range is determined by the insufficiency of heat and, depending on the traits of the strain, extends along the isotherm of active temperature totals of from 1400°–1600 °C.

The planting of spring wheat in dry areas with light snowfall is carried out at mean diurnal temperatures of 5°–6 °C (in the humid regions of the north, at 8°–10 °C). The maturation of grain in the northern reaches of the geographic range takes place at mean diurnal air temperatures of 15°–16 °C (in the south, at 18°–25 °C). A considerable portion of the geographical range for spring wheat is located in arid regions (the steppes of the Soviet Union, Canada, and the United States), where harvest size is determined by spring reserves of soil moisture and by the distribution of precipitation in summer. In these regions, measures aimed at increasing the moisture assurance of young crops are widely used.

Corn is second with respect to geographical distribution range, and third with respect to planting area size (106 million ha), losing out to wheat and rice only. The principal planting areas are situated between 50° northern latitude and 40° southern latitude. However, when growing corn for silage, corn crops in Europe extend as far as 60° north latitude. Climatic conditions in a geographical corn cultivation range are characterized by a great diversity.

Corn is an extremely polymorphous species, made up of an immense number of strains and hybrids, differing very with respect to earliness or heat and moisture consumption. Nevertheless, a comparatively high development temperature is common for a predominant number of the strains. During the germination period, the lower temperature limit is 8°–10 °C; for the period from

Table 1. The relationship between total effective air temperatures (y) and the number of leaves (x) formed in corn strains and hybrids differ with respect to early maturation

No. of equation	Period	Equation of regression	R
(1)	Tasseling	$y = 30.2\,x + 31.8$	0.86
(2)	Young maturity	$y = 35.2\,x + 241.1$	0.83
(3)	Full maturity	$y = 41.6\,x + 290.5$	0.80

planting until maturation, 10 °C. For this reason, the moisture consumption of corn is most precisely expressed in effective temperature totals of more than 10 °C ($t° - 10$ °C). That is to say, 10 °C is subtracted from the value of the positive mean diurnal temperature and the remaining values are totalled. Depending on the early-maturing nature of a strain, the temperature total from planting time until maturation fluctuates over a large range. A basic difference arises between the sprouting and the tasseling periods. A stable relationship exists between the number of leaves formed during this period and a total effective temperature higher than 10 °C, accumulated over this same period.

On the average, a total effective temperature of 30°, $\pm 2°$ ($\sum t > 10$ °C), is reached at the time that the first full leaf appears (for regions where the mean temperature in July does not exceed 20 °C). Thus, under similar conditions, strains which form 20 leaves between the appearance of the third full leaf and the tasseling period (late-maturing strains) reach the tasseling stage having accumulated a total effective temperature twice as high (200 %) as that of early-maturing strains, which form 10 leaves over this same period.

During subsequent periods of development, the difference in total effective temperatures between early-maturing and late-maturing corn strains is considerably lower. For the period between tasseling and young maturity, the difference does not exceed 20 %; for the period between young maturity and full maturity, this difference does not exceed 25 %.

Since the leaf number index characterizes the early-maturing nature of a strain, both during the individual corn development periods and, as a whole, over the entire growth period, the present author has devised equations for the relationship between the number of leaves, which characterizes the early-maturation properties of a strain, and the total effective temperature required for the onset of the basic developmental stages (tasseling, young maturity, full maturity), taking the date of planting into account (Table 1).

The equations presented in Table 1 make it possible to calculate the total effective temperature required by corn strains and hybrids which differ with respect to early maturation, and to compare these totals with heat resources for the purpose of the agrometeorological zoning of corn. These equations may be used for corn-growing regions located north of the 20 °C July isotherm.

For more southern regions, where a significant portion of the growing season takes place in the presence of an air temperature of more than 20 °C, so-called ballast temperatures should be taken into consideration.

It has been established by means of field tests that, during an air temperature increase, the rate of development is accelerated for corn in accordance with an increase of up to approximately 19°–20 °C, in the mean diurnal temperature, after which a subsequent acceleration of development does not occur. For this reason, the mean total effective temperature occurring during the appearance of the first full leaf increases in the southern regions. At a mean air temperature in the 12°–20 °C range, this total is stable, amounting to approximately 30 °C. Thus, at a mean air temperature of 25°, the total reaches 40 °C. Consequently, in the southern corn-growing regions, it is necessary to allow for high temperatures which do not accelerate the development of a corn crop (ballast temperatures) when estimating heat resources.

In order to ascertain more accurately the degree of heat assurance in southern regions over the corn-growing season, and to compare this assurance among the different areas in the geographically cultivation range, the relationship between total effective (ballast-free) temperatures (y) in the 10–20 °C range and an overall total temperature higher than 10 °C was calculated. This relationship, over a total effective temperature range of 600°–1800 °C, has a near-linear nature and is expressed by means of the equation:

$$y = 0.74x + 140; \quad R = 0.94,$$

where $y = -\sum t°$ in the 10°–20 °C range and $x = -\sum t > 10\,°C$.

Using this equation, it is easy to determine effective heat resources in the geographical corn cultivation range.

For the purpose of estimating heat resources in the continental areas of a moderate-climate region, as well as in the subtropics, a correction is introduced over the duration of the frost-free period, which is shorter in this region that the period with a temperature higher than 10 °C.

A correction is also introduced for continental areas in the Soviet Union with respect to the value of the diurnal variation amplitude of the temperature which is reached, on an average, over a month (July), 16°–17 °C, since the amplitude value increases ballast temperatures during the daylight hours. For European countries, where the mean monthly diurnal temperature variation amplitude exceeds 12 °C, this correction is not introduced.

In order to compare heat resources with the total temperatures, which express heat consumption for the development of strains and hybrids that differ with respect to early maturation, Table 2, which demonstrates how the corn development stage will be supplied with heat 80 % of the year, depending on the early-maturing nature of the strain, was devised by Professor Chirkov (USSR) in 1960.

Table 2 was compiled for total effective temperatures of 400°–1200 °C, with allowance for the exclusion of ballast temperatures.

The data presented in Table 2 are used during the agroclimatic zoning of corn strains and hybrids which are different with respect to early maturation.

In order to lay out the zoning, 80 % of the total temperature assurance is calculated for each point, figured from the lower limit of 10 °C. Next, the ballast temperatures are eliminated from these totals in accordance with Eq. (4) and total effective ballast-free temperature isolines are traced in conformance with agrocli-

Table 2. The assurance of heat for the development of corn as a function of the thermal region and early maturation

Total effective temperatures (°C) Assurance 80% of temp.	Development stage	Characteristics of early types and hybrids with no. of characteristic counts number of leaves			
		Medium early 13–14	Medium mature 15–16	Medium late 17–18	Late mature 19–20
400– 500	Tasseling	80– 90	55– 80	45– 70	35– 55
	Young maturity	15– 40	8– 20	3– 10	5
	Full maturity	5– 10	5	–	–
500– 600	Tasseling	90–100	80– 98	70– 90	55– 75
	Young maturity	40– 60	20– 40	8– 25	5– 10
	Full maturity	15– 35	5– 10	5	–
600– 700	Tasseling	100	98–100	90–100	75– 95
	Young maturity	60–100	40– 65	25– 50	10– 30
	Full maturity	35– 60	10– 35	5– 15	5
700– 800	Tasseling	100	100	100	95–100
	Young maturity	80– 95	65– 80	50– 70	30– 55
	Full maturity	60– 80	35– 60	15– 40	5– 15
800– 900	Tasseling	100	100	100	100
	Young maturity	95–100	80– 98	70– 90	55– 75
	Full maturity	80– 95	60– 82	40– 55	15– 40
900–1100	Tasseling	100	100	100	100
	Young maturity	100	100	90–100	75–100
	Full maturity	100	82–100	55– 90	40– 80
1100–1200	Tasseling	100	100	100	100
	Young maturity	100	100	100	100
	Full maturity	100	100	90	80

matic mapping laws. Afterwards, the heat accurance isolines are plotted on the map for the tasseling stage (the use of corn for green fodder and cob silage), young maturity (silage with cobs), and full maturity (using corn as dry grain).

Corn productivity is determined, to a considerable extent, by heat and moisture resources. The climatic conditions which assure the output of large and stable corn harvests for grain and silage are characterized by good soil moisture reserves (140–180 mm of productive moisture in a soil layer 1 m deep) during the major portion of the growing season, in conjunction with optimum temperature conditions (a mean diurnal temperature of 18°–22 °C). When growing corn for grain, heat resources must guarantee the annual maturation of higher yield strains and hybrids: for silage with cobs, at the onset of young maturity; for green fodder and silage without cobs, at the onset of the tasseling stage.

When growing corn (especially for silage), the spacing of plants varies, depending on the early-maturing nature of the strain. Short, early-maturing strains, with a smaller leaf surface than late-maturing strains, are grown with a larger spacing than the latter. Increasing the spacing of young early-maturing strains makes it possible for the crops to form an form an optimum leaf surface area 30–40 thousands of m^2 per ha., as well as allowing the potential for photosynthesis in the early-maturing crops to reach the value of the late-maturing strains.

For this reason, it is necessary to take a plant spacing which is optimum with respect to a particular species, as well as climatic conditions, into account when assessing the climatic assurance of corn productivity.

In regions where heat conditions guarantee the annual maturation of grain in a particular group of strains, productive moisture reserves in the soil during the period from the fourth stage[1] of cob organogenesis to the young maturity stage (the eleventh organogenesis stage) determine corn productivity. With productive moisture reserves of 120–180 mm in a soil layer 1 m deep, climatic conditions which guarantee the highest productivity for a particular group of strains are classified as optimum (5 points). Moisture reserves in the 80–120 mm range are classified as good (4 points), 40–80 mm – satisfactory (3 points), and below 40 mm – poor (2 points). The calculation of 80 % of the moisture reserve assurance, using the above-mentioned scales, and the computation of heat assurance data will characterize climate resources which are applicable to the assessment of corn productivity.

The assessment of moisture assurance for corn productivity may also be carried out by means of a water-variable hydrothermal coefficient.

The modified formula for the hydrothermal coefficient takes the form:

$$K = \frac{0.5\,R_1 + R_2}{0.18\sum t_2}.$$

In order to take moisture resources into account during the growing season, using this equation, precipitation, R_1, over the fall-winter period in included, with a coefficient of 0.5, as well as R_2 over the corn-growing period, while $\sum t_2$ constitutes total temperatures over the same period. The hydrothermal coefficient, calculated for each point in accordance with the specified formula, is compared to grain yield and corn plant mass data for the purpose of determining the closeness of the relationship, as well as devising regression equations.

If there is no reliable perennial data with respect to yield, an assessment of the moisture assurance for corn germination may be carried out in the first approximation by means of a modified hydrothermal coefficient formula, using the following scale:

K	Scale
1.2–1.3	5
1.0–1.1	4
0.8–0.9	3
0.6–0.7	2
≤ 0.5	1

With a three-point moisture assurance estimate, occasional waterings are necessary; with an estimate of one to two points, the cultivation of corn is possible only when employing irrigation.

---

[1] The fourth stage of coborganogenesis stets in when two thirds leaves have appeared (from the total number which during the tasseling stage)

The cultivation of rice is carried out primarily in regions with a monsoon-type climate, where the rainy season is replaced by a dry season. With the onset of the rainy season, rice seeds germinate quickly due to soil humidification, then the lowlands are flooded with rain over the tillering stage and the young crops are submerged until young maturity. Furthermore, the water layer evaporates during a decrease in precipitation and the maturation period takes place without submersion.

At the present time, rice is grown predominantly in fields which are flooded with water, where the water layer is regulated either by means of rain water drainage (in monsoon regions) or through the use of irrigation systems. Rice is a heat-loving crop; thus, heat resources restrict the geographical range for its cultivation. Rice strains which mature earliest, having a growing season approximately 110 days in length, need total active temperatures of approximately 2200 °C. These strains are grown in the Far Eastern regions of the Soviet Union (the maritime district). Strains which mature at an average rate, grown in the southern portion of the Soviet European territory, Bulgaria, and Italy, require total active temperatures of 2800°–3200 °C. In tropical areas, late-maturing strains require active temperatures of 4000 °C over the course of the growing season.

8 The Climate of Trees, Orchards, and Forests

B. PRIMAULT

8.1 Geometric Shape

If one wants to study the effects of climate on different plants, and more specifically on trees, it is essential to consider first of all their geometric shape. Actually, it is on this that the impact of meteorological factors will depend and, therefore, also the importance of the repercussions of climate on their development. This occurs particularly through the effects of radiation (see Sect. 8.2.1).

8.1.1 Isolated Trees

An isolated tree may be compared to a long cylinder of very small diameter, surmounted by a large sphere. This isolated tree (we now think first of all of an apple tree and not, for example, of a conifer) may receive in the same manner, from all sides, not only the direct radiation from the sun and the diffuse radiation from the sky, but also the radiation emitted by the environment and especially by the soil. Thus, with respect to the impact of radiation, it can be stated that it constitutes a perfect receptacle for circumglobal radiation. Such a solid body emits its own radiation in all directions (see Fig. 1A).

In the case of a conifer, the trunk or the cylinder supporting the entire tree loses its importance in relation to the entire tree. The crown, by itself, may be compared to an upright cone. Since the base of the cone, compared with the whole, is only of relative importance, the contribution due to the radiation emitted by the soil is negligible in comparison with that produced by the direct and diffuse radiation. In energy calculations concerning a tree of such shape, it is necessary to consider primarily the global radiation and not the circumglobal radiation as in the preceding case (see Fig. 1B).

Depending on the need, the reference body can be infinitely varied by cutting it into a succession of simple geometric shapes. It will always be, however, a body having a circular horizontal projection (see Fig. 1C).

In the latter two cases, its own radiation is emitted toward some prevailing direction, perpendicular to the largest surface or surfaces.

8.1.2 Orchards

Orchards of Quadrilateral or Quincunx Pattern. In the case of such orchards, we are generally dealing with simple shapes, such as those just described. The only difference between an isolated tree and a tree in such an orchard is the magnitude of the radiation emitted by the other trees in the plantation. Thus, there occurs here an exchange of energy radiated from one tree to the next and vice versa, which makes a considerable contribution compared to that received by an isolated tree.

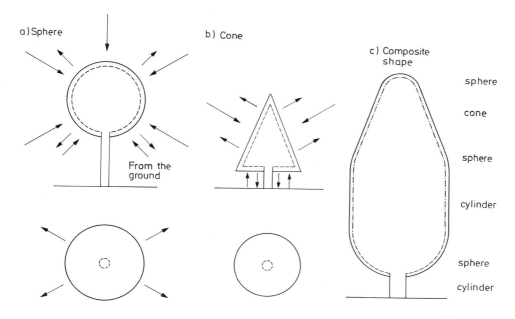

Fig. 1. Geometric shapes related to isolated trees (examples)

With respect to the wind, the series of obstacles that it meets in an orchard and which it can only bypass with difficulty, act as a brake. This is the more effective the larger the orchard and the more closely spaced the trees.

With respect to all the other effects (temperature, precipitation, humidity, etc.), there is practically no difference between an isolated tree and an orchard tree. The production of the entire plantation, therefore, will practically be the sum total of the separate production of isolated trees.

Orchard in Lines. If, instead of placing the trees in geometric alignment, but respecting their individuality, they are planted so that the branches of one touch those of another in a given direction, one is no longer dealing with isolated geometric shapes that can be considered independently of one another (see Sect. 8.3), but with simple geometric shapes formed by several juxtaposed plants. Thus, for example, low hedgerows can be assimilated to a rectangular parallelepiped placed on the ground. Actually, for reasons of antiparasitic treatment, harvesting, and soil cultivation, the present trend is to lower the branches as much as possible and to trim the plants as geometrically as possible to facilitate the passage of the machines through the rows. As a result of this, the effect of soil radiation becomes negligible in comparison with that of total radiation. We are dealing with the sum total of the energy received by a horizontal surface (the top of the hedgerow) and by two large lateral surfaces, while the two ends of the hedgerow (small vertical surfaces) become negligible in comparison with the other three (Fig. 2).

Here also the wind can penetrate inside the orchard and act on evaporation and transpiration inside it; but a careful arrangement of the rows in relation to the

Fig. 2. Geometric shape of an orchard row

prevailing wind makes it possible to slow it down more or less. The effect is maximum if the wind blows perpendicularly to the rows, absent if it blows parallel to them.

8.1.3 The Forest

In contrast with the orchard, the forest is no longer a succession of isolated or integrated geometric shapes, but constitutes a compact whole, which can be assimilated to a deformation in the geographical relief.

With respect to radiation, the receiving surface of the forest (the complex of the treetops) is identical with the surface of the ground. The total impact affects all the trees and only depends on the exposure and the slope of the ground, thus only on orographic features. We shall return later to certain features, such as treetops, edges, clearings, etc.

With respect to the wind, while the forest does not constitute a direct obstacle to its course, the differences in height among the crowns of the trees greatly increase the friction coefficient of the air. This is markedly greater above a forest than on a bare land or at sea. Consequently, the air flow, nearly laminar at sea or on bare land, is disturbed while passing over a forest : there, a marked turbulence is produced. Furthermore, the succession of trees inside the forest prevents the wind from penetrating deeply into it. It follows that the forest as a whole acts upon its own microclimate; decrease in the intensity of the wind, in evaporation and transpiration brings as a consequence, increase in humidity.

The interception of radiation by the crowns of the trees decreases the heating of the ground in the forest, which reduces the temperature in daytime inside the forest. On the other hand, by night, since the radiation from the soil is intercepted by the leaves and to a large extent sent back to the ground, the temperature inside the forest is higher than on open land.

8.2 External Actions on an Isolated Tree

8.2.1 Atmospheric Agents

Radiation. The atmospheric parameter which certainly has the greatest impact on the development of a plant – in this particular case, an isolated tree – is radiation. The plant obtains most of its energy from it to transform carbon dioxide into glucose and to manufacture its other organic components, particularly vitamins and proteins. We have seen in the preceding section the role played by the geometric shape of the tree in the amount of energy received by radiation. Furthermore, the principal difference which appears in the results due to direct radiation and diffuse radiation, on the one hand, and to the radiation emitted by the environment – particularly the soil – on the other, consists in the diversity of frequencies. Thus, global radiation (the sum total of direct radiation and diffuse radiation) is primarily short-wave, that is, from ultraviolet to the near infrared. The radiation emitted by the environment, that is, all solid objects and particularly the soil, is limited to the infrared spectrum. Therefore, in this case it is a long-wave radiation, basically calorific. The use that the plant can make of it depends above all on the type of plant, that is, on its species.

Temperature. The movement of the cell juice of each of the living cells forming the plant is a function of its temperature. The intensity of the life of the cell, including its power of absorption, its transformation possibilities and its capability to divide itself – thereby increasing the number of cells in the whole plant – depends on its temperature; but the temperature of the cells varies from place to place within the tree. Thus, a leaf exposed to the sun will have a higher temperature than one located inside the crown or a cambium cell in the trunk. Therefore, if one wishes to know the productive capacity of the whole plant, these peculiarities must be taken into account. The total activity of the cells forming an isolated tree actually depends on the activity of each one of them, and therefore on their temperature, integrated in relation to the whole plant.

Furthermore, the morphological composition of a leaf exposed to direct radiation is not identical with that of another leaf, shielded from the sun, from its time of sprouting, by its companions. The capabilities for assimilation of the various tissues – particularly of the nervation tissues – depend not only, therefore, on the radiation received by the leaf, but also, or even more, on its morphological and physiological condition.

Generally speaking, it may be assumed that the temperature of the leaves located on the periphery of the crown is a function of the air temperature and of the radiation received. From this energy received from the outside it is necessary to deduct the energy required for transpiration, for the processes of synthesis, and the radiation of the leaf itself. Inside the crown, on the other hand, the temperature of the leaves is very close to the air temperature, as measured in a meteorological shelter. Actually, the periphery of the crown acts as a shutter to intercept radiation, but lets most of the wind pass through, so that an equilibrium is established with the exterior of the covered space, as in the case of the meteorological shelter. However, also in this case, one observes a decrease in temperature as a result of the three losses mentioned above.

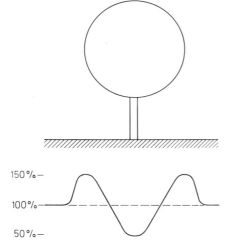

150%—
100%—
50%—

Fig. 3. Precipitation, approximate distribution

Air Humidity. As a result of the diffusion phenomena, the humidity of the air inside the crown of trees tends to reach an equilibrium point with the outside humidity. A gradient is maintained, however, irrespective of environmental conditions, by the fact that the tree in its activity releases into the air present inside the crown considerable amounts of humidity, both through the simple respiration of the leaves and, even more, through the phenomena of transpiration. Consequently, the humidity inside a crown will always be at least slightly higher than the outside humidity. Thus, this gradient, constantly maintained and renewed, produces as a consequence a permanent exchange of water vapor and air between the inside of a tree and the surrounding air, and this irrespective of the environmental conditions (except saturated — 100 % RH-air outside).

Precipitation. The liquid water which reaches the outer surface of a tree – whether it be precipitation water (rain, drizzle, snow, sleet, etc.) or dew – can only in part reach inside the crown. Actually, that water is intercepted by the peripheral leaves and deflected toward the outside of the tree, like the water which falls on the tiles of a roof. Therefore, the precipitation received on an open field is not identical with that which is measured at the foot of a tree. The water which falls on an isolated tree is thus divided quite unequally among the surface of the leaves, the periphery of the cover of the crown and the base of the tree itself (see Fig. 3). The periphery of the crown is thus much better watered that the base of the tree; but a considerable amount of precipitation (approximately 4 mm at the beginning of each shower) is retained at the periphery of the crown and inside it, to water the leaves, the twigs, the branches and the trunk, as well as the flowers or the fruits. This water does not drain down to the ground, but evaporates in place. Therefore, the presence of trees causes an increased evaporation of the water originating from precipitation.

Wind. The action of the wind on an isolated tree may be of two kinds, physiological, and mechanical.

The physiological action has two basic effects: the equilibrium of temperatures with the outside environment, on the one hand, and the increase in transpiration, which means a loss of water and, therefore, a decline in the temperature of the leaves, on the other.

The mechanical action tends to bend the tree along the path of the prevailing wind. This action, however, only results in a deformation of the crown if the prevailing wind blows regularly in the same direction every day. Considerable deformations are then visible in the crown, and this happens on the plains, in valleys with constant winds, on the seashore or on mountain ridges. It is most often a case of strong, regular winds, blowing during the entire vegetative period.

Local breezes, weak by nature and only blowing in some periods of the year, can have similar consequences on the behavior of trees. They are those which appear by night in spring. Actually, they act at the time of the lengthening of the young twigs, by causing them to bend always in the same direction. If, under these particular conditions, the trunk is vertical, the young branches are deflected and the crown as a whole is bent in the direction of the wind. From afar, the tree appears symmetric, because the deviations are minimal, in contrast with what we have already described.

8.2.2 Subterranean Agents

The growth of a tree is affected not only by the meteorological conditions that it encounters in the atmosphere, or by their action on its aerial parts (trunk, branches, twigs, leaves, flowers, fruits), but also, and in certain cases even more, by the conditions determined for it in the soil, that is, at the root level.

Temperature. The principal function of the roots of a tree is not only to anchor it to the ground, but also, or even more, to enable it to absorb the water and mineral salts. The activity of those underground vegetative organs depends, just as much as that of the aerial organs of the plant, on their temperature. Such temperature, however, is no longer affected by radiation or evaporation, but only by the temperature of the environment: the soil temperature. If one desires the establishment of an equilibrium between the capabilities of the root apparatus for assimilation and the activity of the aerial parts of the plant, the temperatures felt by the two groups of cells should be in harmony with each other. Consequently, it is necessary to know the temperature pattern inside the layer penetrated by the roots. That temperature depends in part on the thermal effect of the deep layers, but the greatest fluctuations are caused by the surface losses and gains through radiation, evaporation and, secondarily, through contact with the air which lies over it. The study of the temperature conditions in the soil of an orchard, therefore, is essential to determine whether a balance is achieved between the activity of the underground parts and that of the aerial vegetative organs of each plant.

Water. As a result of the transpiration of the leaves, the tree suffers permanent losses of water to the atmosphere. That water must be replaced by an intake originating from the root apparatus: therefore, there is a constant flow of water from the roots toward the crown; but, for the roots to be able to absorb water, they must not only be kept at a sufficient temperature, but they must also find that water in the soil into which they penetrate.

The study of the water balances through the measurement of precipitation and the calculation of evaporation and transpiration, as well as the study of the movements of water in the soil (capillarity, percolation, etc.) will therefore constitute important elements in the knowledge of the life conditions of the tree.

8.2.3 Synthesis

If one wishes to examine the conditions of the growth of an isolated tree, it is therefore essential to posses basic data concerning the various meteorological parameters both above the ground – that is, in the air – and inside the ground itself. The equilibrium of the plant will depend on the combination of all these particular effects and, without such equilibrium, there is no growth. Thus, even though each air parameter may be favorable to assimilation, if the soil temperature or humidity is insufficient, the plant will be unable to benefite from those atmospheric conditions to produce vegetable matter. The opposite is equally true.

8.3 Transition from the Isolated Tree to the Orchard

As we have seen at the beginning of this chapter, an orchard is composed of a number of isolated trees or groups of trees, and the sum total of the growth of each of them represents the production of the orchard as a whole. Since the environmental parameters do not act in a uniform manner on the aerial parts of the plants, but such action largely depends on the geometric shape of the trees or groups of trees, it is essential to study some special cases.

8.3.1 Orchard of Quadrilateral or Quincunx Pattern

In the case of an orchard with a regular arrangement, the trees actually constitute isolated components. Only the wind, as an independent meteorological parameter, must be considered differently in the case of orchard trees and in that of isolated trees. With respect to all the other elements (radiation, temperature, humidity, precipitation), the trees can be regarded as a succession of isolated trees.

The orchard reacts as a whole in the case of the wind, in the sense that the latter is slowed down by the successive action of each tree. Among the trees in the first row, the wind is modified as if by a juxtaposition of channels separated by columns (thus, the wind is accelerated in the intervals and slowed down in the crowns, see Fig. 4). The wind streams are, however, immediately deflected by the plants in the second row and turned upward (selection of the path of least resistance). The rupture of the wind flow inside the orchard causes as a result the fact that, beginning from the third row of trees, the wind is slowed down by 50% in comparison with that in open spaces.

8.3.2 Orchard Arranged in Rows

In the case of an orchard arranged in rows, the support that the trees give one another radically alters the problem. Actually, the geometric shapes juxtaposed in space are of a quite different nature.

We have seen above (Sect. 8.1.2) the importance of this juxtaposition of the trees without breach of continuity with respect to radiation. However, this meteorological element is not the only one affected: the humidity inside the

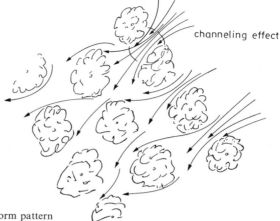

Fig. 4. Wind flow in an orchard of uniform pattern

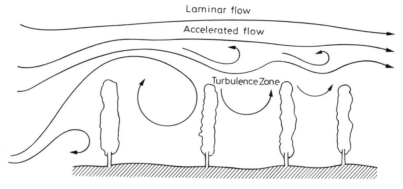

Fig. 5. Wind flow in an orchard arranged in rows (cross-wind)

crowns is also affected, as a result of the absence of lateral diffusion between the crowns which touch each other. Therefore, there is diffusion of humidity toward the environment only through the top and the sides of each row, which creates a quite peculiar microclimate inside the row. The same applies, but to a lesser extent, to temperature. What we have stated concerning the temperature inside the crowns of isolated trees can actually be applied by analogy to the block formed by the succession of crowns in the row.

With respect to the wind, its effect on an orchard arranged in rows basically depends on the angle of attack with reference to the row itself: if the prevailing wind blows in the same direction as the rows in the orchard, neither its force nor its direction will be affected by the rows to any considerable extent, while inside the crowns it will be practically absent. If, instead, it blows perpendicularly to the alignment of the trees in the orchard, it will be intercepted by the first, and eventually by the second row. Later, its action will be greatly decreased and its velocity slowed down to approximately 20% of what it would be in an open field. There appear then an acceleration of the airflow above the orchard and a very turbulent layer between them (see Fig. 5).

8.4 Transition from the Isolated Tree to the Forest

In Sect. 8.1.3, we have shown the action of the forest on the ground. It is then a compact mass which follows the irregularities of the ground and reacts as a uniform whole. Therefore, it is no longer a case of isolated trees, as in the case of the orchard arranged in a quadrilateral or quincunx pattern, or of a juxtaposition of large geometric blocks independent of one another, as in the orchards arranged in rows, but an aggregate mass. Then it is preferable to regard the forest as a whole as a simple geometric shape, as a sort of irregular terrain relief having its own properties. The action of the meteorological parameters on each tree then depends on the position of that tree in relation to the whole or of certain parts of the same tree in relation to the outer surface of the geometric shape formed by the forest.

8.4.1 Boundary Climate

In a certain sense, the edge of a forest can be compared to the first row of an orchard arranged in rows, provided that only its top side and one of its lateral walls are considered. The side of the forest area facing the outside directly receives the effects of the environment. The inner side of the tree should be regarded as a tree inside the forested area. Consequently, the trees which constitute the edge of the forest are asymmetric, developed like the orchard trees in rows insofar as the sector parallel to the edge of the forest is concerned, and like trees inside the forest (see Sect. 8.4.3) insofar as the side facing the forest is concerned. The climatic effects felt by those trees are therefore quite different: they are exposed to the wind from the outside and to the circumglobal radiation on part of their crown, to the global radiation insofar as their tops are concerned, but they are in the shade and shelter by the forest cover for the greater part of their periphery. The effect of the various meteorological parameters on their growth is therefore highly complex.

8.4.2 Climate of the Upper Part of the Crown

The upper parts of the crowns, and particularly their peripheral components, such as the treetops, are subject to climatic conditions approximately similar to those found on bare ground. From the viewpoint of radiation, the peripheral elements feel the global radiation, which reaches the upper part of the forest, and their vegetative organs are strongly heated by the sun. While the humidity there is relatively low and similar to that of the surrounding air, the wind conditions are rather different. In this respect, in contrast with what would happen on bare ground, the wind which circulates above a forest is seriously disturbed by the fact that the surface is not uniform, but consists of a series of obstacles of small size, placed in a dispersed order (see Fig. 5, turbulence zone). Therefore, there occurs an extreme perturbation of the wind flow and the formation of a layer in which motion is very turbulent. Furthermore, one can attribute to this circumstance the fact that a wind gust, initially of a laminar pattern, may, above a forest, be transformed into a miniature cyclone, which will generate forces sufficient to break or uproot certain plants.

From the thermal viewpoint, the effect of the surrounding air is considerably altered by the radiation received and/or by the radiation emitted. Thus, the outer surface of a forest is warmer than the surrounding air by day, colder by night. The daytime heating causes an increased evaporation and transpiration in the upper parts of the tree, and therefore a considerable need for kinetic energy to enable water to rise from the root apparatus toward the top.

8.4.3 Inside Climate

Since the forest forms a compact block, it is not surprising to find even inside the forest a very peculiar microclimate, often necessary for the young trees to sprout and to develop. All the meteorological parameters mentioned above under Sect. 8.2.1. act differently on the plants located inside the forest than on isolated trees.

Radiation. The global radiation which reaches the upper surface of the forest is intercepted by the peripheral plant elements. Consequently, inside the forest, only a limited radiation is apparent, the intensity and the composition of which both largely depend on the nature of the upper plant organs. Thus it happens that certain bands of the global radiation spectrum are completely absorbed by the leaves in the crown and therefore disappear completely from the light which prevails inside the forest. The plants which live in that microclimate generally adapt rather easily to those specific conditions and the morphology of their leaves is thereby modified. Furthermore, for this reason there is a distinction between light leaves and shade leaves, light trees and interior trees.

Temperature. The edge of the forest and the upper part of the crowns act as a meteorological shelter. The thermal fluctuations inside the forest therefore are less marked than outside, and the energy inflow due to radiation is smaller there, if not nonexistent. Consequently, the heating by day and the cooling by night are thereby slowed down, and the daily temperature curve is much less uneven. The only basic difference between the temperatures measured in a meteorological shelter and those originating from inside a forest area consists in the fact that the latter are affected by the transpiration of the plants and the evaporation of water from the ground. Therefore, during the growing season, there is a loss of heat energy due to those two phenomena, a loss which results in a slight decline in the overall temperature.

Humidity. In the preceding paragraph, we have shown that, as a result of plant metabolism, large amounts of water are released by the plants into the atmosphere. Consequently, air humidity is higher inside than outside a forest.

Precipitation. We have seen in Sect. 8.2.1 that an important part of the precipitation which falls on an isolated tree is intercepted by the leaves and that only a very small part of it reached the ground under the tree (see Fig. 3), the latter acting as a roof to deflect the precipitation toward its periphery. In a forest, the crowns of the trees intercept the same amount of water as an isolated tree, insofar as the watering of its leaves and the evaporation of that surface water are concerned. Once watered, however the forest no longer acts as a roof deflecting the precipitation toward its periphery: the trees drip onto each other, so that the amount of water which falls to the ground, from that moment, is practically the same as in the case of open ground. Therefore, a decrease is found in the total amount. If the water falls obliquely (during strong winds), part of it drains along the trunk before reaching the ground. Thus, apart from a certain decrease at the beginning of a shower (of the order of 4–6 mm), the amounts of water collected on the ground in a forest are practically the same as if they fell outside it.

8.4.4 Effect of Exposure

While, in the case of isolated trees or of orchards following a geometric pattern, the slope of the ground only plays a very secondary role in the growth of a plant, the same does not apply to orchards arranged in rows, nor, for even better reasons, to the forest.

In the case of orchards arranged in rows, the angle formed by the successive rows with the line of steepest slope of the ground must be taken into account both with respect to the radiation received (shade produced by a row over the next) and with respect to the flow of cold air and the action of the wind.

In the case of the forest, exposure plays in it a role as important – and applicable by analogy – as in the cultivation of fields. Thus, the angle of incidence of direct radiation on the upper surface of the forest is one of the factors in evaluating the total effect of radiation on the growth of the plants. Consequently, in the study of the climate of a forest, in addition to the traditional meteorological factors, the exposure and the slope of the land will be taken into account.

Literature

Baumgartner, A.: Entwicklungslinien der forstlichen Meteorologie. Forstwissenschaftliches Zentralblatt 86 3, 156–175; 4, 201–220 (1967)

Buys, M.E.L., Kotzé, A.V.: Forecasting the maturity of fruits. The Deciduous Fruit Grower 13, 335–341 (1963)

Flüeler, H., Gysi, C.: Beitrag zur Kenntnis von vier Laubmischwaldstandorten im Lehrwald der ETH. Schw. Z. Forst. 121, 39 (1970)

Hills, G.A.: Regional site research. The Forestry Chronicle 36, 401–423 (1960)

Krivsky, L.: Bestimmung der vorherrschenden Windrichtung aus Windfahnenbäumchen. Meteorol. R. 11, 86–90 (1958)

Masatoschi, M.: Studies on wind-shaped trees: their classification, distribution, and significance as a climatic indicator. Hosei Univ. Tokyo. Climatological Notes 12, 52 (1973)

Miegroet, M.van: Untersuchungen über den Einfluß der waldbaulichen Behandlung und der Umweltfaktoren auf den Aufbau und die morphologischen Eigenschaften von Eschendickungen im schweizerischen Mittelland. Mém. l'Inst. suisse rech. forestières 32, 229–370 (1956)

Runge, F.: Windgeformte Bäume in den Tälern der Zillertaler Alpen. Meteorol. Rdsch. 11, 28–30 (1958)

Schreiber, K.-F., Weller, F., Winter, F.: Natur-, betriebs- und marktgerechter Obstanbau. Obstbau 78, 1–16 (1959)

Utaaker, K.: Approaches in agricultural meteorology on the meso-scale in Norway. Publications de l'OEPP. Série A 57, 17–35 (1970)

Zürcher, U.: Der Wald in der Orts- und Regionalplanung. Schweizerische Zeitschrift für Forstwesen 120, 736–744 (1969)

9 Improving a Climate for Agricultural Purposes

Y. I. CHIRKOV

Contemporary methods of influencing a climate in order to optimize its conditions with respect to agriculture are quite diverse. Basic climate control methods are aimed at regulating water conditions, but in doing so, other climatic components are also changed. The irrigation of arid regions, the drainage of swampy regions, field-protective afforestation, and the construction of reservoirs the most substantial effect on climate.

During the irrigation of dry steppe regions, semiarid zones, and deserts, a substantial change takes place in the water, radiation, and heat balances of the soil. Evaporation values rarely increase. Consequently, the consumption of heat for evaporation increases considerably. This causes a decrease in the temperature of an active surface, as a result of which the difference between the temperature of the air and that of the active surface decreases, and may even produce a temperature inversion in the near-earth air layer. Without irrigation, sunlight is absorbed by the earth's surface for the most part in deserts, due to the small degree of heat consumption for evaporation. For this reason, the temperature of the active surface reaches very high values.

Thus, humidification and temperature conditions are also considerably improved in the upper soil and near-earth air layers by means of irrigating agricultural crops, which produces higher yields. Research conducted in the deserts of Central Asia shows that the mean monthly air temperature was 3.1 °C lower in July and 2.8 °C lower in August under a shelter in a large oasis than in the desert, while water vapor expansibility in July and August was 5.4 mbar higher in the oasis than in the desert. The soil temperature of an irrigated field is 15°–20 °C lower (after watering) than that of an unirrigated field.

An increase in the volume of irrigated areas has already begun to have an effect on climate in a global point of view. The total amount of irrigated land already exceeds 2 million km². During the irrigation of large areas, an air mass transformation takes place over irrigated territories, the incoming dry warm air is cooled and humidified as it moves over an irrigated region. Large reservoirs, constructed in areas heavy in agriculture, also facilitate improvement of the climate in arid regions.

The drainage of swampy territories has an effect on the climate which is opposite to that of irrigation. Evaporation from a soil surface is decreased with a decrease in humidity, while the surface temperature of the upper soil layers is increased. The difference between the temperature of an active surface and that of the air rarely increases; therefore, turbulence and wind activity are intensified.

Conditions which are more favorable to agricultural activity are produced during the irrigation of marshy soils: the soil dries more quickly in spring, which speeds up the start of spring field work; the drainage of excess water, using reclamation systems, improves soil aeration and root system activity. In northern regions, reclaimed soils become warmer, which improves the root nutrition of plants and increases the yield.

The construction of reservoirs in humidified regions has no noticeable effect on the ambient climate. In arid regions, the effect of large reservoirs is felt for several kilometers. In the coastal belt, the air temperature in summer is 2°–3 °C lower than in regions which are distant from the shore, which is caused by the development of fairly strong breezes (up to 3–4 m/s).

Field-protective afforestation is also one of the methods for changing the climate in steppe regions. Of the various forms of timber stands, field-protective strips provide the greatest effect as a means of climate improvement.

Forest strips are located along the edges of agricultural fields, as a result of which a windbreak system is created, reducing wind velocity and turbulent exchange in the near-earth air layer. The weakening of turbulent exchange takes place as a result of the disruption of strong vortex flows as they pass through the forest strip.

Research has shown that the wind-diminishing action of a penetrable forest strip is the most effective. An impenetrable strip acts like a wall against the wind flow. When an air current approaches an impenetrable strip, it passes over the strip with no abatement of the turbulence. The region of calm formed behind such a strip is not very great. During dust storms, soil deposits, which also damage young crops, accumulate in such a region. The intensity of the turbulence behind the region of calm increases rapidly, reaching the same levels found in an unprotected space.

Reducing the intensity of vortex flows in the near-earth air layer has great practical value in that it prevents wind erosion of the soil (dust storms) and diminishes evaporation from young crops, as well as from the soil, thereby keeping the moisture in the soil. A more favorable microclimate and phytoclimate are created in the interstrip space (see Chap. 5).

Many methods of influencing the climate for agricultural purposes are aimed at optimizing the microclimates of fields, meadows, and orchards by means of regulating soil temperature and moisture content. This is discussed in detail in Chaps. 4 and 5.

At the present time, the development of methods and techniques for the improvement of the climate is acquiring an increasingly practical value in agriculture. Only the dissemination of one of the most effective means for optimizing the climate – irrigation – has made agricultural development of more than 12,000,000 ha of land in arid regions of the world possible in the past decade.

10 Greenhouse Climate

J. Seemann

10.1 Radiation Conditions

Solar radiation in the greenhouse is not only of great importance to plant growth, but, in combination with the glass, it produces essentially the origin of the typical special climate, the greenhouse climate. Solar radiation thus becomes climatic element as well as climatic factor.

Solar radiation cannot penetrate into the greenhouse with its total intensity. A certain percentage of the radiation becomes either absorbed or reflected by the transradiative surfaces, glass of corresponding plastic covering. The magnitude of the radiation that is held back from the greenhouse by absorption depends on the chemical composition of the covering material, as well as on its thickness. The glass types and the plastics that are normally used for greenhouses not only cause an overall reduction of the intensity of solar radiation, but they also have a selectivity effect, i.e., on the one hand, they only allow certain spectral ranges to pass and, on the other, they reduce the permeability of individual portions of the radiative spectrum in differing manner. The usual types of glass, to begin with, only permit the penetration of wavelengths between 320 nm and about 2800 nm. Only relatively little of the UV radiation can enter into the greenhouse and this usually only with new glass. Older glass practically does not permit passage of UV. In the case of light, i.e., in the case of wavelengths between 360 and 760 nm, the absorption of the usual types of glass, dependent on their thickness, amount to 1.6 to 2.5%. Glass absorbs between 8 and 10% of the total solar radiation (Völkers, 1954). The absorption of plastic sheets, Plexiglas and polyester plate varies, in part, considerably from glass. Figure 1 shows the spectral transmission of polyester, PVC sheets, and Plexiglas in comparison to the usual silicate glass. Within the light range, the plastics selected here do not differ very much from glass in respect of spectral transmissivity. In the area of short-wavelength heat radiation, however, the absorption is occasionally considerably higher. This is especially true of Plexiglas. In exchange for this, this material allows the transmission of a considerable portion of the UV radiation.

The well-known greenhouse effect is caused by the fact that glass or plastics, in the form of greenhouse coverings, transmit primarily short-wavelength solar radiation, however, absorb very strongly the long-wavelength radiation (heat radiation of the soil, of the plants and of fruit-tree areas under heating). The greenhouse effect is not present to the same degree in individual plastics, as with glass, since they permit the larger portion of the long-wavelength heat-radiation to pass. According to investigations by Hanson (1963), polyester transmits 13%, polystyrene, 37%, and polyethylene even 74% of the radiation between 2500–15,000 nm. Greenhouses that are covered with this material will lose more

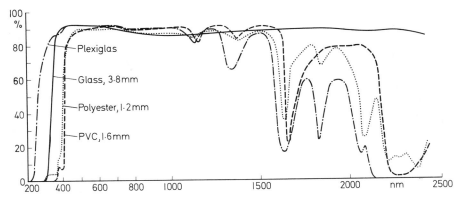

Fig. 1. Spectral transmissivity of Glass, 3.8 mm (——), polyster, 1.2 mm (...), PVC, 1.6 mm (– – –), and Plexiglas (–·–·–). (From Renard, 1969)

heat during the night than greenhouses covered with glass. They generally show higher daily amplitudes of temperatures than do glass-covered greenhouses.

While the radiative reduction by absorption by glass or plastics remains practically the same, the loss of radiative intensity in the greenhouse varies considerably in the course of the year, or even the day, by exterior reflection. The magnitude of the reflection is determined, along with the surface condition of the glass or the plastics, also by the angle of incidence of the solar radiation, i.e., the angle which the radiation forms with the glassing surface. In this way, reflection becomes dependent on the position of the sun, the degree of slope of the roof, and the upright walls, as well as on the geographical direction in which the greenhouses are built. In the case of window glass and with an angle of incidence of 0°–40°, this amounts to 7–8% for light. The values increase considerably with increasing angle of incidence of radiation.

In horticultural practice, this fact automatically leads to the question of the most suitable slope angle of the greenhouse roofs and the direction of the structure. The question as to whether the light intensity can be increased primarily in winter and the heating by the sun can be reduced during the summer is dependent on this. In general, one can state that a steeper roof would be more suitable in winter, while a flatter roof would be preferable in summer. Since such differentiations are no longer possible today, considering full-year cultivation, one will generally select a roof slope that is optimal from both standpoints, radiation and construction. A roof slope of 26.5° has been established on this basis for the standard greenhouse. For the direction of the building, East-West location is preferred because of the increased light availability in winter. In corresponding investigations in the vicinity of London, Whittle and Lawrence (1959) determined that light incidence in winter can be higher by about 12% in an East-West structure than in a greenhouse in North-South direction. In view of the daily course of carbon dioxide assimilation of the plants, Stocker (1949) points out that the North-South position is the more natural one. In all considerations that are made in respect of the problem of building direction and roofangle, one should not forget that marked differences can be expected only in direct solar radiation.

Other factors besides those mentioned also affect the radiation conditions in greenhouses. To begin with, of course, opaque structural parts do not transmit solar radiation. The reduction of the total radiation can, dependent on the form of construction of the greenhouse, amount to about 10% in the most unsuitable case. An additional factor is presented by soiling of the glass or plastic covering. This can negatively influence the radiation conditions and have quite a detrimental effect on greenhouse cultures, especially at times when the light is weak. The intensity of soiling depends on the distance from the source, the type of soiling, and the weather conditions. The accumulation of dirt particles is much greater in dry weather then in rainy weather conditions. According to investigations by Seemann (1951), an average of 4% of the light was absorbed during a period of ten months by deposited dust and soot particles in horticultural operations that were removed from industry, in an operation suffering considerably from soiling, up to 56% of the light was absorbed.

Two types of glass are being used in horticultural operations for greenhouse glassing, bright glass and clear garden glass. While the bright glass has a smooth surface like window glass on both sides, clear garden glass is "nubbed" on one side (the inside, when applied to the greenhouse). Along with the surface structure and the differing transparency, these two glass types also differentiate themselves by the fact that clear garden glass exerts a stronger scattering effect on the passing solar radiation.

10.2 Artificial Light

In a large part of the world, the light intensity from natural sunlight is generally so small in winter that difficulties can develop for the cultivation of plants in the greenhouse. The growth is usually considerably inhibited, and the quality of the plants leaves much to be desired. For this reason, efforts have been made for many years to eliminate as well as possible the difficulties that arise from the unsuitable light conditions by means of artificial illumination. If artificial illumination is to be successful in application, one must understand what can be expected from it and how it must be used.

Two important goals are pursued in horticulture with artificial illumination. Both sufficient CO_2-assimilation and photoperiodic effects are to be attained. For a general increase of photosynthetic output (CO_2-assimilation) of the plants, artificial light is applied either before daylight or after dark or as additional light on dull winter days. When light is added to increase photosynthesis output of plants, one must be concerned with both light intensity, and spectral composition of the light sources.

An example for the differing influences of illumination strength on CO_2-assimilation of cyclamen and head lettuce under otherwise optimal temperature- and humidity conditions, is given in Fig. 2. It represents the dependence of CO_2-consumption – i.e., assimilation output – of these two plants in its light-dependence as percent of the maximally possible consumption. One can see that maximum assimilation output is attained by about 5000 lx for that variety of head lettuce and by about 10,000 lx for cyclamen. An increase of the light intensity

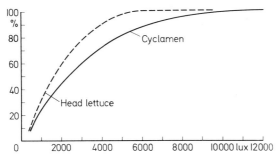

Fig. 2. Carbon dioxide consumption in percent of the maximally possible consumption by cyclamen (——) and head lettuce (---) in its dependence on light intensity. (From Seemann, 1973)

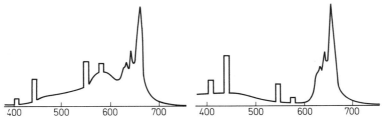

Fig. 3. Spectral light density distribution of two different fluorescent lamps, left 40 W/32 and right 40 W/77 (L-Fluroa)

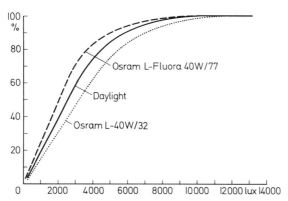

Fig. 4. CO_2-assimilation of a cyclamen plant in percent of the maximally possible CO_2-consumption in daylight and with fluorescent lamps 40 W/32 and 40 W/77

beyond this is of no practical importance to photosynthesis. The curves also show that an increase of the lower light intensities produce a greater effect than does the same increase at higher illuminations. At 1500 lx, the lettuce already uses 50% of the possible CO_2-amount, assimilable, the 50% value is reached by cyclamen at about 2500 lx. For practical application, this indicates that an increase of light intensity at higher illuminations only produces slight additional effects in CO_2-assimilation. In the range of higher light intensities, artificial illumination – perhaps as additional light over daylight – will, therefore, not be economical.

In artificial illumination for an increase of carbon dioxide assimilation, spectral composition of the light must be noted together with illumination intensity. It is known from plant-physiological investigations that the individual color ranges of the light have different effects on plant growth. Two spectral ranges of the light are especially important, the wavelengths about 660 nm (red) and between 400 and 450 nm (blue). The red component supports all growth processes with the exception of the synthesis of the bio-growth materials. Red is most strongly utilized by the plant for CO_2-assimilation. Blue has both an inhibiting as well as a supporting effect. It stimulates assimilation and supports biogrowth materials, but it inhibits too strong a longitudinal growth. Blue cancels the etoil element. The wavelengths between 470 and 550 nm, probably, only have a secondary role in plant development.

The horticulturists' selection of light sources in respect of their light composition is limited to those available on the market. The simplest and best-known artificial light source is the light bulb. It contains practically all wavelengths as solar radiation, however with a differing intensity distribution. Its energy maximum is located at 950 nm (solar at 500 nm). The light portion of the total radiative amount of the light bulb only amounts to about 8 %. (Of this, 57 % are in the red and 6 % in the blue.) The remainder of 92 % is infrared and heat radiation. The relatively low light-yield and the high heat radiation make the light bulb uneconomical for artificial illumination. Especially great importance, on the other hand, must be assigned to all lamp types with high light yield. This is the case, especially with fluorescent lamps. Figure 3 presents the spectral radiation density of three fluorescent lamps that are often used in horticultural practice. The differing effect of this spectral distribution on CO_2-assimilation can be seen from Fig. 4. Fluorescent tubes radiate almost exclusively in the visible range and, therefore, release almost the total energy in the form of light. The excessive heat radiation that often interferes in plant illumination, such as occurs in the use of light bulbs, mercury vapor lamps, or mixed-light lamps, does not occur with fluorescent lamps. The relatively high energy consumption that is necessary for a noticeable increase of assimilation by artificial illumination limits the economy of this measure in commercial horticulture. For this reason, artificial illumination is used almost exclusively for plant propagation during light-poor periods. In this, it offers considerable advantages to the gardener. Propagation time is considerably shortened and cultures experience better young development, something that later expresses itself in yield and quality.

The economics of artificial illumination can be increased considerably when it is controlled, i.e., if the light installation is switched on only when a certain illuminative strength from natural light is not available and is shut off again when a certain light intensity is exceeded. Such a control for artificial illumination is best handled with photoelectric switches and a timer. As already mentioned, artificial illumination serves, aside of the increase of assimilation, also the production of photoperiodic effects (long-day – short-day effect). It is known that certain plants only begin to bloom when, during a certain time period, they are exposed to a short day with light of 10 or less hours. They are called short-day plants. Long-day plants, on the other hand, require 14 h of light or more during certain times. A third group of plants is indifferent to the length of day. With the aid of artificial illumination, the gardener is enabled to determine, himself, the time of flowering within given limits in short-day and long-day plants. The most interesting example for this is chrysanthemum culture in respect of a long-day – short-day treatment. In order to be able to raise several cultures in a single year, the rooted chrysanthemum plants are held for about 30 days under long-day. Since the chrysanthemum is a definite short-day plant, this supports the vegetative growth considerably. Subsequently for 60 days and until flowering, the plants receive a short-day treatment. If the short-day period falls into summer, the plants must, naturally, be darkened for an appropriate time. This is the case for locations between the 42° and 58° latitude, between approximately May 1 and the middle of September, i.e., periods with more than 14 h of daylight.

It is necessary that the night (darkness phase) take a continuous course on short-day treatment. An interruption of the night with appropriate amounts of light leads to long-day effects. The magnitude of the effect depends on the total amount of the interfering light. Approximately the following is valid within certain limits: 2 min with 10,000 lx or 200 min with 10 lx. This effect is being used in horticulture today in photoperiodically controlled cultures in order to obtain a long-day effect. This process has proven to be more economical than a continuous extension of the day. The regulation of the illumination periods for long-day – short-day treatments takes place with a timer.

10.3 Heat Balance in the Greenhouse

Heat conversion in the greenhouse is quite differentiated. Figure 5 presents processes without quantitative evaluation of the individual heat currents in simplified form. Natural heating of the greenhouse, as already mentioned, takes place by solar radiation. Solar radiation of the wavelength between 360 and 2800 nm after passing through the glass, impinge upon the ground of the greenhouse – the surface of the soil, the plants and the remaining items. Of these solar rays, a portion is reflected by the surfaces and, to a certain percentage, again leaves the greenhouse through the glass. In this process, again, a portion is re-reflected back into the greenhouse by the glass surfaces. The remainder of the incident solar radiation is converted into heat on the surfaces. By means of convection, the heat is passed on from the surfaces into the greenhouse air. The soil surface conducts heat into the soil. This heat flux is inverted when radiation decreases, or at night. Thus, the heat that has, in a manner of speaking, been stored in the greenhouse soil, becomes again available for the overall heating of the greenhouse. In an unheated greenhouse, about nine tenths of the total heat release of the greenhouse is covered during the night by the heat stored in the soil; but the surfaces of the greenhouse also radiate heat. This concerns heat radiations in the long-wavelength region (above 2800 nm), this radiation can, therefore, not pass through the glass. Instead, it is absorbed. This heat, by way of heating the glass and through reflection at the glass surface, remains within the greenhouse to the largest extent.

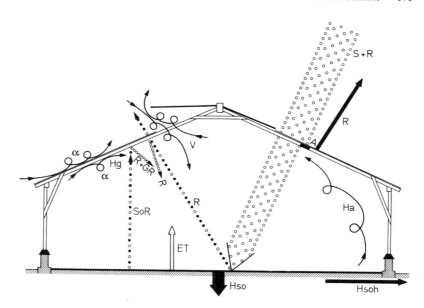

Fig. 5. Radiation and heat balance in the greenhouse (by Seemann, 1957). $S+R$, Sun radiation; R Reflection; A, Absorption; Hso, Heat flux into the soil; $Hsoh$, Horizontal heat flux; Ha, Heat flux into the air; V, Heat exchange and ventilation; SoR, Soil heat radiation; GR, Glass radiation; ET, Evapotranspiration

During the colder period of the year, when the energy from solar radiation is insufficient for the necessary warming, greenhouse heating represents the added heat source. The most important role in this is played by warm water radiators and warm air heating systems. Warm-air heating represents pure convective heating in which the heat is added directly to that of the greenhouse air. In contrast, radiator heating adds heat radiation to convective heat transfer.

Along with heating for the warming of the greenhouse air, soil heating is of great importance. It should generally be considered to be a type of additional heating that produces better heating of the soil than could be attained with normal greenhouse heating.

A relatively large portion of the energy from solar radiation, or of the heat added to the greenhouse by way of heating, is consumed for vaporization. While a part of this energy remains in the greenhouse in the form of latent heat and is liberated again by condensation of water vapor on the inside of the glassing and the structural parts, the greenhouse loses heat by conduction through the glassing, the structural parts and the soil as well as by air exchange through leaks and airing. The magnitude of the heat losses by conduction depends on the magnitude of radiation from the outer surfaces, the difference between interior and exterior temperature transit number and the wind velocity that (controls) losses by air exchange, dependent on the tightness of the greenhouse, the type and sort of airing, and also the wind velocity and temperature difference between interior and exterior air.

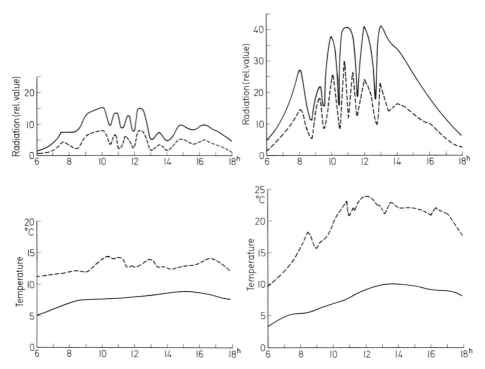

Fig. 6. Radiation intensity and temperature conditions in a greenhouse (———) and in open country on a dull day (*left*) and a bright day (*right*). (According to Seemann, 1957)

10.4 Temperature Conditions

The air temperature conditions in the greenhouse differ, in part, considerably from those in open country. The cultivation of horticultural plants is, after all, displaced into the greenhouse because of these specific temperature conditions. In the closed, unheated greenhouse it is always during the day, and generally during the night, warmer than it is in the open country. To what degree it will be warmer in the unheated greenhouse than in the open country depends on the level of solar radiation, or the nightly radiations. The dependence of the temperature conditions on the amount of solar irradiation is clearly shown in Fig. 6. In this case, the greenhouse air was heated, in a closed greenhouse with a ground surface of 75 m^2, on a sunny day (22 April, 1955), in contrast to a poor day (21 April, 1955) with a radiative intensity that was higher by 36% and with approximately the same exterior temperature, more strongly by 47%. The amount of warming of the air in a greenhouse under the influence of solar radiation, and the resultant daily temperature fluctuations, also depend on the size of the greenhouse, or its air volume. The greatest temperature fluctuations between day and night occur in hotbeds because of the relatively small air volume.

In respect of vertical temperature distribution, in comparison to open country, opposing conditions result in the unheated or insufficiently ventilated greenhouse. While in the open country and under sunshine, the temperature of the lower air layer is reduced with the height, it increases with the height in the greenhouse. The air, then, is the warmest in the vicinity of the roof of the greenhouse. The amount of the gradient of the temperature depends on the size of the structure and on the weather in the open country. In a greenhouse with good roof ventilation, the vertical temperature conditions largely correspond to those of the open country. At night, the air temperature in cloud-free weather or in relatively low outside temperature is colder in the roof area than close to the ground.

The course of the temperature in the heated greenhouse is controlled by the temperature control, the temperature distribution by the technical conditions. King (1969) has shown that temperature gradients of more than 5 °C can occur in the heated greenhouse at 20 cm above ground with stagnant air. These great differences can, however, be largely eliminated by the suitable application of ventilators.

The special conditions of the heat conversion in the greenhouse and, especially the fact that direct heat radiation from the soil through the glass is prevented, lead to the soils in the greenhouses having higher temperatures than in the open country. Initially, it becomes apparent upon a general consideration of the course of temperature in the soil that the heat transport in the ground of the greenhouse and of the country takes place in approximately similar direction; but the soil in the greenhouse is, in all depths, considerably warmer than in the open country. The temperature difference between 5 and 50 cm depths and the daily amplitude in the upper soil layers is, in compensation, higher in the open country than in the greenhouse.

While, in summer, the soil over practically the entire greenhouse surface warms up equally, a very pronounced horizontal temperature gradient between the center and the edge develops in winter. In larger greenhouses, the temperature reduction towards the edge amounts to about 0.5 °C/m. In smaller houses, the temperature in the upper soil layer can drop up to 3 °C/m. The plant temperature and, especially here, the leaf temperature, generally runs parallel to the air temperature in the greenhouse, as well as in the open country. Through the energy supply from direct solar radiation and other radiations (heating tubes or electrical lamps with high temperature radiation), the plants attain temperatures above those of the surrounding air. The leaf temperature drops below the air temperature in situations of small amounts of radiation or high transpiration.

Figure 7 presents an example for the course of temperature of a leaf of *Ficus elastica* in the greenhouse and in open country in fine weather. The most important part of this is in the fact that the leaf temperatures in the greenhouse are generally higher than in the open country, in the case of direct comparison as well as in comparison with air temperature. This is not really surprising, if one considers that the value of the leaf temperature is dependent on the total heat conversion and that, in the greenhouse, the convective heat transport is smaller than in open country. Added to this is the fact that long-wavelength radiation cannot radiate directly through the glass or the plastic, and the plant is therefore in a radiative exchange only with the greenhouse glassing.

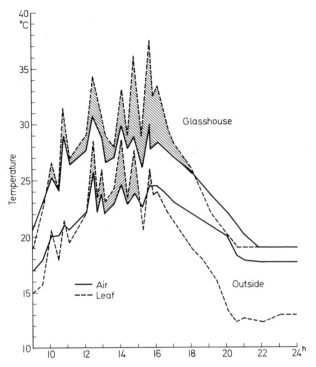

Fig. 7. Course of the temperature on leaves in the greenhouse and in open country. (From Seemann, 1973)

The leaf temperature will be below the air temperature in cloudless sky at night and even sometimes in the later afternoon hours in negative net radiation, earlier in the open country than in the greenhouse. This also can be seen from Fig. 7.

10.5 Temperature Control in the Greenhouse

We must differentiate between the control methods in the heated greenhouse and the measures for temperature control in the greenhouse during the summer in considering the control of greenhouse temperatures.

In the heated greenhouse, heat regulation is conditioned to the control of the air temperature and, hence, the regulation of the heating. Not very long ago, the horticulturist strove for temperatures that were as equal as possible in his greenhouse. Today, we know from a series of scientific investigations (Went, 1950; Seemann, 1956) that an essentially constant course of temperature is in no way advantageous to plants. If one includes knowledge on plant physiology, photosynthesis, and respiration in one's considerations, it must be obvious from the start that a certain differentiation of the course of temperature, at least between day and night, must be advantageous. The recognition that the consumption of structural materials that have been gained is reduced by lower nighttime temperatures and that better plant development can be attained in this manner has become common among gardeners. Nightly temperature reduction is being carried out today in almost all modern horticular operations.

Under consideration of the photosynthetic output of the plants, it is useful, during the daylight hours, to adapt the temperature to the prevailing light conditions in the heated greenhouse. High temperatures will be applied in the heated greenhouse only when a relatively high light intensity is available for strong CO_2-assimilation. High temperature with low light availability is detrimental to plant development and, additionally, means unnecessary energy consumption for heating. In the same way, too low a temperature with good light conditions is disadvantageous to plant growth, because, in that case, the light that is available in sufficient amounts cannot be fully utilized for photosynthesis.

Such heat regulation cannot be carried out in the heated greenhouse by manual control. This problem, that has been a subject of scientific and practical research for many years (Seemann, 1956), led among others to the development of an electrical control device that makes light-dependent heat control as well as appropriate temperature reduction automatically possible during the night.

This type of heat regulation in the heated greenhouse is the only proper one from ecological points of view and, additionally, the most economical. It permits an optimal development of the plants that corresponds to winter conditions, and leads to a considerable savings of fuels for greenhouse heating.

While temperature control in the heated greenhouse is relatively simple, provided there is a flexible heating installation and appropriate control, temperature control during the summer months produces some difficulties. In any case, technical expenditure is greater for this. It is important in summer to avoid overheating of the greenhouse atmosphere and, especially, of the plants. This is especially important when there is direct sunlight. This produces an increased heat exchange in the greenhouse, as well as on the plants. Special importance must be ascribed to climate control during the summer months for the increased heat exchange of the plant and the leaves. The importance of an increased heat exchange becomes understandable, in respect of the plant leaves, if one considers that increased added heat from direct solar radiation on the plants leads to an increase of the transpiration. If the water household of the plant is strongly taxed by increased transpiration, the plant reacts to this by closing of the stomata and other physiological processes and limits the transpiration. Connected to these processes is an increase of the diffusion resistance for CO_2 from the atmosphere into the leaf. This results in a reduction of the photosynthetic output of the plant. Control of the temperature conditions in the greenhouse is possible by the following measures during the warmer season with solar radiation:

1) By reduction of the net radiation. In practice, this is attained by shading the greenhouse (Seemann, 1973; Simon, 1967).

2) By increasing the air exchange and the heat transfer number in respect of the heat exchange of the plant. Airing of the greenhouse contributes to this (Seemann, 1957).

3) By the "destruction" of solar energy (heat) by way of increased evaporation. For this purpose, for-spray processes or appropriate evaporative installations (so-called wet walls) have been applied (Kreutz, 1949; Seemann, 1962; Stein, 1961).

In greenhouses that are used for scientific purposes, direct air-conditioning plants are often installed.

10.6 Humidity

Along with light and temperature, humidity is one of the most important elements in the greenhouse climate.

The following reasons may be decisive, in general, for a deviation of humidity in the greenhouse in contrast to open land:

1) The varying exchange conditions of the air volume at an equal or higher water vapor supply by way of evaporation of the water that is commonly plentifully available in the upper soil layers in the greenhouse.

2) The temperature conditions that deviate from the open country in the greenhouse by which relative humidity is considerably influenced.

The following can be stated in summary according to the investigations of Seemann (1957) in respect of the humidity conditions in greenhouses: the course of absolute moisture in unaired greenhouses is in close relationship to soil temperature, it increases when the soil warms (increased evaporation) and is reduced when the soil cools. Absolute moisture is lowered by ventilation even with steady soil temperature, water re-supply, however, is increased by evaporation. In a constant course of temperature, an equilibrium between water vapor re-supply in the greenhouse and the absolute moisture in the open country (on the basis of air exchange) is established, absolute and relative humidity, then, are constant in the greenhouse.

"Dry" air (low relative humidity) can be attained on the one hand in the greenhouse practice by increasing the air temperature and, on the other, by ventilation. Moist air can be attained by increasing evaporation from the soil in the closed greenhouse with relatively low air temperature. The latter case will especially occur when radiative heat or floor heat is being used, presuming, of course, that the soil is sufficiently moist. In the case of hot-air heating, the air will generally be dryer than with other heating systems, because the soil usually remains cooler.

10.7 Evaporation and Water Consumption

With sufficient water supply to the soil in the greenhouse, generally potential evapotranspiration will predominate, i.e., evaporation of the water on soil surfaces and on the surfaces of leaves will largely correspond to that of a moist surface.

Plants transpire through the leaves and, even though only to a limited degree, through other parts of the shoots. One differentiates between stomatous (by way of the stomata) and cuticular transpiration (by way of the cuticula). While the latter constitutes only a small percentage of the total evaporation and can take place day and night, the degree of stomatous transpiration depends, among other things, on the width of the open stomata. To any special degree, stomatous transpiration takes place only during the day, since in most plants the stomata are closed during the night.

The evaporation formula according to Dalton provides a good impression on the physico-meteorological conditions for the degree of evaporative intensity.

According to this and in modern presentation, evaporation is:

$$V = \frac{w \cdot 0.623d}{p \cdot c_p} \alpha(E' - e),$$

where $(E' - e)$ is the saturation deficit of the water vapor, e is the water vapor pressure of the air and E' the maximum water vapor tension (saturation vapor pressure) – both in mm mercury – in the border layer of the evaporating surface, relative to its temperature.

α is the heat transfer number, d the heat of evaporation of the water (approximately 597 cal/g water), p the air pressure and cp. the specific heat at constant pressure. w represents the degree of water-covering of the evaporating surface. Relative to the evaporating leaf, w indicates the degree to which the stomatous openings permit a diffusion of water vapor from the interior of the leaf. An entire series of important concepts can be obtained from this model in respect to evaporation and water consumption in the greenhause. Since α cannot be very great, considering the small amount of air movement, evaporation is primarily controlled by the saturation deficit. The value of the saturation deficit is determined other than by the actual water vapor content of the air, which is also influenced by the water vapor content of the outside air because of the air exchange, and especially through solar irradiation, because E' is a function of the surface temperature. Related specifically to the plant, this means that with increasing solar radiation, the transpiration and hence the water consumption of the plant increases. The saturation deficit is increased in sunny weather, in addition, by the fact that ventilation of the greenhouse, dryer exterior air leads to a lower water content (e) of the greenhouse atmosphere. The transpiration of the plants cannot be raised in an unlimited manner in opposition to the evaporation of the moist soil surface. If the water household of the plant is exposed beyond a certain messure, the plant limits the water transport, something that takes place, among others, by the closing of the stomata. This reduces the water-covering factor and the transpiration is reduced.

Morris et al. (1963) and Morris and Lake (1962) have concerned themselves to an especially great degree with the transpiration of plants in the greenhouse. They were able to prove that the course of transpiration and hence the water consumption of the plant is, to a special measure, a function of solar radiation. According to Nicolaison and Fritz (1954), the water consumption in the greenhouse amounts to about 2571 for tomatoes and 871 with beans per 1 kg of sprouted matter and vegetative period.

Literature

Hanson, K.J.: The radiation effectiveness of plastic films for greenhouses. J. Appl. Meteorol. 2 (1963)
King, E.: Contributions to the greenhouse climate. Erwerbsgärtn. 24 (1969)
Kreutz, W.: Meteorological observations in the greenhouse. Holzminden 1949
Morris, L.G., Lake, J.V.: The water loss from plants and soil under glass, Vol. III. Hortic. Cong. Nice, 1958. Pergamon 1962
Morris, L.G., Postlethaite, J.D., Edwards, R.J., Neale, F.E.: The dependence of water-requirements of glasshouse crops on the total incoming solar radiation. N.I.A.E. Tech. Memo 86 (1953)

Nicolaison, W., Fritz, L.: Investigations on the water consumption of our vegetable types. Gartenbauwissenschaft *1* (1954)

Renard, W.: Plastics in the greenhouse and their light-technical properties. Tech. Commun. Int. Soc. Hortic. *2* (1965)

Seemann, J.: Investigations on the soiling of glass roofs. XXIII. Act. Rep. Gärtn. Versuchsanst. Friesdorf 1951

Seemann, J.: Heat control in the heated greenhouse according to ecological considerations. Gartenbauwissenschaft *22* (1956)

Seemann, J.: Climate and climatic control in the greenhouse. Bayer. Landw. Verl. 1957

Seemann, J.: A contribution to the problem of cooling of greenhouses during summer with simple methods. Gartenbauwissenschaft *1* (1962)

Seemann, J.: Climate under glass. Tech. Note. World Met. Org. (1973)

Seemann, J., Horney, G.: A contribution to the application of luminescent lamps for artificial illumination in horticulture and their effect on CO_2-assimilation. Arbeitsgem. Elektriz. Landwirtsch. (AEL) Issue *3* (1972)

Simon, J.: Shading of greenhouses with special consideration of plant-physiological basics, climatic conditions and technical possibilities. Diss. Hannover 1967

Stein, J.: Mat-cooling for greenhouses. Tech. Gartenbau. Bundesgartenschau 1961 (1961)

Stocker, O.: Basics of natural greenhouse culture. Stuttgart: Ulmer 1949

Völkers, O.: Plateglass data. Fachverbd. Fensterglasind. Frankfurt 1954

Went, F.W.: The earhart plant research laboratory. Chron. Bot. *12* (1950)

Whittle, R.M., Lawrence, W.J.C.: The climatology of greenhouses. I. Natural illumination. J. Agric. Eng. Res. *4* (1959)

11 Open Fields and Shade

J. Seemann

The ecological differences in open fields and in shade are of decisive importance to the development of natural plant communities. General distinction is made between plants that require light and those that prefer shade. In their natural locations, these show a different morphological structure. In natural plant communities, plants of both types have established themselves in direct vicinity, so that the shade plants practically live in the shade of the sun-loving plants. In plant culture technology, this principle is often used to advantage by the establishment of so-called mixed cultures. However, shade is also provided with technical means where shade-plants are raised in single culture, such as e.g., in horticulture (Seemann, 1974; Simon, 1967). Greenhouses are shaded during the summer. Shaded areas are constructed for perennials that must be kept available over longer periods of time, and certain seedlings, as well as plants in their early stages, are often provided with shade.

The most decisive influence on differing development of plants in open fields and in the shade is exerted by solar radiation. Here it is important to consider separately the ecological effects of total radiation and light. In the shade, the plant is exposed to less available light and a reduced net radiation; but this reduction of radiation is not the only meteorological influence on the plants in shade. As a result of lower net radiation, the air- and soil temperatures are generally lower during the daylight hours, while during the night they may well be higher than in open country, especially in weather with little wind (Salati et al., 1966). In addition, there also results a reduction of the evapotranspiration since the total energy conversion is reduced. The decisive ecological factor in the comparison of sunny and shaded locations is, and remains, radiation.

As already mentioned, light and total radiation must be considered separately in this comparison as individual ecological factors. The reduction of the light intensity in shaded locations generally and necessarily produces a reduction of the photosynthetic output when the light intensity drops below a certain level. Pronounced shade plants have, however, a lower CO_2-assimilation level to begin with than do sun-loving plants. Bohning and Brunside (1956) determined, for typical shade plants, such as philodendron and oxalis rubra, maximal illumination strengths of 4000–11,000 lx, with a compensation point at 500 lx. In the case of sun-loving plants, such as *Coleus, Glycine, Heliatus annuus, Phaseolis vulgaris* etc., the same authors determined assimilation peaks at 21,000–27,000 lx, where the compensation point was located at 1000–16000 lx. Shade plants have weaker respiration and, for this reason, have a relatively more advantageous material balance than do sun-loving plants. According to Ruge (1956), the assimilation output in sun-loving plants is 10–15 mg, while it is up to 20 mg in cultured plants; in shade plants, however, it is only 3 to 5 mg of CO_2 $dm^{-2}h$. In sun-loving plants, the intensity of respiration is also higher with its 1.5–3.5 mg in contrast to only 0.15–1.5 mg CO_2 $dm^{-2}h$ in shade plants.

The effect of total radiation in the shade-free location is characterized by increased heat conversion on the leaves of plants. In intensive solar radiation, i.e., with high net radiation, it is primarily the water household of the plant that is exposed to greater stress. This can, perhaps, be shown most easily by means of the heat household equation in its following, extended manner of writing:

$$0 = Q - \alpha_t(t_L - t_A) - \frac{w\,0.632\,d}{cp \cdot P} \alpha_e(E' - e),$$

where net radiation (Q) – sensible heat (α_t = heat transition number, t_L = leaf temperature, t_A = air temperature) – Transpiration (expressed by Dalton's evaporation equation, where w = water-covering factor, α_e = water vapor transport value, d = evaporation enthalpy, p = specific heat at constant pressure p and (E' − e) = saturation deficit (E' here refers to the temperature of the leaf) is equal to zero. One can do without the consideration of a heat storage in the leaf, since these values are very small.

It can be readily seen with the aid of the heat household equation that a high saturation deficit is produced with high net radiation and that therefore transpiration must be very intense. If this condition continues over an extended period of time, then the plant reacts to too high a water-loss by narrowing or closing the stomata and through cell-physiological processes. Transpiration, at least the stomatous, is limited (w becomes <1), or ceases essentially altogether (the transpiration member of the equation becomes approximately = zero). Through this, then, the energy that has been made available in the form of latent heat will also contribute to the heating of the leaf. This means that respiration is increased at the same time. The closing of stomata and the other cell-physiological processes increase the diffision resistance to CO_2. The consequence of this is, again, a reduction of the CO_2-assimilation to below the compensation point.

The processes described here, that take place because of excessive heat conversion on the leaves in plants in shade-free locations probably represent the most important factors in the problem of shade- or sun-loving-plants. It is known that a photosynthetic depression can also occur in sun-loving plants under the indicated conditions. However, on the basis of its morphological structure, they are more resistant to these stresses than are shade-plants (Bannister, 1971). Shading has proven to be advantageous during periods with high net radiation in the cultivation of plants that must be considered sun-loving plants, for the above-mentioned reasons. An experiment of this author showed, however, that shade-plants can also be cultivated in sunny locations under certain circumstances (Seemann, 1958). Here, cyclamen and gloxinias were cultivated successfully during the summer in a greenhouse without shading, as long as a sufficient removal of heat took place with appropriate technical means.

In contrast to this, Klenert (1974) could show that sun-loving plants, such as the grapevine, develop very disadvantageously when they are cultivated in a shady location. The reduction of the light intensity that is connected to this apparently exerts a very negative effect on the quantity of grapes as well as the quality of the

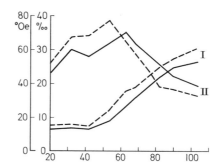

Fig. 1. Effect of shade on the quality of the young wine from grapes (———) shade, (– – –) open location; *I* new wine weight in °Oe; *II* acid in %$_{00}$. (From Klenert, 1974)

new wines. Figure 1 shows, for example, how the ripening process in shaded grapes takes a much more disadvantageous course than in open locations. In unshaded grapes, not only the new wine weights (sugar content!) increases more rapidly, but the acid reduction takes place earlier and more intensely than in grapes that obtain only reduced solar radiation as a result of continuous shading.

Literature

Bannister, P.: The water relations of heath plants from open and shaded habitats. J. Ecol (Oxford) *1* (1971)

Böhning, R.H., Burnside, C.A.: The effect of light intensity on the rate of apparent photosynthesis in leaves of sun and shade plants. Am. J. Bot. *43* (1956)

Klenert, M.: Artificial change of the meteorological conditions in grapevines and the effect on the fertility of the grape and the growth of the berries. Diss. Giessen Univ. 1972 (see also Z. Vitis, Vol. 13, 1974)

Ruge, U.: In: Maatsch, R. (ed.), Pareys illustrated horticultural encyclopaedia. Hamburg: 1956 (5th Ed.) Parey Pub.

Salati, E., et al.: Influentia do sombramento na temperatura do sol em cafegal. Serv. Meteorol. (Rio de Janiero) Bol. Tech. *2* (1966)

Seemann, J.: Temperature course in greenhouses with one-sided turn-hatch ventilation and with fan ventilation. Dtsch. Gartenbauwissenschaft *4* (1958)

Seemann, J.: Climate under glass. WMO Tech. Note *131* (1974)

Simon, J.: Shading of greenhouses with special consideration of plant-physiological basics, climatic conditions, and technical possibilities. Diss. Hannover Univ. (1967)

12 Optimum Climate for Animals

B. PRIMAULT

12.1 Definition of Optimum Climate

Large and small cattle, as well as courtyard animals, all belong to the so-called "warm-blooded" animals. These animals thus have a physiological system – called thermoregulation – the purpose of which is to maintain their body temperature within a very narrow range, within which alone life is possible. Either a lowering of the body temperature below a certain limit, or a rise above an upper threshold causes the death of the individual. Those thresholds vary from species to species, and even from age to age within the same species. Each individual may nevertheless endure either a cooling or a heating of the body, but only for rather short periods of time.

Since the external temperature is not constant, in the case of cold weather the animal compensates for the heat losses by means of an internal combustion system which gives the body the necessary calories to maintain a uniform internal temperature. In hot weather, on the other hand, and above their ideal body temperature, warm-blooded animals have various cooling mechanisms. On the one hand, they consist of an increased vascularity of the subcutaneous areas, and, on the oother hand, of the evaporation of a certain amount of water, either from the outer surface of the body (perspiration), or inside the body, more particularly in the respiratory organs. In the two latter cases, the energy required to transform the water from the liquid phase to the gaseous phase is removed from the body to be released into the environment; but it is also necessary that the air in the environment be capable of absorbing that additional moisture. If its temperature is higher than the ideal body temperature and its humidity is very high (close to saturation, for example), these cooling mechanisms cannot function satisfactorily, and the body of the animal becomes warmer until death follows.

Therefore, for each category of domestic animals, five different thermal zones are defined:

a) hypothermia, a range within which the animal, utilizing all its heat-producing resources, cannot supply sufficient heat to maintain its ideal temperature;

b) a broad range within which the body temperature is at its optimum level, but the animal must compensate for more or less considerable heat losses toward the outside; the animal is subjected to cold stress;

c) a zone of thermal indifference, which means that the normal metabolism of the animal supplies the number of calories required to maintain its temperature at the optimum level, but without any surplus;

d) a neutral range, within which, however, conditions are no longer ideal, because the temperature control system of the animal must already produce some cooling; the animal is thus subjected to thermal stress from heat;

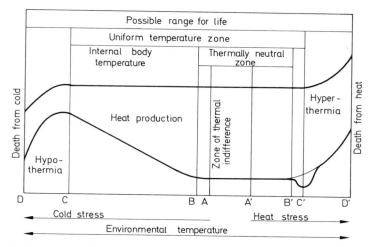

Fig. 1. Temperatures and critical ranges. (From Bianca, 1968)

e) finally, a zone of hyperthermia, where the temperature control mechanisms no longer produce a sufficient cooling of the body temperature to maintain it at its normal level (Fig. 1).

Depending on their original climate, the various species of domestic animals are more or less adapted to heat and cold. Thus, the limits of the uniform temperature range are only applicable each time to a well-defined species or beed.

Similar differences, or of even greater amplitude, are also found in the same individual at different ages. Thus, for example, the leghorn hen has a range of thermal indifference located between 18° and 28 °C, while its chick requires approximately 34 °C to feel comfortable. The ideal temperatures for the pig range between 0° and 15 °C, while the sucking pig requires 33 °C. The cow requires a temperature ranging between 0° and 15 °C, while the calf requires a temperature ranging between 13° and 25 °C. The sheep easily tolerates temperatures ranging from −3° to +20 °C, while the lamb only feels comfortable between 29° and 30 °C. We could give many more such examples (see Fig. 2).

These few examples show the great variations in the temperature ranges comfortable for adult animals, while newborn animals have much more narrowly defined ranges of comfort, and generally those are at levels considerably higher than those of their parents.

What we have just described concerning the temperature applies equally, by analogy, to humidity, because the latter contributes to the well-being of the animals, through the operation of the temperature control systems.

Consequently, rational breeding must tend to provide the animals with temperature and humidity levels corresponding to their range of comfort, which means located within their zone of thermal indifference, and hydric indifference, respectively (see Chap. 13).

We have seen that warm-blooded animals, by means of certain internal mechanisms, were able to control their body temperature: to supply calories to

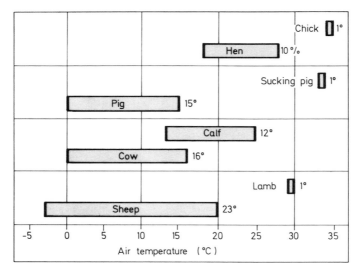

Fig. 2. Zones of thermal indifference. (From Bianca, 1968)

compensate for losses or to cool their organism, overheated by the environment. Therefore, by analogy, they must also have, but to a lesser extent, the ability to adapt to environmental conditions different from those of their original environment. Such ability to adapt varies greatly from species to species. It is also noted, in studies of selection, that certain individuals by nature have zones of thermal indifference different from those of their race of origin. By selection and cross-breeding, it is then possible to isolate lineages whose zone of thermal indifference is

Table 1. Optimum living conditions for cattle

Type	Temperature range (°C)	Humidity (%)
Calves for breeding	5–20	50–80
Calves while fattening	18→12[a]	50–60
Young breeding cattle	5–20	50–80
Young cattle while fattening	10–20	50–80
Milk cows	0–15	50–80
Sucking pigs (microclimate for newborn animals)	33→22	50–80
Young pigs and pigs for slaughter	22→15	50–80
Sows (pregnant and lactating), boars	5–15	50–80
Lambs	12–16	50–80
Sheep for slaughter or wool	5–15	50–80
Horses (riding, racing or draft)	8–15	50–80
Chicks (microclimate)	34→21	50–70
Female chicks and capons	17–21	50–80
Egg-laying hens	15–22	50–80

[a] → means that the temperature must be gradually lowered as the animals grow in age or gain weight

located below or above the levels normally assigned to their species. With the subsequent generations, these become a sort of genetic trait of the breed, and one obtains cattle which will feel comfortable under climatic conditions different from those prevailing in their area of origin.

By analogy, a transplantation of certain lineages may, after some generations, lead to a kind of mutation of the genetic trait defining the zone of thermal indifference. This is the so-called "acclimatation" of the breed.

12.2 Bovine Cattle*

Generally speaking, the zone of thermal indifference varies with the age of the animal: the older it grows, the more it moves toward lower temperatures. Thus, in order to be comfortable, a newborn calf will need higher environmental temperatures than its mother or father. Furthermore, depending on the purpose of the breeder (production of meat or of milk), the limits of the zone of thermal indifference of his cattle will be different.

12.2.1 Calves for Slaughter

As the weight of the animal increases, the optimum environmental temperature decreases. Thus, for example, a 50-kg calf will feel comfortable at 18 °C, while a 200-kg calf will begin to pant at the same temperature, showing signs of heat stress. Actually, a temperature of only 12 °C will be best for it.

12.2.2 Breeding Cattle

As in the case of the calf for slaughter, breeding or stock cattle needs lower and lower temperatures as its weight increases. Beginning at 20 °C for a weight of 50 kg, the ideal temperature for development at 500 kg is found to be only 5 °C. Such cattle is intended for reproduction, rather than for the production of meat or milk.

As stated above, bovine cattle can adapt rather easily to environmental conditions quite different from the optimum ones. Thus breeding cattle may be left in the pasture, under considerably varying temperature conditions, not only from day to day, but also from day to night. In this respect, we point out that the effect of radiation on the surface temperature of cattle, in case of high absorption of heat through the hair, may play a not insignificant role in determining the possibility of keeping breeding cattle on open ground.

12.2.3 Beef Cattle

While fattening steers, care will be taken to ensure that the losses due to temperature control are not too great, otherwise the yield will become doubtful. Consequently, care will be taken to provide cattle being fattened with environmental temperatures higher than those for the cattle intended for breeding.

---

* The temperatures shown here apply to most breeds in the temperate zone: Durham, Charolaise, Friesian, Franc-Comtoise, Alpine brown, etc.

Thus, in the case of 200-kg steers, one will attempt to maintain a temperature of 20 °C; the latter will be gradually lowered to 10 °C for 500-kg animals.

12.2.4 Milk Cattle

It has been found that low temperatures had a favorable effect on milk production. This is less a question of the amount of milk secreted than of its quality, that is, its fat content. For this reason, in the case of milk cows, lower temperatures than those suggested for bovine beef cattle will be preferred. Irrespective of the weight of milk cows, one will try to keep them at temperatures ranging from 0° to 15 °C, the latter temperature constituting an upper tolerable limit, beyond which fat production declines very rapidly.

12.3 Goats

Goats, the descendants of certain varieties of antelopes, live in their natural state in the most diverse climates. Therefore, it is not surprising to find breeds of goats well adapted to the temperature conditions either of the high mountains or of the arctic regions, or of subtropical, or even tropical, regions. Consequently, it is very difficult to set ideal temperatures for breeding goats, the more so because such breeding generally takes place in the open air in natural pastures.

12.4 Ovines

What we stated concerning goats is applicable to sheep. Actually, this type of cattle originates from very different areas, so that its zone of thermal indifference varies enormously from breed to breed. Thus, the European breeds have a zone of thermal indifference located between 16° and 12 °C for lambs, and between 15° and 5 °C for sheep used in the production of meat or wool. The comment made with respect to milk cows (see 12.2.4) also applies here to the small cattle raised for its milk production: the lower the temperatures, the higher the fat content of the milk will be. This fact is important in the production of cheese from sheep's milk.

12.5 Pigs

There appears to be at the present time a marked trend toward the consumption of pork meat. Consequently, on the one hand, small-scale breeding (three to six animals per farm) by farmers tends to disappear, while, on the other hand, large-scale production enterprises are developing (300–800 animals, or even more). The larger the enterprise, the more desirable it is to adapt temperature and humidity to the requirements of the species, in order to assure the profitability of the enterprise. Since those ideal conditions vary with the age and the weight of the animals within wide margins, the only profitable solution is a partition of the installations which will place the animals in different groups, depending on their temperature and humidity requirements.

12.5.1 Pregnant and Lactating Sows

Irrespective of the weight of pregnant sows (generally between 150 and 350 kg), the most adequate temperature for them lies between 8° and 15 °C. This temperature range will be maintained after giving birth and for the entire duration of location. Since, however, the young animals require quite different environmental conditions to develop properly (see 11.5.2), it will be essential to have separate stalls for the sucking pigs and their mothers. The two groups will only meet at the temperature boundary for sucking.

12.5.2 Sucking Pigs

At birth, a sucking pig requires relatively high temperatures, up to approximately 33 °C. If it is not possible to feed artificially, it will then be necessary to have special boxes where the young pigs can go to warm themselves up, while they will leave them to reach the part of the pigsty where the mother is kept, which is maintained at 8° to 15 °C. As the sucking pig gains weight, it tolerates temperatures that tend to decrease. Thus, at 20 kg, its ideal temperature has already been lowered to 22 °C. Consequently, depending on the weight gained by the pigs, the temperature of their stall can be lowered without subjecting them to thermal stress; on the contrary, they will only feel better.

12.5.3 Young Pigs

As in the case of sucking pigs, the pigs intended for fattening or breeding have a lower ideal temperature as they fatten. Thus, a 20-kg pig will feel most comfortable at 22 °C, while a 60-kg pig will preferably move to areas at a temperature close to 15 °C.

12.5.4 Pigs for Fattening

Beginning at a weight of 60 kg and up to the end of the fattening period (approximately 100 kg per head), the ideal temperature does not vary at all. If one wishes the metabolism of the animal to compensate rather accurately for the heat losses due to a temperature gradient existing between the body temperature of the animal and that of the air in the environment, it will be kept in an environment close to 15 °C. In this manner, through its natural metabolism, the body supplies the calories required to maintain the internal temperature at the ideal level without supplying too great amounts of heat or forcing the animal to struggle against the heat; but during fattening pigs are very sensitive to the humidity levels in the environment. They will be kept within a narrower humidity range than other fullydeveloped or breeding pigs. This involves a relative humidity ranging from 50 to 70%. Only under this condition will the respiratory mechanisms release in the most balanced manner the calories due to the general metabolism.

12.5.5 Boars

In order to ensure the best semen for the reproduction of the species, boars will be kept under temperature conditions similar to those required by pregnant sows (see 12.5.1).

12.6 Other Domestic Animals (Camels, Yaks, Llamas, etc.)

While the ideal conditions for most domestic animals tratiditionally raised in the temperate zone are well known, unfortunately this does not apply to breeds such as camels, yaks, llamas, etc. The latter are kept under natural conditions, which means in pastures and not in closed or air-conditioned stables. Until now, the need to know their zone of thermal indifference has therefore been less pressing. For this reason, research carried out in this field is practically nonexistent. We shall only state that, depending on the areas of origin of the breeds raised, the zones of thermal indifference are adapted by natural selection: those animals are acclimatized to their habitat and to their breeding areas, respectively.

12.7 Fowls

As in the case of the other domestic animals mentioned under 12.6, most species of fowls are raised in the open or in cages without air conditioning. The temperature and microclimate conditions ideal for them therefore have not yet been determined. This, however, does not apply to hens: there is rather accurate information concerning the most appropriate temperature conditions for their development.

12.7.1 Chicks

In nature, generally chicks are born in a nest, where they are protected from the effects of cold or heat by the body and the feathers of the mother. Therefore, it is not surprising that, in order to survive, chicks require a relatively high temperature. Upon hatching, it is 34 °C, and gradually it decreases by approximately 2 °C each week of their life. When the chicken reaches a weight of 300 g, its ideal temperature is approximately 21 °C.

More than any other domestic animal, in order to develop properly, chicks require a rather low relative humidity. This is set between 50 and 70 %, preferably close to 55–60 %. These figures, by themselves, already show how important is the microclimate in which the chicks develop. Actually, they can struggle against excessive temperatures only with great difficulty. Thus, for example, 300-g chicks, kept at 30 °C, will mostly die of exhaustion from loss of breath. On the other hand, very young chicks raised at about 25 °C will practically not grow at all, because all the food ingested will be used to keep their body at its normal temperature.

12.7.2 Broilers and Egg-Laying Hens

These two types of poultry have a zone of thermal indifference located between 17° and 21 °C, but their metabolism is in perfect balance at approximately 18 °C. As in the case of the chicks, the relative humidity should be kept close to 70 %. Actually, humidity is the worst enemy of the plumage of hens and, if their feathers become sticky as a result of excessive humidity, the thermal screen that should form between the body of the animal and the air in the environment loses its effectiveness and the heat losses are very great. They must then be compensated by the ingestion of additional grain, causing a decrease in the profitability of the farm.

12.7.3 Other Fowls

At the beginning of this paragraph, we remarked that, as a result of the general conditions of poultry production, there was no very accurate information available concerning the zones of thermal indifference of fowls other than hens (for example, ducks, geese, turkeys, etc.).

We shall note, however, a general feature applicable to all birds: their need for a very low relative humidity in order to develop properly. Consequently, poultry farms, whatever they may be, will be kept sheltered from the rain and especially will be located on well-drained land. Even ducks, geese, and other water fowls require dry ground to rest and build their nests. Of course, in addition to that, they must also have available water surfaces to play and search for their food.

Literature

Bianca, W.: Der Einfluß der Alpung auf die Körperentwicklung und die Blutbeschaffenheit von Jungziegen. Schw. Landwirtsch. Mheft 30, 5, 197–213; 6, 241–254 (1952)

Bianca, W.: Heat tolerance in cattle. Int. J. Biometeor. 1, 5–30 (1961)

Bianca, W.: Die Anpassung des Haustieres an seine klimatische Umgebung. Schw. Landw. Forsch. 10, 155–205 (1971)

Johnson, H.D.: Environmental temperature and lactation. Int. J. Biometeorol. 9, 103–116 (1965)

Primault, B.: Conditions atmosphériques limites pour le transport des animaux domestiques, p. 4. Rapport de travail Inst. suisse Météorol. 15 (1971)

Rist, M.: Soll-Temperaturen in Schweineställen. Die Grüne 85, 621–622 (1957)

Schram, K., Thams, J.C.: Die Temperatur eines frei aufgestellten Körpers als Maß für die Erwärmung und Abkühlung von Bauwerken. Schw. Bl. Heizung und Lüftung 4, 3–12 (1967)

Weber, F.: Warmes und kaltes Klima als begrenzende Faktoren für die Leistungen unserer Haustiere. Schw. Landwirtsch. Mkeft 33, 223–234 (1955)

Winn, P.N., Godfrey, F.F.: The effects of humidity on growth and feed conversion of broiler chickens. Int. J. Biometeor. 11, 39–50 (1967)

13 The Climate Inside Animal Shelters

B. PRIMAULT

13.1 The Effect of External Factors on the Walls and the Roof

Any body, even if placed inside a building, is affected to a more or less considerable extent by the meteorological factors of the external environment. The animals who live in shelters, themselves part of the house, the barn, or built separately, constitute no exception to this rule.

If it is considered desirable to create for domestic animals, and therefore inside the shelters, the optimum microclimatological conditions for their development, in their construction and installation one should keep in mind the possibilities offered by the conditions prevailing outside of affecting the climate inside. This influence is a direct function of the meteorological elements properly so called, but also, or even more, of the construction itself (shape, exposure, color, material used, etc.).

13.1.1 Radiation

As shown above (see Chap. 3), the temperature of any solid body is directly affected by the radiation received. On the other hand, the body itself radiates thermal energy toward the outside. If one wishes to create inside a building temperature conditions ideal for the surface of an animal, it is essential to start from this radiation effect and to consider it seriously. There are formulas which make enable the calculation of a precise balance sheet of radiation both on the outside walls and on a body placed in a shelter. In the former case, this effect depends on the exposure of the wall, that is, on its azimuth, on the height of the sun, which varies in the course of the year, and on the surface ratio between the outer walls and the roof.

It is possible to prevent excessive heating due to the radiation received by building porches facing south (shade falling on the main wall), or by placing reserve fodder or straw above the shelter (barn).

At the present time, modern construction techniques make it possible to use reflecting materials and to install ventilated double walls in order to prevent strong heating of the shelter in daytime (in certain regions, such materials are forbidden, connected with the preservation of the integrity of natural sites).

In warm regions (Mediterranean, subtropical, and tropical climates), experience has demonstrated the beneficial effect of painting roofs and walls white, which reflects a large part of the radiation received.

While the radiation received raises the overall temperature of the building during the day, the radiation emitted by it, on the other hand, lowers its temperature by night. Therefore, there is an alternating action of radiation on the overall temperature of the building.

13.1.2 Temperature

Radiation is not the only meteorological factor which acts on the overall temperature of a building. By being in contact with the walls and the roof, the temperature of the surrounding air also affects that of the building. Thus, even if the radiation received is low (with overcast sky, for example), the temperature of a building is affected by the magnitude and the direction of the thermal gradient existing between its outer surface and the environment.

Furthermore, the temperature of the surrounding air must be considered when one wishes to control the temperature inside by means of forced draft (see Sect. 13.4).

The importance of the outside temperature, however, is less than that of radiation, and it is easy effectively to prevent the undesirable effects of this factor by adjusting the thermal conductivity coefficients of the walls.

13.1.3 Wind

The effect of the outside temperature on the overall temperature of a building through the contact of the air with the walls and the roof is enhanced by the wind. Actually, in calm air, a zone of thermal balance is formed around a building, and the loss or increase of heat through the outside air is thereby greatly reduced.

An air flow, on the other hand, alters that zone of balance, so that the outside air can act directly and continuously on the walls and the roof. The stronger the wind, the greater the effect of this disturbing action.

13.1.4 Precipitation

The shelter must protect the animals not only against the harmful effects of temperature and radiation, but also against those of precipitation, liquid or solid. Furthermore, precipitation has a direct effect on the temperature of the building as a whole. Actually, generally it causes a lowering of the temperature of the roof and, in the case of strong winds. of the walls, Such lowering is felt not only during the precipitation itself (because rain water and, above all, snow have a lower temperature than that of the building), but also after it has stopped, because the water remaining on the building will evaporate; the evaporation heat will remove from the building large amounts of calories. This phenomenon of cooling through evaporation begins with the first drops and only ceases when the last particle of water remaining on the walls or the roof has evaporated. The cooling effect of evaporation can also be used to cooling the shelters by heavy radiation or high outdoor temperature.

13.1.5 Snow Cover

In vast areas of our globe, precipitation does not fall only in the form of rain, but also of snow. The snow then accumulates on the roofs of the shelters for domestic animals and remains there for several weeks, or even several months. The layer thus formed has particular thermal properties, because it forms a screen against the effects of the outside radiation and temperature (see Sect. 3.1.1 and 3.1.2). Thanks to its very great insulating and reflecting properties, the presence of a layer of snow reduces the temperature fluctuations to which the roof is subjected as a result of the radiation received and of the air temperature. Thus, the temperature changes recorded on the surface of the snow only affect the first few millimeters or, at the most, a few centimeters (in the case of old snow) in the thickness of the layer, and therefore generally do not reach the outer surface of the building.

In the case of snow accumulated by the wind along the walls of the building, similar thermal effects can be observed. In the latter case, however, and as a result of the heating of the wall from the inside (see Sect. 3.3.1), the snow melts rather rapidly when in contact with the vertical walls, and a cavity is formed between the mass of snow and the building itself. If this cavity is closed in all directions, the thermal screen effect of the snow mass is fully preserved. If, on the other hand, the cavity is open at one of its two ends, or upward, and enables the outside air to circulate freely within it, the snow mass only protects the wall (and most often only in part) against radiation (see Sect. 13.1.1) and no longer against the thermal effects due to contact with the outside air (see Sect. 13.1.2).

13.2 Effect of the Temperature of the Walls on the Climate Inside

While the meteorological and climatic conditions of the environment surrounding the building have an important effect on the temperature of its walls, the microclimate prevailing inside the building itself acts on their temperature in an identical manner. Everything stated in the preceding paragraph, therefore, is also applicable in this case, with the only limitation that the radiation received is no longer due to the sun, but to the bodies of the domestic animals, or of any other object placed in the shelter.

The effect of the microclimate prevailing inside, on the walls, however, is not the only factor to be considered. The effect of the walls on the climate inside the building will also be taken into account. Therefore, there is a close interaction between the one and the other.

13.2.1 Radiation

The warmer a body, the more energy is radiated from its surface into the environment. In this particular case, the more the walls are heated, the greater will be their radiated heat, and this will be directed toward the outside as well as the inside of the shelter. This latter radiation in part is directly absorbed by the animals, whose temperature is raised. On the other hand, if the walls are cold, they radiate less and, therefore, give the animals a cold feeling, because, while constantly losing heat through radiation, they only receive little of it in exchange. The ideal conditions for the development of domestic animals therefore require that the heat losses inherently related to the radiation from their bodies toward the walls be compensated for, at least to a large extent, by the radiation from the walls toward the animals, so that the balance of those two effects produces the ideal conditions of the general metabolism of the animals.

13.2.2 Humidity

When one enters an animal shelter of the old type, either in the cold regions of the world, or during the season of bad weather, one is often impressed by the moisture which oozes from the walls or from the ceiling. This oozing – added to the effect of the gases mixed with the air – may go as far as to cause chemical changes on the surface of the walls (formation of saltpeter, for example, due to the action of ammonia on limestone and lime, in the presence of water).

Such a tendency of the walls to form a deposit of moisture always indicates insufficient insulation; but let us attempt to explain its cause:

Any animal constantly releases into its environment large amounts of water ($352 \, g \cdot h^{-1}$ for a 700-kg milk cow, for example). That water originates from sweating, if the temperature is high, but mainly from respiration, irrespective of temperature. Actually, the air contained in the lungs must be saturated with moisture to enable the osmotic exchanges of gases (CO_2/O_2) to take place normally within the alveoli of the lungs, which otherwise would dry out very rapidly. The saturated air is then exhaled toward the outside and thereby it raises the moisture content in the building. If the overall temperature of the building is close to the dew point, the humidity cannot be absorbed by the air and it condenses directly in the form of fog, the droplets of which are deposited on all solid objects, be it the bodies of the animals or the walls of the building.

These condensation phenomena depend first of all on the temperature and the dew point of the air inside the building. However, it is quite possible for the air contained in the building to have a sufficiently high temperature, and a sufficiently low humidity, respectively, to absorb all that released by the animals. The relative humidity will rise, of course, little by little, but will not necessarily reach the saturation point. If, under such conditions, the temperature of the walls, of the ceiling, or of the floor is lower than the dew point, water will condense on those surfaces. Therefore, in the selection of the site of a building and installation of a shelter, great attention will be paid to these condensation phenomena. Should they take place in spite of this, that will prove that a mistake was made in the evaluation of the thermal conditions of the walls, and it would be desirable to remedy it as soon as possible.

13.2.3 Temperature

We have seen in the preceding paragraph that the temperature of the walls affects the humidity in the building, because any condensation on the walls is accompanied by a lowering of the air humidity (in $g \, m^{-3}$); but the water which is thus deposited on the walls has a not inconsiderable effect on the durability of the building itself.

Furthermore, the temperature of the walls is not only of primary importance in determining the humidity in the building: it also directly affects the air temperature as a result of the contact between the air and the wall. The phenomena described above (13.1.2), therefore, are applicable to this case as well.

13.3 Insulating Properties of the Walls

In Chapt. 11, we have seen that the optimum climate for the development of domestic animals often is not that offered by nature in the open. Therefore, if we leave our cattle in the open air, its development potential is decreased and its profitability becomes questionable. For this reason shelters are built, to make possible the artificial creation of more favorable microclimatic conditions.

In certain cases, the sole purpose of the shelters is to protect the cattle from an excessively strong action of direct and diffuse radiation, in order to avoid excessive heating of the skin surface. For this purpose, it will not be essential to install closed shelters, but simple roofs, leaving the space reserved for the animals open to all winds.

Radiation is not the only factor to be considered in the study of the living conditions and development of cattle: the temperature and humidity of the environment also play an important role, and the shelter must primarily insulate farm cattle from the harmful effects of those two factors.

13.3.1 Thermal Insulation

If one compares the temperature of the air, and more precisely of the outside air, and the ideal temperature mentioned above (see Chap. 11), depending on the kind and age of the cattle, an important difference is often found. The walls which shelter the cattle from the outside meteorological elements must be capable of decreasing, or even eliminating, this gradient. Therefore, the thermal insulating properties of the building must be calculated as a function of the gradient.

It is not necessary, however, in the selection of the materials and in the calculations preceding construction, to consider the extreme conditions shown by a climate analysis, because the wall neither heats nor cools off instantaneously. Thus, in addition to its insulating role, it plays the role of a temperature-control device. Furthermore, the complete absorption of all external temperature fluctuations by a wall would require the construction of disproportionate shelters, the profitability of which would then be questionable. Also, a preliminary climatological analysis should show which is the gradient which occurs at least three times in five years (the three times may occur in the same year) and during a period exceeding four consecutive days. That constitutes the basis on which it will be profitable to make the calculations concerning thermal insulation.

At the present time, materials and methods of construction are available which are quite varied and meet all needs. However, it should not be forgotten that, as one tends toward an increasingly effective insulation, construction costs increase. Before taking the decision to build, therefore, it is essential to have a climatological study done, showing the maximum temperature gradient to be expected. From that figure, an economic study of the market will be performed, aiming at determining whether the expenditure required by an insulation which meets the local conditions would still be bearable, in economic terms.

13.3.2 Insulation Against Humidity

In the study of the microclimatic conditions capable of producing optimum well-being for the animals, we have emphasized the importance of humidity. The walls of cattle shelters therefore must be insulated not only from the thermal viewpoint, but also from the viewpoint of humidity. The condensation phenomena mentioned under 13.2.2 often let water penetrate inside the walls, which deteriorate very rapidly. Such condensation may take place inside as well as outside the shelter. Consequently, the waterproof coating which plays the role of insulator against humidity on the walls must be applied on both sides of them on the surfaces in direct contact with the air. This hydraulic insulation must prevent any penetration of water into the wall mass. Consequently, special care will be taken in sealing all the openings in that coating (water pipes or electric cables passing through the walls, installation of door or window frames, etc.).

13.4 Ventilation

While the simple fact of placing cattle inside a building insulates it from the undesirable effects of external meteorological elements, such confinement is not sufficient, by itself, to create the ideal microclimatic conditions for the development of the animal. Actually, the animals release both heat and water vapor as a result of their internal metabolism. As a consequence, temperature and humidity will increase very rapidly inside the shelter, and the microclimatic conditions inside it will very soon become unsuitable for life. In order to prevent such a situation, it is necessary to add to the protective action of the shelter a microclimate control effect, by means of natural or forced ventilation.

13.4.1 Heating

If, as a result of a negative radiation balance, the shelter cools beyond tolerable conditions, sometimes it is possible to introduce into it warmer air from the outside, which will compensate for the decrease in temperature. In most cases, however, the air outside is colder than inside, and therefore it cannot heat the shelter.

13.4.2 Cooling

The opposite situation occurs much more often, which means that the air inside the shelter is heated beyond tolerable limits as a result of the heat released by the metabolism of the animals. Then it is necessary to eliminate the excees heat. If the temperature gradient between the inside air and the outside air is sufficient, it is possible to reestablish the balance by letting the latter circulate in the shelter. Standards and ratios have been determined and graphs are available which show, by head of cattle and category, the amounts of air required to reestablish optimum conditions (see Fig. 1).

13.4.3 Elimination of Water Vapor

The general metabolism of the animals involves a considerable release of water. The water is released in the form of vapor into the environment – that is, into the air inside the stables – by perspiration and respiration. Therefore, it tends to accumulate there. A change of the air inside will make it possible to eliminate the water vapor. The volume to be introduced will depend on the absolute humidity of the outside air, that is, on its capacity to hold additional water vapor in suspension, and on the amount of water vapor released by the cattle. The

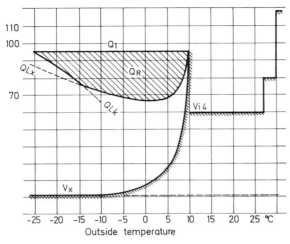

Fig. 1. Ventilation curve for pigs. V_i Ventilation rate depending on heat, V_x ventilation rate depending on water vapor, Q_T heat released per head (100 kg), Q_{Lx} losses depending on water vapor, Q_{Lk} losses depending on carbon dioxide, Q_R residual heat

difference between the outside and the inside dew point temperatures will be the determining factor in these calculations.

As in the preceding case, tables will facilitate the calculation of the amounts of air to be introduced in order to eliminate the water vapor (see Fig. 1).

13.4.4 Elimination of Harmful Gases (Carbon Dioxide, Methane, Ammonia, etc.)

The chemical reactions which compose the metabolism of individuals, as well as those which occur in the stable litter, cause a considerable release of harmful gases. As in the case of water vapor, they tend to accumulate inside the stables. Their concentration may reach a level dangerous to the life of domestic animals. Therefore, care will be taken to eliminate them permanently by proceeding to an adequate ventilation of the stables. Such ventilation will depend on the volume of harmful gases produced, particularly carbon dioxide (the release of methane and ammonia may be decreased by preventive measures, such as the choice of stable litter, the prompt evacuation of fetal matter and their separation from the urine, etc.).

As in the case of the elimination of water vapor, there are tables which make possible a rapid calculation of the amounts of air to be moved in order to eliminate the carbon dioxide produced by the respiration of the animals (see Fig. 1).

13.5 Artificial Climate Control

In most cases, external conditions do not make it possible to obtain an optimum microclimate inside animal shelters by the simple adaptation of construction (exposure and shape of buildings, materials forming walls and roof) and ventilation. It is then necessary to resort to artificial climate control.

However, all artificial climate control processes are very expensive, both in their installation and, especially, in their operation. Heating and humidification of buildings require energy and a well-designed installation.

However, the cooling and drying of the air are the particular operations which consume often surprising amounts of energy (one will recall, in this connection, the blackouts in the American electric network, during the summer, following the simultaneous switching on of a very large number of room air conditioners).

13.5.1 Heating

In the polar and subpolar regions of the world, in the mountains, or in the temperate zones in winter, the walls are not sufficient to insulate cattle adequately from the outside air temperature. Furthermore, as a result of the essential ventilation, the temperature to which the cattle is subjected becomes too low to assure an optimum metabolism. Therefore, it is necessary to supply the missing calories.

The optimum microclimate for the development of certain kinds of animals (sucking pigs, chicks, etc.) requires high temperature, which even constitute an essential prerequisite for survival.

Consequently, in order to create such optimum living conditions and thus the highest possible profitability for the farm, it is essential to supply additional calories. This can be done either by thermosiphon, or by hot-water heating, or again by heating the air introduced by the ventilation systems. The calorific energy will be of vegetal (wood), mineral (coal, oil) or electric origin.

13.5.2 Cooling

During the summer, and especially in the Mediterranean, subtropical and tropical zones, radiation is intense, so that the buildings are greatly heated by it. The outside temperature, generally very high, does not make it possible to lower the inside temperature to create an optimum climate there. Therefore, it is necessary to provide for this artificially. Unfortunately, the installation of all the cooling systems known at the present time is very expensive. Furthermore, they consume great amounts of energy. Consequently, the artificial cooling of an animal shelter, in order to create optimum microclimatic conditions inside it, is most often irrational, in economic terms. Actually, the expenses involved in doing so exceed by far the profit obtained by an increased productivity of the cattle. For this reason, before proceeding to the installation of such systems, it is necessary to perform very accurate calculations of the amount of cooling required, of the likely financial profits due to the increased productivity, and of the general profitability of the enterprise.

Apart from a few exceptions, the artificial cooling of animal shelters is not profitable.

13.5.3 Humidification

It has been shown in Chap. 11 that the humidity of the air breathed plays a very important role in the well-being of animals. Such humidity must be relatively low for the majority of domestic animals. Nevertheless, in many desert or semi-desert areas, the natural humidity of the air is insufficient to provide the desired well-being for the animals. Consequently, water will be vaporized in the ventilation system, thereby adding the water vapor which is lacking in nature. Such systems will be carefully studied in order to avoid an excessive consumption of water. They are generally highly profitable in practice.

13.5.4 Air Drying

In contrast with what we have just stated, the external atmospheric humidity conditions are favorable in most of the regions where intensive cattle breeding is practiced. Therefore, it is sufficient to ventilate the stables adequately in order to provide humidity conditions ensuring an optimum development of the animals.

In certain areas with a marked maritime climate, the outside air is so saturated with humidity that its absorption properties are almost nonexistent. Then it is not possible to eliminate the water vapor by simple ventilation; it will be necessary to dry the air before blowing it in, thereby creating a more favorable type of microclimate. This artificial drying is generally performed by applying the principles of air conditioning: the fresh air flow is brought to a temperature markedly lower that the dew point outside. A large part of the water vapor then condenses on the pipes forming the radiator, and the water thus produced must be eliminated outside the building. The dried air can then be blown into the stable. Since the process described above lowers the air temperature, however, sometimes it will be necessary to reheat it later in order to obtain the optimum temperature conditions.

The greatest drawback of such a system is its cost, because, as in the case of air conditioning (see Sect. 3.5.2), the installations required to dry the air, as well as their operation, are very expensive, which makes their profitability highly questionable.

13.6 Open Stables

By carefully studying the climatic conditions prevailing in certain cattle breeding areas (subarctic or mountain areas), it often appears that the optimum conditions for the cattle occur naturally during a great part of the year. Consequently, in those regions, it would be wrong to insist in building closed shelters for cattle at any price. While radiation requires that cattle, particularly bovines, be taken back to the stable when the temperature decreases to a level lower than that comfortable for man, it is in contrast with our scientific knowledge (see Chap. 11). It would often be preferable to leave the cattle exposed to the weather outside rather than to confine it in stables where temperature and especially humidity are higher and, generally, less favorable to its comfort.

It might be concluded from this that the best solution would be to let bovine cattle roam on the pastures. Old research, which however remains fully valid, has demonstrated that the hair of many breeds of bovines absorbs a great amount of heat radiation. If this cattle is exposed to the sun, its skin temperature often rises above the tolerable threshold. Actually, bovines eliminate most of the calories produced by their metabolism through an intense blood flow in the subcutaneous tissues. If the latter are greatly heated from the outside – as in the example just mentioned – the heat exchange cannot take place, because the temperature gradient is reversed (the temperature of the subcutaneous tissues is higher than that of the body of the animal). This results then in "heat strokes", which can be fatal.

Furthermore, cattle should not be exposed to precipitation.

Consequently, wherever temperature and humidity are naturally close to the optimum conditions for breeding, the closed stables will be advantageously replaced by open stables, to which the cattle has access according to its needs and where it can circulate freely. It is then protected from radiation and precipitation, but directly enjoys the temperature and humidity conditions of the local climate. Great installation and maintenance costs will be avoided in this manner.

In all the decisions involving the establishment or maintenance of the particular microclimatic conditions for cattle breeding, the well-being of the animal is to be considered first, and not that of the personnel in charge of its care.

Literature

Bianca, W.: Das Stallklima und das Tier. Schw. Landwirtsch. Mheft *48*, 263–274 (1970)

Cena, M., Courvoisier, P.: Untersuchungen über die physikalischen Faktoren des Stallklimas unter besonderer Berücksichtigung der Abkühlungsgröße. Schw. Arch. Tierheilk. *41*, 303–336 (1949)

Gysel, A.: Erhebungen über Temperatur und Luftfeuchtigkeit in Schweineställen samt einigen Unterlagen zur Gestaltung des Stallklimas auf Grund ausländischer Angaben. Schw. Landwirtsch. Mheft *40*, 331–334 (1962)

Hader, F.: Indexzahlen der Witterungsbewertung bei Bauarbeiten. Wetter und Leben *12*, 143–147 (1960)

Ober, J.: Der Wärmehaushalt der Viehställe und der konstruktive Wärmeschutz der Stallbauteile. Mitteilungen für Tierhaltung *73*, 12–20 (1961)

Oosterlee, C.C.: The influence of climatic factors on the heat balance and milk yield of the cow in relation to the design of farm building in the Netherlands. Int. J. Bioclimatol. Biometeorol. *III*, Sect. C (1959)

Piccot, F.: Climatisation des étables. Rev. suisses d'Agr. *IV*, 209–218 (1972)

Primault, B.: Le risque d'enneigement dans les étables ouvertes. Rech. agron. en Suisse. *3*, 406–416 (1964)

Raeuber, A.: Meteorologische Vergleichsmessungen zwischen Schuppenstall and Freiland in Groß-Lüsewitz. Angew. Meteorol. *2*, 217–222 (1956)

Smith, C.V.: Ventilation and associated patterns of air flow. Agr. Meteorol. *1*, 30–41 (1964)

14 Plant Climate (Heating and Cooling)

B. PRIMAULT

In the present chapter, we shall only deal with changes in the microclimate in open fields, and therefore we shall ignore from the beginning the special climate of greenhouses, which is covered by a separate chapter.

14.1 Physiological Effects of Extreme Temperatures

14.1.1 Definitions

In order to develop properly, a plant must be able to live in an environment in which the temperature exceeds a certain level, depending on its phenological condition. From that threshold, called "zero vegetation point", all other external conditions remaining sufficient, or even optimum, the quantitative development of the plant is practically proportional to the increase in temperature. This increased development, however, is not without limits. Each phenological condition of each plant is limited at its highest level by a maximum tolerable temperature. From this second threshold, growth decreases very rapidly as the temperature rises and then falls to zero, at which point the plant has died as a result of the heat.

Below the zero vegetation point, the plant can survive at relatively low temperatures without suffering great harm. If, however, the temperature decreases beyond a new threshold (natural resistance to frost), the plant also dies.

Consequently, for each variety and for each phenological stage, four temperature thresholds must be considered: the natural resistance to frost, the zero vegetation point, the maximum tolerable, and the absolute maximum (Fig. 1).

Thus, between the point of natural resistance to frost and the zero vegetation point, the vegetative organs remain in a quiescent state, while between the maximum tolerable and the abolute maximum their growth decreases very rapidly as the temperature rises. Between the zero vegetation point and the maximum tolerable, plant development is, as a first approximation, in linear proportion to the temperature.

It is therefore very important, when introducing new crops, to know those thresholds and to compare them with the climatological data in order to see whether they might be exceeded during the various phenological stages through which the plant proceeds during its annual development. Otherwise, it will be necessary to devise artificial measures to protect the plants from the harmful effects of such conditions.

14.1.2 Frost

We have seen above that one of the important thresholds is the natural resistance to frost. When frost is mentioned, one generally thinks of the transition of water from the liquid state to the solid state, and, implicitly, of the corresponding

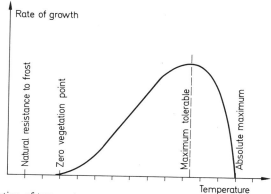

Fig. 1. Diagram of plant growth as a function of temperature

temperature, which is 0 °C or 32 °F. Plants, however, generally do not suffer any harm when the temperature of the air crosses that dramatic threshold of zero degrees. Some already "freeze" at temperatures well above zero, others resist perfectly well at temperatures which may each -6 °C, or even -10 °C or more, depending on their species and on their phenological stage.

The notion "plant frost", therefore, must be dissociated from the notion of "water freezing". The same applies to its climatological definition. In this case, it is a purely physiological concept closely related to the plant itself or, more precisely, to some of its organs, because it is extremely rare for a plant to freeze completely.

Mechanics of Frost. Plant tissues are composed of agglomerations of cells whose membranes do not touch one another in a continuous pattern, but leaves among them spaces full of an aqueous liquid which is called intercellular water (see Fig. 2). This intercellular water is not pure, but always contains dissolved salts of mineral and vegetal origin, that is, inorganic and organic substances.

When the frozen tissue, prepared after having been previously defrosted, is examined under a microscope, it is surprising to find that most of the cells have burst toward the outside. If, instead, one proceeds to the same examination of nondefrosted tissue, it is surprising to see that the number of burst cells is extremely small, and that the edges of the cell membrane are then folded toward the inside of the cell and not toward the outside (Fig. 3).

It has long been believed that the damage caused by frost was the result of the bursting of the cells. This would be due to the transition of the cell juice from the liquid state to the solid state, meaning from water into ice. Since, in this case, the volume of the mass suddenly increases, it might be assumed that this would produce a mechanical pressure sufficient to cause the bursting of the cell membranes. It would have been the case of a phenomenon similar to the bursting of a bottle full of water exposed to intense cold.

The second procedure (examination of cells not previously defrosted), makes it possible, on the other hand, to suspect entirely different phenomena. It is no longer a case of the water contained in the cell which would pass from the liquid state to the solid state, but of the water located in the intercellular spaces. This water, when solidifying, also increases in volume. However, in relation to the water remaining

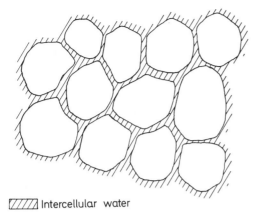

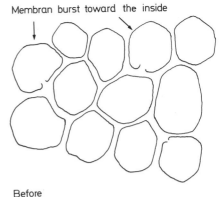

Membran burst toward the inside

Before

Defrosting

After

⬚ Intercellular water

Fig. 2

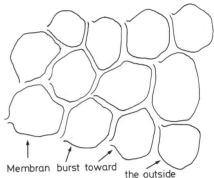

Membran burst toward the outside

Fig. 2. Diagram of parenchymatous tissue

Fig. 3. Direction of cell bursting. *Above* tissues still "frozen", *below* tissues examined after natural defrosting

Fig. 3

liquid inside the cells, and as a result of the fact that the cell membranes are always somewhat plastic, the mechanical damage is much less serious than one might expect on the basis of the first experiment.

At the onset of frost, which means when the temperature decreases, the aqueous solution which fills the intercellular spaces – and which is much less concentrated than the cell juice itself – may pass into the solid state. However, since this is not pure water, the transition only occurs at a temperature below zero degrees. Therefore, the more concentrated the intercellular solution, the lower its freezing temperature. The fact that this aqueous solution freezes does not by itself cause damage to the surrounding tissue. In exceptional cases, that is, when the foreign body thus formed – the ice crystal – becomes rather large in size, it may break one or more of the cell walls which surround it.

After the ice crystal is formed and as a result of the vapor pressure gradient which is immediately established between the water and the ice, it grows by coalescence. The water composing the cell juice crosses the semipermeable cell wall by a phenomenon of osmosis. As a result of this, the cell juice thickens and the risk of its freezing decreases proportionally. However, as a result of the drying which takes place inside the cell, there occurs a rapid transformation of the proteins, particularly those which form the nucleus. These transformations are generally irreversible.

When the temperature rises again, the water attracted outside the cells by the ice crystals returns to the cell juice. Actually, the difference in concentration causes an osmotic exchange between the fluids separated by a semipermeable wall. The proteins, however, cannot be regenerated and the cell rapidly dies, even though, from the mechanical viewpoint, it has not been damaged.

The fact that previously defrosted tissues, or tissues subjected to a rapid reheating due to sunshine, show a very large number of burst cells is explained by this return of the water which inordinately inflates the cells, whose juice is highly concentrated. Under natural conditions, the defrosting of the plant usually takes place rather slowly, so that such lesions are rather rare.

Special Cases. In the preceding paragraphs, we have always emphasized that it was primarily a transition of the water from the liquid state to the solid state. It appears, therefore, that it is necessary for the temperature to decrease to at least below zero before a plant can suffer frost damage. This is actually not true at all. Actually, in such plants as banana trees, coffee, or tea plants, for example, as soon as the temperature decreases below a certain threshold, the cells begin to release water. It is thus a case of loss through exudation. This exudation, however, is not due to a passive osmotic transfer on the part of the cell wall, but an active rejection. The phenomena of the drying up of the proteins and their consequent irreversible transformations are then of exactly the same nature as in the case of the formation of ice crystals in the intercellular spaces.

Therefore, when one speaks of frost damage, one should rather say that it consists in a drying out of the cells due to the cold.

14.1.3 Thermal Shock

Even if one does not find damage such as that described above, typically due to extreme temperatures, a rapid change in environmental conditions – particularly temperature – may cause temporary changes in the cell components. This does not necessarily involve the proteins, as in the case of frost, but also the sugars and fats.

Thus, a sudden decrease in temperature, as well as a sudden rise of it, causes a physiological trauma from which the plants recover with difficulty, although its traces cannot be detected in the shape or structure of the tissues. Their metabolism alone is affected, and for several days, or even weeks, the rate of growth no longer follows the curve shown in Fig. 1.

In most cases, the respiration and evaporation-transpiration phenomena are not affected by such thermal shocks. Only assimilation is affected by them.

14.1.4 Physiological Effects of Heat

We have seen that, as the temperature rises, the activity of the cells also increases. Such activity requires ever-increasing amounts of water in order to take place under optimum conditions. Therefore, it is not surprising to assume that at a certain point the water supply of the cells no longer takes place under optimum conditions. The plant suffers a stress from dryness, even though the soil is sufficiently supplied with water, and the surrounding atmosphere is sufficiently humid not to cause excessive evaporation and transpiration.

Actually, in most cases, the heating of plant tissues is caused by intense radiation, which, simultaneously with the rise in the temperature of the environment (thus, of the air), activates the assimilative functions of the cells (this mainly in the leaves).

However, such activation of the assimilative functions requires an increased consumption of mineral salts, of vitamins, etc. Simultaneously, the amount of the products of assimilation (particularly sugars and their derivatives) increases in the leaves. Then it is easy to detect the point at which a physiological reaction of the plant occurs. It consists first in a gradual thickening of the cell juice, which proportionally reduces assimilation as a result of its increasing opacity. Later, phenomena of drying out intervene. They are similar to those which we have mentioned in the case of frost. In this case again they may to the death of cells, or even of entire tissues.

14.1.5 Scorching

In some plants, and more specifically in cereals, during the ripening of the grains, particular physiological phenomena occur if the temperature exceeds a certain threshold. As in the case of frost or of thermal shock, the level of those thresholds varies from plant to plant. The phenomenon is called scorching.

Grain is subject to this accident at the lacteous or pasty stages of ripening. When the temperature exceeds the critical thresholds, there occurs an extremely rapid (in the case of wheat, only two to three days) drying out of the soft mass and an almost instantaneous transformation of the sugars in starches, without, however, thereby affecting the germ. However, the grain can no longer grow. If this accident occurs at the beginning of the lacteous stage, the grain which will be produced from it will be extremely small and the weight of one thousand grains will therefore be low. Since the germ is not affected, the germinating properties of such grain are not thereby decreased. It should be pointed out here, however, that the starch reserve contained in each grain is smaller than that of an average grain. Therefore, the young plant growing from it will not have as large a reserve of food to develop before the first leaf and the young root replace that reserve in feeding the whole plant.

In the case of scorching, the sugars are not the only substances affected. The grain proteins are also affected. Thus, grain subjected to high temperatures (according to Geslin, three consecutive days above $30\,°C$) practically contains no more gluten. Its baking properties are therefore severely impaired, or even destroyed (soft grains).

Consequently, in considering microclimatic changes, it will be necessary to take into account these peculiar phenomena in the growth of cereals.

14.2 The Struggle Against Low Temperatures

In this respect, we shall also refer to Chap. 18. Furthermore, in all these considerations, it must be remembered that the important thing is not to affect the temperature of the surrounding air, but that of the plants themselves.

14.2.1 The Passive Struggle

Selection of Species. We have seen in Sect. 14.1.2 that the natural resistance to frost (or, more accurately, to temperatures below the critical threshold) may vary quite considerably from type to type, and even from species to species within the same type. Therefore, geneticists have selected fruit trees, for example apple trees, whose resistance to frost is greater than that of others. A study of the climatic conditions prevailing in a given place, and particularly the study of the frequency tables of the occurrence of temperatures lower than the threshold of natural resistance to frost, will make possible to determine which species – and within the species, which variety – will be preferable to cultivate. The selection of species and varieties is the first step in the passive struggle: the climate is accepted as it is, without attempting to modify it, even in detail, and only what can develop under those naturally occurring conditions is planted.

Cold Air Pockets. It has been known for a long time that some places are more exposed to frost than others, which means that cold air collects there. In the case of heat losses through radiation, the plants in such places will not be reheated by the air in their environment.

A first appropriate measure to struggle against exceedingly low temperatures is to avoid planting sensitive varieties in those places.

Since cold air – heavier than the air in the environment – flows toward the lower areas in the relief, it is possible to deflect this creeping movement of the air through plants sensitive to the cold by raising a wall, a barrier or a live hedge. In this manner, the field will be protected without requiring man to act at a specific time of the day or year. We shall not here dwell on a detailed explanation of those movements of cold air, referring the reader to the vast literature published on this subject.

14.2.2 The Active Struggle

Heating. Two physical phenomena may combine in lowering the temperature of plants: the inflow of cold air or frost by advection and the cooling of the plant by radiation.

In the former case, in order to prevent a lowering of the temperature of the plants below the threshold of natural resistance to frost, it would be sensible to heat the entire air mass. This can be accomplished by means of high-power burners acting on an air flow generated by a fan. The warm air thus produced will be led into the field themselves through large plastic pipes. Holes pierced in the pipes will enable the air to spread among the plants, thereby heating the entire mass. Thus, one cannot properly speak of heating the plants, but of heating the air which surrounds them.

A similar effect of heating the air mass locked inside a cultivated area may be attained by means of a rotary fan sucking in the air heated by a central burner. In this latter case, the losses will be much greater than in the former, as a result of the very high temperature gradient at the outlet of the fan, which gradient will have the effect of deflecting the hot air flow upward, and therefore outside the cultivated area.

Sprinkling. When 1 g of water passes from the liquid state to the solid state it releases 80 calories (freezing heat). It has thus been thought of making use of those 80 calories to maintain the plant organs at a temperature higher than their threshold of natural resistance to frost. They are then enclosed in an ice dome, and thereby kept at a temperature below zero degrees. (This process, therefore, can only be used if the threshold of natural resistance to frost is below 0 °C.)

In order to do so, water is sprayed as uniformly as possible on the plants by means of sprinklers. The water freezes at the moment of its impact on the plant, if the temperature of the latter is below 0 °C. Each gram thus frozen supplies its environment with 80 calories. Part of those calories is lost in the surrounding air, while part (the greater part) is intercepted by the plant and the ice dome which covers it, and they are heated. When the temperature of the plant is measured, it is found that at each passage of the jet of water the temperature rises, and then gradually decreases until the next passage. If the minimum reached is still above the threshold of natural resistance, no damage will be found, although all this heat movement takes place below the freezing point of the water.

Such a process can only be effective if the threshold of resistance is below 0 °C to a significant extent (at least approximately 1°). To protect plants whose threshold of natural resistance to frost is above zero (for example, bananas and coffee), such a process is ineffective by its very nature.

Let us also point out that the water thus supplied must remain on the plant. In the case of vine shoots "in nursery" or at the "green points" stage, the wool which surrounds the shoots is water-repellent. Therefore, it rejects the water, which does not remain attached to it, and consequently cannot produce the dome whose formation is required for their protection. Thus, it is not really the ice which protects the plant, but the fact that water freezes almost continuously on the plant organs, releasing calories as it changes its state, and the majority of those calories are absorbed by the plant.

In protecting by sprinkling, attention will be paid to the fact that the same amount of calories released at the time of freezing is necessary when the ice melts. Consequently, if the operation is interrupted too early, the melting ice will absorb a great number of calories from the plant, with a risk of causing damage. Before interrupting the sprinkling, it will be necessary to wait for sufficient radiation to cause the ice around the plant organs to melt, so that the ice, formed during the night, will fall by itself (see also Sect. 14.3.3 and Chap. 18.2.3).

Mixing of Air. In Sect. 14.2.2, we have shown how it is possible to prevent the cooling of the plants following an inflow of cold air by replacing the latter with warmer air. In the case of a lowering of temperature due to radiation, the same procedures can also be used to maintain the plant tissues above the threshold of natural resistance to frost.

In this case, the lowering of the temperature of the plants is due to a loss of heat which is not balanced by radiation from the environment. The air which surrounds the plants or which lies on the ground cools off by contact. If the plantation is located on sloping ground, the cold air flows downward and, therefore, is replaced by warmer air originating from the upper layers of the atmosphere. The plant is then constantly surrounded by relatively mild air, and thus less exposed to the dangers of frost.

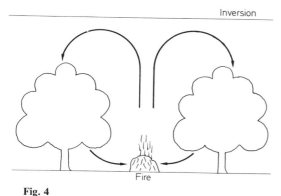

Inversion

Fire

Fig. 4

Fig. 5

Fig. 4. Air movements in the protection of plants by heating under a low temperature inversion

Fig. 5. Fog appearing very thick to the eye, but without any strong protective effect

In large plains (for example, in California), this flow does not occur. Thus, a pocket of cold air, very wide but of limited thickness, forms on the plain. Above that cold air, there is relatively warmer air. It has therefore been thought of warming the plants by drawing that warmer air into the plantations, thereby mixing the two masses.

This is generally done by means of powerful propellers placed on top of masts dominating the plantations and powered by internal combustion or electric motors.

Such a process can only be profitably used if the height of the cold air pocket does not exceed four meters.

Generally speaking, as the altitude rises, the temperature becomes lower. In the case of a clear night and as a result of either an accumulation of cold air behind an obstacle or its stagnation on a wide plain, a different process occurs: on the ground the air is colder than above. The transition is generally sudden, and this level is called "temperature inversion".

If a very marked temperature inversion (at least 6 °C) is repeatedly detected at a few decimeters (at the most, 3 m) above the plantation, it is possible to cause a general motion or mixing of the air between the ground and the temperature inversion. In order to do so, a great number of small fires or small heaters are set up in the plantation (Fig. 4). The mixing does not affect the inversion as in the case of the propellers.

Such meteorological conditions, however, are rare, and therefore such procedure will be used in practice only in very limited areas.

The Struggle by Counter-Radiation (Not to be Confused with the Heating in 14.2.2). Under 14.2.2, we have seen that one of the frequent causes of the lowering of the temperature of plants is radiation. In order to prevent this lowering of the temperature from crossing the threshold of natural resistance to frost of the plants, it is possible to compensate for it by counter-radiation.

The origin of counter-radiation most frequently is natural. It is caused by the cloud cover. It is known that the risk of frost is much smaller under an overcast than under a clear sky. This is not due to the fact that the loss of heat through radiation from the plants is thereby diminished. In this case, however, this radiation is intercepted by the cloud cover, the lower surface of which is then heated. It radiates proportionally more calorific energy downward, that is, toward the plants. The latter intercept the radiation and consequently become warmer. Thus, there is a constant exchange of radiation between the plants and the ground, on the one hand, and the cloud cover, on the other.

Perfectly identical phenomena take place inside a fog layer. Thus, if the dew point of an air mass is above the threshold of natural resistance of the plant, there is no need to fear frost damage. Actually, if, as a result of radiation, the temperature of the plants, and therefore of the air in contact with them, goes below the dew point, a fog will form and the exchanges of radiation will take place between the plants, on the one hand, and the fine droplets of water composing the fog, on the other.

Attempts have often been made to replace clouds and fog of natural origin by artificial layers; but in this case it is important to know certain physical phenomena of radiation. The latter cannot be intercepted by bodies in suspension in the air (fog or smoke), unless the diameter of those particles is greater than the wavelength of the radiation produced. Since heat radiation is a radiation whose wavelength is greater than that of visible light (infrared), the droplets or particles to intercept it must be much larger than those which impair visibility. Therefore, a fog or smoke appearing very thick to the eye will not necessarily be effective in preventing a loss of infrared radiation.

Most, if not all, of the products offered on the market to replace a lack of natural fog by artificial fog or smoke are ineffective; because the particles which compose them are too small. Visual demonstration could not persuade an informed practitioner. We have often had the opportunity of attending such demonstrations, in which visibility was reduced to 1 m by a fog which, from the viewpoint of radiation, was totally useless (Fig. 5).

So far, we have dealt mainly with radiation exchanges between the plant and its environment. However, while it is not always possible to intercept and to send back radiation produced naturally (see also 14.2.2), it is possible to compensate for the losses of radiation by an artificial supply. This principle constitutes the basis for most of the so-called "heating" systems used in practice.

The purpose of the use of small heaters, fuel oil burners, briquette fires, and paraffin candles generally is not to warm the atmosphere (with the exception of the mixing mentioned under 14.2.2). In the case of the Californian heaters, of the fuel oil burners, even of some gas burners, the fuel being consumed heats a metal cover which will radiate heat energy toward the plants. In the case of briquette fires or paraffin candles, the radiation from the flame itself or the red heat of the embers will produce the desired emission of radiated heat.

Covers. Straw covers to protect the young plants from the grip of the frost have been used for many years in the vineyards of southern Germany, Switzerland, and Austria.

Many tests have demonstrated that there was practically no heat gain under the covers in comparison with the air surrounding the unprotected vines. However, experience has shown the usefulness of those covers.

The small heat gain produced by the covers is due to an exchange of radiation between the plant (eventually its prop) and the straw stalks forming the cover. The difference in temperature as compared to a freely radiating plant is generally less than 0.5 °C. It reaches a maximum of 0.8 °C. Therefore, it cannot act as a thermal insulation.

We have seen above (see 14.1.2) that frost damage was caused by the transformation of the proteins due to a drying out of the cell juice as a result of the formation of ice crystals in the intercellular spaces of the plant organs. Such ice crystal formation is often initiated or at least greatly activated by the presence of hoarfrost on the surface of those organs (on the wax layer or on the cuticle). By a chain reaction, those ice crystals outside the plant cause the mass of intercellular water to pass from the liquid state to the solid state. Thus, if, as a result of radiation, the temperature of a plant organ goes below the dew point of the air mass surrounding it and that dew point is below 0 °C, the condensation of the water from the atmosphere on the plant will take place in the form of hoarfrost. If we replace the plant with a mass of straw, the temperature will go below the dew point on the periphery of that straw mass. The hoarfrost will then form on the first stalks of straw located on the surface of the cover. As a result of the vapor pressure differential between the water and the ice, the moisture in the air contained inside the cover – that is, surrounding the plant to be protected – will be attracted by the peripheral ice crystals. Consequently, we will have inside the cover a much drier air than outside the cover. As a result of the phenomena of transpiration and diffusion of water vapor inside the plant tissues, the concentration of the aqueous intercellular mass will increase, its freezing point will be lower, and therefore the threshold of natural resistance of the entire plant will be proportionally lower. It is then a case not of thermal protection, but of water protection.

With the new methods of planting and trimming vineyards, the use of individual straw covers became impossible. An attempt was therefore made to replace them with plastic tents, believing that the same effect would be obtained. However, two physical considerations prevented such a replacement: the size of the masses present and the inpermeability of plastics. The plastic tent has a considerably smaller mass than the straw cover, for the same outer surface. The exchanges of radiation between the plant and the new protection will be much more unfavorable, the more so because the plastic will not absorb the entire radiation emitted by the plant, but only a minimal part. Furthermore, the exchanges of vapor between the air mass contained under the cover and the hoarfrost crystals which form on its outer surface cannot take place because of its impermeability. The air mass enclosed under a plastic sheet therefore does not dry as in the case of a straw cover.

Even by piercing many holes in the plastic sheet, it is not possible to eliminate this latter drawback, while the former is completely unaffected.

The straw cover is an individual protection for plants. Therefore, it can only be used if the plantation is composed of isolated plants which can be protected (vineyards in isolation, tomatoes, etc.). On the other hand, when the plants form a carpet (for example, strawberries), the cover must also assume a continuous shape. The same arguments, however, can be used, which means that it is possible effectively to reduce an excessive radiation from low plants forming a carpet by spreading armfuls of straw or by laying reed screens on frames.

In recent tests, an attempt was made to replace the reed screens or straw spreads with aluminum foil. While the latter presents the same drawbacks as plastic with respect to the exchanges of water vapor between the enclosed mass and the ice crystals forming on the outer surface of the protective mat, it offers the great advantage of reflecting the radiation from the plant directly and practically completely, without absorbing it. Therefore, we have a different physical phenomenon. It is no longer a case of receiving radiation and transforming it into heat activating the radiation of the protective material, but of returning such radiation by simple reflection to the plant which produces it.

14.3 The Struggle Against Heat

We have seen under 14.1.4 and 14.1.5 that high temperatures may also be harmful, or even more harmful than low temperatures. Thus, for example, in California, grapes do not ripen satisfactorily (sugar/acidity ratio) because radiation is too intense, causing an increase in the production of sugar.

Similar findings have been obtained in Israel with respect to citrus fruit orchards. Furthermore, it has been found, especially in the Moselle region, that a high daytime temperature impaired the quality of the wine. Consequently, in these three particular cases, attempts were made to lower the temperature of the plants by appropriate means.

14.3.1 Covers

In order to avoid a strong radiation on a plantation, the first idea which comes to mind is to cover it with a sort of sunshade. Thus, burlap or reed screens will be laid on supports set horizontally above the plants to be protected. Since the sun cannot penetrate into the plants, it will not heat them so much any more. The temperature of the plant organs will thereby be kept within acceptable limits.

Such a practice requires the investment of considerable funds. It is necessary to set up frames in the plantation to support the burlap or reed screens. The latter material must be available and stored so that it can be laid out or folded as necessary. By so doing, direct sunshine will be permitted or prevented at the most appropriate time for the growth of the plants.

14.3.2 Reflecting Particles

In order to reduce costs, other solutions have been studied. This led to spraying the plantations with a powder containing very fine particles of aluminum. As a result, the leaves are covered on their outer surface – therefore, on the part exposed to sunshine – by a reflecting agent. The reflecting surface sends the radiation back into the environment, without letting it heat the plant organs located below to any considerable extent.

(It has been found that this treatment was also effective in the struggle against the cold, due to the very strong albedo of the aluminum particles. However, it can only be recommended if the frequency of frost is low.)

14.3.3 Spraying

While 1 g of water releases 80 calories in passing from the liquid state to the solid state, it absorbs approximately 590 calories to pass from the liquid state to the gaseous state. Thus, the evaporation of water causes a considerable lowering of the temperature both in the environment and in the solid body on which the evaporation takes place.

(Evaporation is often the cause of frost damage, when the methods of protection by spraying are used in an atmosphere with a relative humidity of less than 80%. In the struggle against frost by spraying, therefore, it is essential to know the humidity of the air before operating the protective equipment.)

This consumption of caloric energy is used to lower the temperature of the vegetal tissues and of the surrounding air when direct solar radiation produces too intense a heating for the proper development of the plants. The water sprayed into the air in part evaporates there, and the remainder, falling on the ground and the plants, evaporates on them, thereby causing a new lowering of the temperature. Thus, a double moderating effect is obtained.

The drawback of this system is the production of a considerable increase in the humidity of the air, and therefore an increase in the risk of cryptogamic diseases.

15 Droughts, Dry Winds, Dust Storms, and Hail

Y. I. CHIRKOV

Drought is a complex phenomenon which results from the prolonged absence of precipitation in conjunction with high evaporativity, which causes dehydration of the root zone of the soil and upsets the water supply of plants. As a result, the plant productivity is sharply reduced.

Drought is a normal climatic phenomenon in semiarid and desert regions. It develops in steppe and forest-steppe zones as a result of the inadequate replenishment of moisture reserves in spring, insufficient precipitation (less than normal) during the spring-summer period, and elevated atmospheric temperature.

A productive moisture reserve of less than 60 mm during the spring in a soil layer 0–100 cm deep characterizes poor moisture assurance over the course of the growing season. According to the findings of M. S. Kulik, a reduction in productive moisture reserves within the range of 10–19 mm in the arable soil layer (0 – 20 cm) characterizes a dry season. An arid season, or drought, sets in when productivemoisture reserves are less than 10 mm.

Dry wind is a meteorological phenomenon which is characterized by low relative humidity and high temperatures in the near-earth air layer, and by winds which sometimes reach great intensity. This process results in high evaporativity, causing a disturbance in plant water balance, damage to plant organs, and, in some cases, loss of young crops. The harmful action of the dry wind on plants is intensified when wind velocity increases. The formation of dry winds in an area with a temperate climate is a result of the transformation of air masses (usually on the periphery of an expansive anticyclone). Dry winds in the European territory usually occur when a high *anticyclone* settles above the central regions and over the southeast. Air masses arriving from the north and characterized by comparatively low absolute humidity are transformed. In the process, the relative humidity of the atmosphere is reduced sharply as a result of the warming of the air mass. The transfer of warm air from the south and the southeast has somewhat of an effect on the drying phenomenon, but its manifestation is primarily local.

It is necessary to know the ratio of moisture intake to potential expense (evaporativity) in order to determine the degree of dryness of a given region. This ratio was first used by V. V. Dokuchayev (at the end of the 19th century). Subsequently, many formulas which characterize dryness have been suggested by Russian and foreign researchers.

N. V. Bova (in 1941) advanced the following formula:

$$K = \frac{10(H+Q)}{\sum t},$$

where K is the coefficient of dryness; H is the productive moisture reserve in spring in a soil layer 0–100 cm deep; Q is the amount of precipitation from spring to the

Table 1. Agrometeorological criteria of dry winds (Ye. A. Tsuberbiller)

Types of dry winds	Evaporativity, in mm per day	Humidity deficit at a station for 13 h, in mm wind velocity, in $m \cdot s^{-1}$	
		<10	$\geqq 10$
Mild	3–5	15–24	10–20
Moderator intensity	5–6	25–29	20–24
Severe	6–8	30–39	25–34
Very severe	>8	$\geqq 40$	$\geqq 35$

time of the calculation; and $\sum t$ is the mean total diurnal air temperatures from the date of the transition through $0\,°C$. When the value of the K coefficient is 1.5 or less, this indicates the onset of a dry period. This value corresponds to the beginning of drought damage to spring wheat in the southeastern section of the Soviet European territory.

Soil humidity data, which take into account the level of agricultural technology, particularly measures for reducing nonproductive evaporation from the soil surface, are a reliable index of dryness. If, during the tillering and young maturity stages, there have been three ten-day dry periods (a productive moisture reserve of less than 10 mm in a soil layer 0–20 cm deep), this indicates a moderate drought. If there have been 4–5 ten-day dry periods, this signifies a severe drought. If, in this case, the productive moisture reserves in a soil layer 20–100 cm deep were lower than 60 mm at the beginning of the stage, this would indicate a very severe drought, toward the end of which the near-complete dehydration of the soil 1 m deep and a yield reduction of approximately 70–80 % on the average for a given region would occur.

Agrometeorological indices of the intensity of a dry wind take into account the magnitude of evaporativity, the relative atmospheric humidity and temperature, or the magnitude of the atmospheric humidity deficit, as well as wind velocity.

According to Ye. A. Tsuberbiller of the Soviet Union, who has studied the agrometeorological nature of dry winds in the most detail, a mild dry wind begins when the atmospheric humidity deficit is 15 mm or more (Table 1).

The wind itself is the intensification factor in a dry wind complex, since even a moderate wind ($3–7\,m \cdot s^{-1}$) blows continually through a grass stand, intensifying the interchange of air in the plants and the consumption of moisture by the crops. High evaporativity in very strong dry winds results is an evaporativity of more than 8 mm per day, which is equivalent to a consumption of more than 80 t of water from 1 ha.

The effect of dry wind on crops varies according to soil humidity. In a strong dry wind, and with productive moisture reserves of 15–20 mm in the arable soil layer, leaves lose their turgor during the daylight hours. With a productive moisture reserve of less than 10 mm in the arable soil layer, plants wilt during the very first hours, and do not regain their turgor completely during the night. In this case, photosynthetic productivity is reduced to zero and a sharp decrease in potential yield occurs.

When the productive moisture reserves in the root zone of the soil are maintained at a level which ensures total evaporation from the crops (not lower than 80% of the field water capacity), damage to the plants by dry winds does not usually occur. According to the data of Ye. A. Tsuberbiller, the productive moisture reserves required for 13 h in a soil layer 0–20 cm deep, relative to atmospheric humidity deficit, are: 25–30 mm of productive moisture at a deficit of 20–30 mm and 30–40 mm at a deficit of 30–39 mm.

It is known that optimum irrigation of farm crops ensures high yields under conditions wherein the plants are subjected, almost annually, to the effect of dry winds in summer.

The struggle against drought and dry winds ins carried out in three directions. First, drought-resistant crop strains like Bezostay I, for example, which has gained recognition not only in the Soviet Union, but also in many foreign countries, are created.

Second, farming methods aimed at increasing the moisture assurance of plants are used. Among these are all types of irrigation, the retention of snow-melt water, snow retention, clean fallow, timely arrest of dampness in the spring, furrowless plowing, field-protective forest strips, and other techniques which improve the phytoclimate of crops.

Third, the most efficient territorial placement of farm crops, based on the suitability of their ecological features to climatic conditions, is effected.

Dust storms are one of the most dangerous of meteorological phenomana for farm crops. They occur as a result of natural, as well as anthropogenic, factors, and are often associated with modes of farming which are unsuitable to a given climatic region. The appearance and development of dust storms are caused by a complex of agrometeorological factors, among which are strong winds (greater than $10 \, \text{m} \cdot \text{s}^{-1}$), dehydration, the dusty condition of the upper soil layer, the absence or poor development of plant cover on fields, and the presence of vast open spaces. Dust storms are usually observed when the relative atmospheric humidity is lower than 50%. The absence of snow cover and ice crusts, as well as poor cementation of the soil and its shallow freezing during the winter, are among the factors under examination.

The complex of factors mentioned is found in steppe, semiarid, and desert regions. The northern limit of the occurrence of dust storms essentially coincides with the boundary of the steppe zone.

Durst storms occur mostly in the spring, when the wind intensifies and the fields are plowed up or their vegetation is still poorly developed. Severe storms are observed in the steppe at the end of the summer, when the soil is drying out and the fields are beginning to be plowed up after the spring crop harvest. Winter dust storms are a comparatively rare phenomenon.

Upper soil layers begin to be carried away at a wind velocity as low as $8–10 \, \text{m} \cdot \text{s}^{-1}$. Under the action of the wind, soil particles break away from the surface, the lighter of these being carried great distances in the form of dust. The heavier particles, when they fall, dislodge other soil particles, which are drawn into motion. As a result of this, the process of soil blow-off acquires the nature of a chain reaction. The intensity of the soil blow-off is proportional to the cube of the wind velocity. For example, when the wind velocity rises from $12 \, m \cdot s^{-1}$ to $15 \, ms \cdot s^{-1}$, the intensity of the soil blow-off (erosion) increases almost twofold.

With an abatement of the wind at barriers (forst strips and structures), the heavier particles drop out, forming earthen accumulations. The lighter soil particles remain suspended in the atmosphere for a long period of time. Consequently, visibility and illumination worsen during dust storms. The sun barely shines through the howling curtain.

It is necessary to take into account the degree of development of the plants in assessing the effect of dust storms on young agricultural crops. The inspection of young winter crops exposed to dust storms reveals that slightly developed seedlings, not having reached the tillering stage, are the most severely damaged. Well bushed-out seedlings create a substantial barrier to the wind flow at the field surface and diminish soil blow-off. Deposition of airborne soil particulates is observed in the presence of well bushed-out seedlings more often than is blow-off. Slightly developed seedlings are extremely susceptible to blow-off, the principal cause of damage to them. Analysis of the conditions surrounding the development of dust storms and data from inspections of the damage caused by them point to the need for measures which assist in reducing wind force at the soil surface and which increase the cohesion of soil particles.

Among such measures are systems of openwork forest strips and windbreak rows which diminish the force of the wind, stubble retention, furrowless plowing, and the use of chemical preparations which enhance cohesion of soil particles.

The development of measures for controlling dust storms requires making allowances for the direction of the prevailing winds, relief, microclimatic features of fields, and soil features.

Hail develops during the warm season, when thermal convection (rising air currents) is very intense and massive cumulonimbi form. A zone develops in these clouds at a rate greater than the maximum velocity of the rising currents (roughly higher than the mid-point of the cloud) in which large drops condense and accumulate. Hail can form in such clouds if the maximum velocity of the rising currents is more than $10 \, m \cdot s^{-1}$ above the zero isotherm level and if the cloud peak has a temperature of $-20°$ through $-25°C$ and is located above the crystallization altitude.

Lifted by the rising air currents to the upper portion of the cloud, large drops freeze and form hailstone nuclei, which grow rapidly as other supercooled drops settle on them. The portion of the cloud in which the principal growth of hail occurs is called the hail focus.

The growth of hailstones from 2–3 to 20–30 mm occurs very rapidly, within 4–6 min. The higher the velocity of the rising currents and the longer they rise, the larger the hailstones become. Hail falling in the Amur region on July 24, 1968, for example, reached a size of 6–7 cm with weights from 62–70 g. Some hailstones weighed 500 g.

Areas where the most serious hail damage occurs are located in piedmont and mountainous regions, where, on hot summers days, thermal convection is especially intense, due to the great disparity in the heat of various relief forms. Hail damage causes enormous losses in agriculture.

In recent decades, the struggle against hail damage has been pursued by means of actively influencing hail formation processes within the clouds.

Suppression of the development of large hailstones forms the basis of the practice of hail cloud control. To achieve this, hail clouds are bombarded within 15–20 min after they begin to form with special rockets and anti-aircraft charges, carrying a reagent (silver iodide, lead iodide).

The introduction of the reagent creates an immense number of crystallization nuclei (10^{12} nuclei result from 1 g of reagent), onto which the reserves of supercooled water localize. This impedes hail formation.

The reagent must be delivered to the portion of the cloud housing the hail focus in order to achieve a significant effect. The seat of the hail focus is determined by radar.

The Hydrometeorological Service of the Soviet Union is currently conducting hail-control operations which protect young crops covering an area of over 5 million ha in Moldavia, the southern Ukraine, the Caucasus, and Central Asia. Losses from hail damage in the protected area have been reduced by 80–90 %.

16 Irrigation

J. Lomas and J. Levin

16.1 Introduction

The use of climatic data in crop husbandry can be found in the most ancient human societies, no doubt because of the close connection between the supply of water and food in the early centers of civilization.

The first published description of rainfall measurements from India dates back to 300 B.C. and contains the following reference to the agroclimatic use of such data: "According as the rainfall is more or less, the superintendent shall sow the seed which require either more or less water" (Kantilya, quoted by Stanhill, 1973).

While the strong dependence of agriculture on climate has been recognized by the most ancient farmers, climatic information is being used to a limited extent only in long-term or day-to-day agricultural management even in the most advanced countries, in spite of the fact that the volume of literature published on the subject has been growing at an exponential rate.

In the words of Austin-Bourke (1968) "the task of the agrometeorologist is to apply every relevant meteorological skill to helping the farmer to make the most efficient use of his physical environment, with the prime aim of improving agricultural production, both in quantity and quality".

We have endeavored, in this publication, to provide guidelines on how the existing climatic information can be utilized in a semiarid climate for an optimal utilization of the available water resources in agriculture. While the main purpose here is to supply some procedures by which a central authority could determine the optimum water allocation between different regions and different crops within the framework of climatic and weather constraints, the two last chapters of this publication deal with the problem of integrating weather information in the decision-making process of a specific agricultural enterprise.

16.2 Prediction of Irrigation Needs

These irrigation needs determined for a given crop are not universally valid for a variety on environmental conditions. Even within a given arid or semiarid zone, variations in climatic conditions are so large that differences in crop behavior and evapotranspiration are considerable. This chapter is concerned with the irrigation of a wide range of crops which can generally be cultivated in arid and semiarid regions having a long, dry summer season. Considerable research has been conducted in Israel in order to achieve maximum water use efficiency in the irrigation of various crops. The information presented here is compiled from various publications[1].

---

1 See *Irrigation in Arid Zones.* Yaron, B., Danfors, E., Vaadia, Y. (eds.). Volcani Institute, Bet Dagan, 1969

16.2.1 The Purpose of Irrigation

The main purpose of irrigation is to supply the water needed for the economically optimal production of a given crop.

In certain situations the water applications may be determined by other than purely economical factors. For example, irrigation schedules may be modified to overcome such problems as slow penetration in heavy soils, leaching salts (requiring heavy irrigation) or the interrelations between fertilizer application and irrigation needs.

While in general, the main problem will be the maximization of crop production with the minimum outlay of water, careful consideration must be given to the various factors in the planning stage.

Various indicators are used to assess irrigation needs and determine irrigation scheduling. The main ones that are used are:
- soil water availability
- climatic indices based on the ratio between the evapotranspiration of the commercially optimal crop, and the evaporation of a pan
- the timing of the most efficient irrigation, i.e., the knowledge or the period within the growth season, where irrigation produces the highest increase in yield.

16.2.2 Soil Water Availability

Two important concepts were developed as a result of early work relating soil water to plant response, the *field capacity* and the *wilting point*. They are now generally accepted, although with reservations, as being respectively the upper and lower limits of water availability for plants. The availability of water within these limits has been the subject of controversy for some decades, with different schools of thought opposing each other. In recent years, however, some degree of consensus between research workers has been reached, and the use of an electrical analog to describe water flow in the soil–plant–atmosphere continuum has helped to clarify matters.

The following three equations may be used to describe water fluxes in the various segments of the soil–plant–atmosphere system:

$$Q_{S-R} = - \frac{\psi_R - \psi_S}{R_S + R_R}, \tag{1}$$

$$Q_{R-L} = - \frac{\psi_L - \psi_R}{R_R + R_L}, \tag{2}$$

$$Q_{L-A} = - \frac{\psi_A - \psi_L}{R_L + R_A}, \tag{3}$$

where Q denotes flux (expressed as a volume per area time), ψ denotes water potential (expressed as force per area), and R denotes resistances in the soil (S), roots (R), leaves (L), and surrounding free air (A) (Stanhill, 1962).

It should be noted that the minus signs are assigned to the right-hand part of the equation because the flow from soil to roots and from shoots to atmosphere is considered to have a positive value.

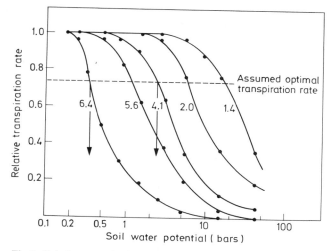

Fig. 1. Relative transpiration rate as a function of soil water potential and evaporative demand. (After Denmead and Shaw, 1962)

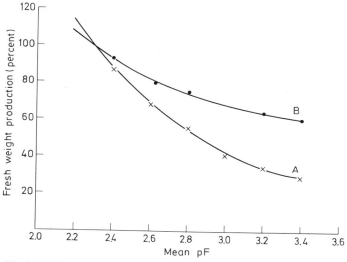

Fig. 2. Relative fresh weight production of various crops (lettuce, radish, and spinach) for different soil water potentials and evaporative demand. (After Bierhuizen and Vos, 1958)

Equations (1)–(3) can be used to explain the difficulty in providing a general definition of soil water availability. With low rates of evaporative demand and hence low flow values, plant water potential ψ_P need not be much different from soil water potential ψ_S, even if the resistance to water flow in the soil R_S is great. Conversely, under conditions of high evaporative demand, ψ_P may be much greater than ψ_S, even if resistance to water flow in the soil is low.

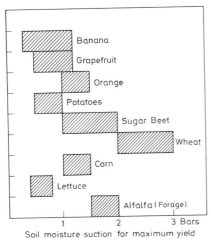

Fig. 3

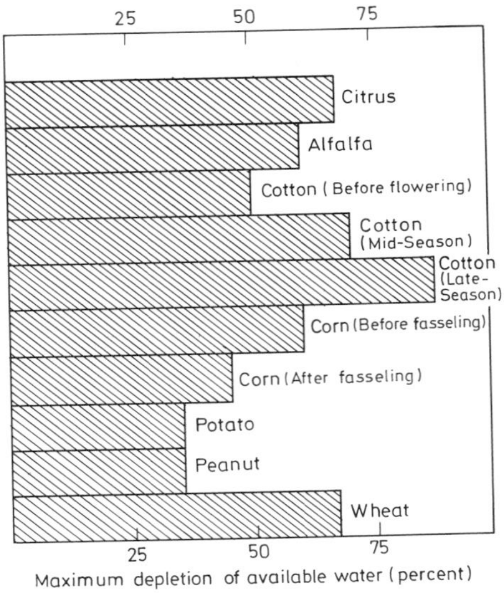

Fig. 4

Fig. 3. Soil suction ranges for maximum yields in different crops

Fig. 4. Maximum admissible depletion of available water for various crops

These conclusions are illustrated in Fig. 1 for transpiration and Fig. 2 for growth. Figure 1 shows that under low evaporative demand the transpiration rate was not affected by increasing soil water potential up to values of 10 bars. Under conditions of high evaporative demand the transpiration rate was reduced at soil water potential of less than 1 bar.

It has been shown that under semiarid conditions, optimal crop production is obtained when the relative transpiration rate (or the evapotranspiration ratio) is maintained at a given value. Figure 1 illustrates the fact that the maintenance of such an optimal relative transpiration value will be obtained at different soil water potentials, under varying climatic conditions.

Similar results are shown in Fig. 2 for three vegetable crops growing on two soil types under two levels of evaporative demand (Bierhuizen and de Vos, 1958). Under conditions of high evaporative demand (line A), the relative yield decreased from 100% at a soil potential of 0.2 bars to 30–40% for a soil water potential of 2.5 bars. Under low evaporative conditions (line B), the decrease was less, and for comparable soil water potentials the yield was only reduced to 60–70%.

It is therefore difficult to determine which is the precise soil moisture suction or potential that is to be chosen for optimum growth, in view of the shifting of the optimal potential with a change in climatic conditions.

The knowledge of the range of water potential at which maximum yield can be obtained may, however, provide a useful tool for scheduling irrigation. Soil suction ranges for maximum yield for different crops are shown in Fig. 3.

A similar indicator of irrigation requirements is the maximum permissible soil water deficit prior to irrigation, i.e., the maximum soil water depletion in the main

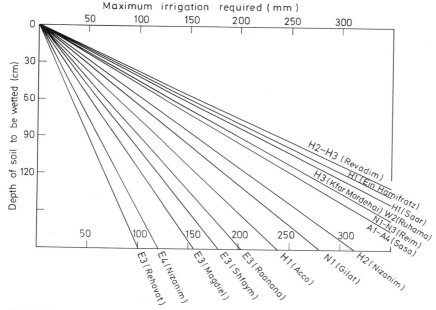

Fig. 5. The relation between water required, depth to be wetted and type of soil

root zone (see below) that can occur without causing significant decrease in the final yield. In deciding on the value of this element, as indeed with the previous one, primary consideration must be given to the critical growth stages of the given crop, such as jointing, tasseling and silking in corn, flowering and boll set in cotton.

Figure 4 summarizes information on the maximum admissible soil water depletion in various crops, the information being based on data collected by various research workers in Israel (Stanhill and Vaadia, 1967; Bielorai, 1969; Mantell, 1969; Shimski, 1969).

When either the maximum soil suction or the maximum soil water depletion have been reached, and irrigation is required. By means of the nomogram presented in Fig. 5 the amount required to wet a certain depth of soil can be obtained for different soil types in Israel.

The letters and numbers appearing at the bottom right-hand side of the nomogram are the soil classification symbols used in the Soil Association Map of Israel[2].

The nomogram illustrates the advantage of an intimate knowledge of the physical characteristics of the land one has to cultivate. The locations in Rehovot, Magdiel, and Raanana where soil samples were collected are all classified as *Hamra* (E 3) but the variation in water storage capacity between them is quite considerable.

It should be emphasized that the nomogram indicates the maximum irrigation to be given, i.e., when the soil water depletion in the given layer has reached the

2 Soil Assoc. Map of Israel, 1 : 250,000, Dan, J., Raz, Z. (ed.). Ministry of Agriculture, 1970

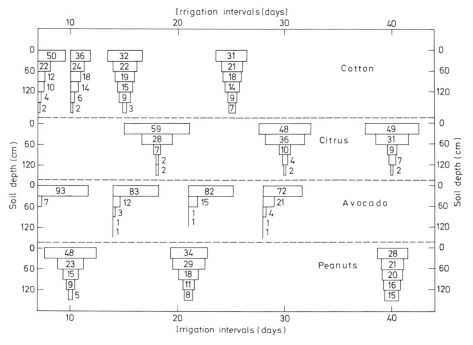

Fig. 6. Amount of water extracted from crop roots (expressed as percent of the total) from various soil layers, as a function of irrigation frequency

wilting point. The quantity of water applied should be reduced accordingly, if the soil moisture before irrigation is above that point.

A point that should be taken into consideration, when assessing water availability to crops is the rooting depth of the crops. Rooting depths vary not only according to species, but also according to irrigation treatment. Irrigation prior to seeding will provide a certain moisture reserve for later root development. Plant roots will grow deeper in soils that are kept wet but not saturated, lack of aeration in wet soils hindering root development.

The main root zone, which may be defined as the depth of soil in which 80–90% of the water extraction occurs, varies in depth with the irrigation treatments, and water availability to a crop must therefore be judged according to the past soil water balance of that crop. One can see from Fig. 6, for example, that peanuts irrigated every 40 days may extract water in considerable quantities up to a depth of 120–150 cm, while the same crop irrigated every 10 days will obtain most of the water from the top 90 cm.

In an extreme case, one can consider that an avocado plantation irrigated every week will not be able to utilize water located below a 60 cm depth, its root system not being developed below that depth. Water availability is therefore strongly dependent on root development, which is itself a function of past irrigation treatments.

16.2.3 Evapotranspiration

Potential Evapotranspiration. Numerous methods have been devised to calculate the potential evapotranspiration (i.e., the evapotranspiration of a short crop completely covering the soil surface and never suffering from lack of water) on the basis of climatic data.

We will only mention the methods of Thornthwaite (1948), Blaney and Criddle (1950), Turc, Thornthwaite-Holzman, Dalton, Energy Balance, and combination methods, the most popular version of the latter being the Penman method (Penman, 1948).

All these methods are based on the dependence of free water evaporation (or the transpiration of a freely evaporating crop surface) on a number of meteorological parameters, mainly net radiation flux, wind speed, and relative humidity of the air[3].

From a practical point of view, the Penman method appears to be the most useful, the calculations involved being relatively simple (tables have been presented by Slatyer and McIlroy, 1961) and the accuracy is relatively high.

The Canadian Department of Agriculture (Russelo et al., 1974) has also published tables giving the potential evapotranspiration as a function of total solar radiation and minimum and maximum air temperature. The main drawback of their approach from a practical point of view is that radiation measurements are not made on a routine basis, and the procedure is therefore of limited utility in agricultural practice.

Techniques have been devised to measure evaporation directly, i.e., by means of evaporation pans, the most commonly used being the U.S. Weather Bureau Class A Pan, circular, 121 cm in diameter, 25.5 cm deep, made generally of galvanized iron and mounted on a wooden platform. The pan is filled to within 5 cm of the rim, and evaporation measurements are made by hook gauge or by refilling to a fixed point. The procedure is simple enough that the farmer can set up the instrument on his land and use it to determine the water requirement of the crop.

Actual Evapotranspiration. The pan evaporation is very close to being the potential evapotranspiration, which is about that of a crop growing constantly at field capacity. This, however, is different from the situation in actual agricultural practice where the crop grows most of the time at soil moistures that are much below field capacity. What is their evapotranspiration then?

The simplest answer, and one that has been shown experimentally to be quite correct, is that "commercial" crops, i.e., crops that are not maintained at field capacity, but are irrigated according to existing agricultural know-how, evaporate at a rate that is proportional to the evaporation of a Class A Pan under the same conditions. The ratio of proportionality depends on the nature of the crop, its phenological age (if it is not perennial) and the irrigation technique used (the evapotranspiration of a crop under trickle irrigation may be as much as 40% lower than that of one under sprinkler irrigation, for the same yield).

---

3 A review of 12 of the more known methods, their limitations and their suitability in different climatic and geographical regions can be found in: Lecarpentier, C. (1975) *L'Evaporation potentielle et ses implications geographiques.* Ann. Geogr. No. 464: 385–413

Table 1a. Ratio between actual evapotranspiration and Class A pan evaporation for various crops

Crop	Location	Pan ratio				Source
		Young	Mature	Ripening	Average	
Grass	Caesarea, Israel				0.94	Stanhill (1964)
Lucerne	Bet Dagan, Israel				0.70	Lomas et al. (1972)
Sorghum	Thorsley, Alabama	0.30	1.15			Doss et al. (1964)
Sorghum	Gilat, Israel	0.30	0.88			Stanhill (1962)
Cotton	Gilat, Israel	0.20	0.85	0.40–0.10		Stanhill (1962)
Maize	Ames, Iowa	0.27	0.90	0.40		Fritschen et al. (1961)
Maize	Gilat, Israel	0.25	0.72			Stanhill (1962)
Peanuts	Bet Dagan, Israel	0.30	0.56			Stanhill (1962)
Vine	Even Sapir, Israel	0.30	0.58			Stanhill (1962)
Citrus	see Sect. 3					

Table 1b. The evapotranspiration ratio as a function of time under sprinkler irrigation[a]

Crop	Location	Month						
		April	May	June	July	August	September	October
Sorghum	Gilat	0.18	0.30	0.40	0.66	0.30		
Maize	Bet Dagan	0.40	0.60	0.95	0.45			
Cotton	Gan Simuel	0.30	0.30	0.37	0.80	0.77	0.41	
Peanuts	Gilat		0.22	0.60	0.74	0.82	0.60	
Peanuts	Bet Dagan		0.20	0.42	0.70	0.72		
Apples	Upper Galilee		0.85	1.10	1.25	0.90	0.58	
Bananas	Coastal Plain	0.30	0.50	0.60	0.80	0.90	0.90	0.70

[a] Source: Shalhevet, J., Mantell, A., Bielorai, H., Shimahi, D.: Water requirements of field and orchard crops Pamphlet No. 156, Agricultural Research Organization, Institute of Soilds and Water. Bet Degan (1976)

The knowledge of that relationship coupled with the measurement of the Class A Pan evaporation, will allow the determination of the actual evapotranspiration of its crop. Table 1 presents these relationships for a number of crops, based mainly on experiments carried out in Israel during a number of years.

In most crops the evapotranspiration rate varies however with the life of the crop. In annual crops the greatest water loss during the early stages is by evaporation from the bare soil. In the absence of rain or irrigation, and under intense drying conditions, such water loss is quickly reduced by the formation of a surface mulch of dry soil. During the late growth stages, evapotranspiration is

Table 1c. Water applications as a fractions of Class A pan evaporation for crops under trickle irrigation[a]

Crop	Location	Soil	Water applied	Remarks
Tomatoes	Arava, Israel	Sandy to loamy sand	0.8–1.0	Very good response
Tomatoes	Sinai Desert, Israel	Coarse sandy soil	o.8	Good response using short intervals between irrigations
Tomatoes	Ein Gedi, Israel	Not given	0.7	Response about 30% better than with spray which was applied to 0.9 E_{pan}
Tomatoes	Nahal Kalin, Israel	Loany sand	0.6 0.8 1.0	0.8 E_{pan} gave 5–10% better results than 0.6 and 1.0
Tomatoes	Griffith H.S.W. Australia	Clay loam	0.7 0.35	Better yields using 0.7 E_{pan}
Tomatoes	San Diego, USA	Heavy clay	0.6	Marketed yield: 27.3 t/ha
Peppers	Arava, Israel	Sandy to loamy sand	0.8–1.0	Very good response
Green Peppers	Havat Eden, Israel	Not given	0.8	Yield: 15.3 t/ha
Cucumbers	Arava, Israel	Sandy to loamy sand	0.8–1.0	Very good response
Sweet corn	Arava, Israel	Sandy to loamy sand	0.8–1.0	Very good response
Vines	Bnei Atarot, Israel	Sandy clay with clay at 80 cm	0.4–0.5 at the beginning of the season but up to 0.8 to 1.0 E_{pan} by the end of the summer	Average application 0.5–0.6 E_{pan} but peak application 0.9–1.0 E_{pan}
Table grapes	Yetvata, Israel	Sandy loam	0.7–0.8	Yield of 12.9 kg/vine
Vines	Griffith N.S.W. Australia	Banna sand	0.5 0.25	20% higher yield using irrigation at 0.5 E_{pan}
Apple trees	Scores by, Vic Australia	Shallow perseable soil with clay subsoil	0.4 at the beginning of the season incrissing 0.6 at the end	
Apricot	Hatearim, Israel	Silty loam	0.7–1.0	
Citrus	Kfar Hayarkon, Israel	Not given	0.3–0.75	No adverse effects with 0.3 E_{pan}
Bananas	Kfar Hayarkon, Israel	Not given	0.45–0.90	Best results with 0.45 E_{pan}

[a] Source: Jobling, G.A., *Trickle irrigation design manual* (60137), Part I. New Zealand Agr. Eng. Inst., Lincoln College, Cantebury (1974)

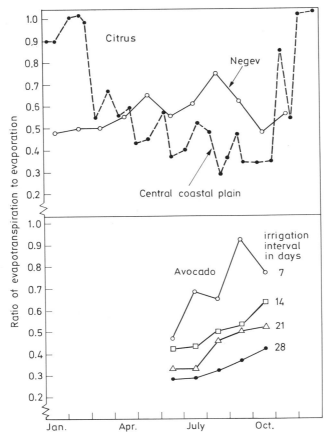

Fig. 7. Changing relationship in relative evaporation in perennial crops with time and irrigation treatment

reduced by foliage senescence and the interruption of irrigation application and subsequent drying out of the soil. With perennial deciduous crops a similar seasonal fluctuation is to be expected with the development of new foliage in the spring and the onset of the resting stage in the fall. It is only for perennial evergreen crops with a complete or constant-sized ground cover that a linear relationship between evapotranspiration and evaporation from a water surface can be expected throughout the year. Even that cannot be taken for granted, as can be seen from Figure 7, which presents the changing relationship in relative evapotranspiration during the life of a few crops.

While the evapotranspiration ratio of citrus in the Northern Negev appears to remain relatively constant throughout the year (Bielorai, 1969), measurements made on a citrus orchard situated in the Central Coastal Plain show that the ratio tends to increase in the winter and to decrease during the irrigation period of the dry summer months (Kalma, 1970). Figure 7 illustrates also the main and inherent weakness of irrigation control by measurement of evaporation.

One can observe that the evaporation ratio of the avocado crop varies with the irrigation treatment. The decision as to which is the treatment to be chosen is a prerequisite to the use of evaporation pan in irrigation control. When no information is available as to the effect on yield or crop quality of irrigation reduction or change, irrigation control by pan evaporation can at best maintain the existing irrigation efficiency, but cannot improve on it.

The crop parameter that is generally considered first when assessing irrigation treatments is the total yield, but the effect on other elements, such as fruit quality, or fruit size, may be investigated.

The knowledge of the response of crop to irrigation quantified into *response curves*, or *production functions*, coupled with data of evapotranspiration ratios of selected irrigation treatments [for a detailed and recent review of research data see *Irrigation of field and orchard crops in semi arid regions*, Shalevet, Moreshet, Bielorai, Shimshi (eds.), International Irrigation Centre P.O.B. 49, Bet Dagan, 1976] will permit an increase in irrigation efficiency, either used as tools by themselves, or integrated in automatic systems of irrigation monitoring.

16.3 Assessment of Irrigation Requirements by Nomograms

16.3.1 Introduction

The objectives of agricultural policies, to cover the local requirements or increase agricultural exports (or a combination of both), have generally been pursued in advanced countries by an emphasis on higher productivity and a shift from less to more remunerative crops. In arid and semiarid countries, where the water requirements of most agricultural crops are not met by nature, these changes can be achieved only be extending the area under irrigation. In Israel, for example, the irrigated area increased sixfold, from 30,000 ha in 1948 to 180,000 ha in 1972. This trend, combined with the fact that the amount of water consumed by agricultural crops has exceeded 80 % of the total water consumption of the country for the last ten years, highlights the economic importance of a sound methodology in irrigation water distribution and allocation planning. Such planning has, however, too often been rules by tradition and empirical evaluation, while using only to a limited extent the large body of information gathered in recent years on crop water requirements.

The purpose of this chapter is to show how the gap in information flow, between investigation on one side and planning and extension on the other, can be bridged to a certain extent by incorporating the data obtained by research into simple graphical procedures and nomograms, providing thereby a more accurate assessment of crop water requirements and of the climatic and soil effects on these requirements.

The present analysis will be limited to the citrus crop which consumes 30 % of the water used in Israel's agriculture, and plays therefore a very important role in its agricultural hydrology. However, the approach expounded here can easily be applied to other orchard crops under irrigation.

Table 2. Ratio of mean daily evaporation from citrus plantations (ET) to mean daily screened Class A pan evaporation (Eo)

Country	Variety and age	ET/Eo	Source[a]
1. Arizona	Valencia, 17–19 years	0.56	Hilgeman and Van Horn (1964)
2. Arizona	Valencia, mature	0.66	Van Bavel et al. (1967)
3. Arizona	?, mature	0.52	Hilgeman and Rodney (1961)
4. California	Navel, mostly mature	0.40	Van Bavel et al. (1967)
5. Israel	Shamouti, mature	0.48	Goldberg and Gornat (1968)
6. Israel	Shamouti, 35 years	0.52	Kalma (1970)
7. Israel	Shamouti, 7–10 years	0.30–0.40	Kalmar et al. (1973)
8. Israel	Shamouti, 8 years	0.52	Bresler et al. (1965a)
9. Israel	Marsh grapefruit, 10 years	0.33–0.59	Bielorai and Levi (1970)

[a] First six data sources as quoted in (Kalma, 1970)

16.3.2 Required Data

In order to assess the irrigation water requirements, the value of the following elements must be determined: the evapotranspiration of the crop, the physical characteristics of the soil and the climatic conditions.

Crop Evapotranspiration. Data on evapotranspiration rates from citrus plantations in Israel and elsewhere are available from a number of investigations (see Kalma, 1970). Water requirements of citrus can, therefore, be assessed from the evapotranspiration – pan evaporation ratio (ET/Eo; see Table 2) if reliable Class A pan data from the considered region are to be obtained. When the investigations include different irrigation treatments, the estimate of ET/Eo can even be improved.

Figure 8, for example, presents the grapefruit yields obtained in the northern Negev region of Israel, as a function of the evapotranspiration pan evaporation ratio during the irrigation period (Bielorai, 1969; Bielorai and Levi, 1970). The results indicate that above the threshold value of 0.50, yields are not affected by the seasonal ET/Eo value. Similar conclusions can be drawn from the data summarized in Fig. 9, which originates from eight seasons (1964–1972) of measurements made in a Shamouti orange orchard of the Central Coastal plain of Israel (Mantell and Goell, 1973). According to these data, decrease in yield is observed when the ET/Eo decreases below 0.45–0.50. While a minimum value for ET/Eo can therefore be admitted, seasonal variations should also be taken into consideration. Kalma (1970), while reporting an overall average value of 0.52 for the ET/Eo in an orange orchard at Rehovot, Israel, found mean rainless season (summer) values of 0.45–0.46. Bielorai and Levy (1970) have similarly observed that rainy season (winter) values are higher than the values observed, made during the summer irrigation season. The hot and dry winds (hamsin) that blow during the spring appear to be able to depress the ET/Eo value, as they also blow during the night, increasing pan–evaporation, while having little or no effect on evapotranspiration, the stomata being closed (Baruch Reuveni, oral communication).

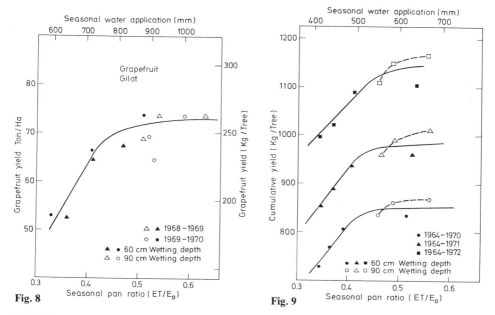

Fig. 8. Grapefruit yield as a function of seasonal pan ratio, and water application

Fig. 9. Cumulative citrus yields as a function of seasonal pan ratio and water application

A word should also be said about the effect of the irrigation technique. Bresler et al. (1965) found an ET/Eo ratio of 0.44 when the irrigation was supplied in alternate rows of trees, as compared with 0.69 when every row was irrigated. Heller et al. (1973) reported similarly a reduction in water use of 18% for alternate row irrigation as compared with every row irrigation for equivalent yields.

Depth of Water Stored in the Soil. The amount of water that can be stored in a layer of soil, expressed as equivalent depth of water, can be computed from the formula:

$$d_w = \frac{(FC - WC) \times ds \times D}{100},$$

where d_w = depth of water stored in the soil in cm, FC, WP = field capacity and wilting point, respectively in % dry weight, ds = bulk density of the soil, D = layer of soil considered in cm.

Banin and Amiel (1969, 1970) have shown that field capacity, available water and wilting point can be calculated with a relatively high accuracy, if one of the three variables is known (see Table 3).

The extension service or planning body in charge of determining water allocation does not always dispose of all the required information, and the difficulty can be obviated in the following ways:

1) The addition to the Soil Association's map of the considered region or country, of tables indicating the main physical characteristics of the classified soils, would enable the planning body, after pinpointing the location of the region considered, to extract the required information from the tables.

Table 3. Relation between physical properties
of some soils in Israel. (After Banin and Amiel,
1970)

Regression equation	Coefficient of determination $r^2 \times 100$
$W_{15} = 0.582 \, W_{1/3} - 2.06$	90.5
$W_A = 0.553 \, W_{15} + 6.05$	54.4
$W_A = 0.418 \, W_{1/3} + 2.06$	82.9

W_{15} = Soil water content at wilting point
$W_{1/3}$ = Soil water content at field capacity
W_A = Available water

2) In the absence of such detailed maps or tables, the general case is that only one or two of the required characteristics are known, and an estimate of the others can be made. For example:

2a) *Density unknown:* Most soils have a bulk density varying between 1.30 and 1.60, and by fixing the value of the unknown density at 1.45, the error in the estimate of water content should not exceed 10–12%.

2b) *Field capacity or wilting point unknown:* A definite relationship exists generally between these two characteristics for different soil types. Figure 10 shows the relation between field capacity and wilting point for 26 different soils of Israel. Wilting point can, therefore, be expressed as a function of field capacity:

$$Y = -0.39 + 1.057 \, X, \tag{4}$$

$X = \log_{10}$ of field capacity; $Y = \log_{10}$ of wilting point, both in % dr.wt.

It follows that once an empirical relation has been established between wilting point and field capacity for the region considered, the knowledge of one of the two characteristics may supply a relatively accurate estimate of the other.

2c) *Depth of the soil to be irrigated:* An irrigation is generally considered adequate when the main root zone, which in citrus is the 0–90 cm soil layer (Bielorai and Levi, 1970; Shmueli et al., 1973), has been brought to field capacity. Very satisfactory results have, however, been obtained when only a fraction of the main zone has been brought to saturation (see for example Figs. 8 and 9; Shmueli et al., 1972). An added advantage of applying water amounts which do not replenish the deficit in the entire rootzone, is that as there is no surplus for leaching and drainage below the root zone, any such losses are, therefore, nonexistent or at least considerably reduced.

Soil Moisture Tension. Soil moisture stress or soil moisture tension is a factor expressing the water status of a soil more adequately than either soil water content or available water. Most crops show a higher sensitivity to the former than to either latter factors. Recent experiments have shown, for example, that citrus yield decreases are proportional to the mean soil moisture tension increase above 1 atmosphere (Heller et al., 1973) or to the length of time that the mean soil moisture is maintained at or above 1 atmosphere (Bielorai and Levi, 1971).

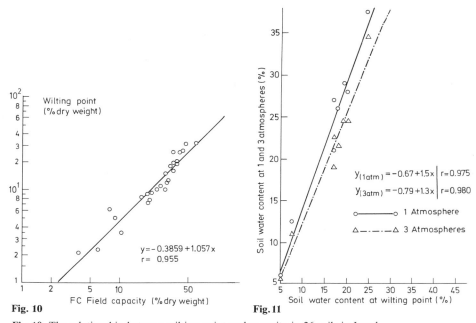

Fig. 10

Fig. 11

Fig. 10. The relationship between wilting point and capacity in 26 soils in Israel

Fig. 11. Soil water content at one and three atmospheres as a function of soil water content at wilting point

Shmueli et al. (1973) summarizing the citrus water requirement experiments conducted in Israel during the 1960's, observe that citrus seems to differ in sensitivity to soil moisture tension at various times during the season, with a high sensitivity (threshold 1–2 atm) during the period of flowering and fruit-set (beginning of June) and a lower sensitivity during the rest of the season (threshold 2–4 atm). The use of this information in water allocation planning is restricted by the fact that soil water retention curves, which give the soil water tension as a function of soil water content, are even less common and harder to obtain than the other soil data. It can be observed, however, that these curves are generally quite similar in outline, and can be characterized by the range of soil water content that they span or even by one of the starting values, the soil water content at wilting point or field capacity. Figure 11, based on data from Rawitz (1965) and Heller et al. (1973) indicates that a high correlation can be obtained between the soil water content at wilting point and the soil water content at some other water tension value (1 and 3 atm, for example). Soil moisture content at 1 or 3 atm tension can, therefore, be calculated by either Eqs. (5) or (6):

$$Y(1\,atm) = -0.67 + 1.5\,X, \tag{5}$$

$$Y(3\,atm) = -0.79 + 1.3\,X, \tag{6}$$

where X = soil moisture content at wilting point, Y = soil moisture content at 1 and 3 atm soil moisture tension.

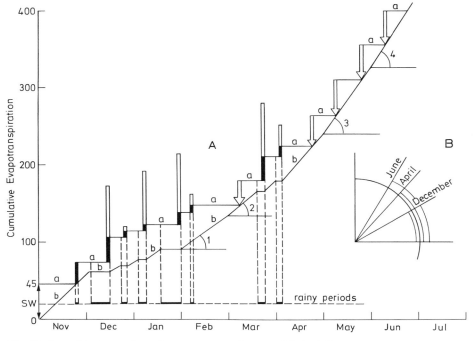

Fig. 12. A graphical procedure of crop water requirements (Method A)

From a known value of the wilting point, the soil water stress threshold in terms of soil water content can, therefore, easily be calculated.

Effective Rainfall. The question of what part of the total rainfall during a given period is effective and consumed by the crop, and what part is lost by runoff, evaporation etc., is a complex one. In the procedure presently exposed, we limited ourselves to the following criteria, empirically chosen:

1) The rainfall of an isolated day (i.e., without rainfall before or after) was considered effective if it was equal or higher than twice the daily evaporation rate (Class A pan evaporation).

2) The rainfall of a rainy period of two or more days was considered effective if the total rainfall amount of the period was higher than twice the daily evaporation rate.

3) Any rainfall amount that caused the root-zone to exceed field capacity was considered lost by deep drainage.

16.3.3 Graphical Method for Water Requirement Evaluation

Method A. The method is similar to the procedure suggested by Halkias (1963) for the evaluation of water supply systems.

Basically the irrigation need is determined by the intersection of two straight lines, one representing the amount of water stored in the soil (a), and the other the cumulative soil water depletion induced by evapotranspiration (b) (see Fig. 12).

Table 4. Climatic conditions over a citrus orchard in Rehovot during 1968–1969. [After Kalma, J.D.: Some aspects of the water balance of an irregated orange plantation. The Volcani Institute of Agric. Res., Rehovot (1970)]

Month	Rainfall periods	Effective rainfall (mm)	Mean daily pan evaporation (mm)	Crop evaporation (mm)
November	25.11.68	27	2.7	1.9
December	4.12.68–17.12.68	99	1.6	1.1
	23.12.68–26.12.68	15		
January	4. 1.69–8.1.69	79	2.2	1.5
	18. 1.69–30.1.69	92		
February	7. 2.69–8.2.69	25	2.2	1.5
March	18. 3.69–23.3.69	117	3.8	1.7
April	1. 4.69–4.4.69	42	5.1	2.3
May			6.3	2.8
June			7.5	3.4
July			7.4	3.4

The procedure will be best understood by its application in an actual case: Citrus orchards in Rehovot, in the central coastal plain of Israel, are situated on red-brown Mediterranean soil (with a field capacity and wilting point of 5.9 and 2.3% by weight, respectively, and a bulk density of $1.6 \, g \cdot cm^{-3}$) and were submitted during 1968–1969 to climatic conditions as shown in Table 4. The effective rainfall as tabulated in Column 2, was determined according to the criteria presented in 2.4. Column 4 was computed by assuming a pan ratio of 0.45 during the rainless period and 0.70 during the rainy period (see 2.1).

Until June, the soil water tension threshold is 1 atm. Assuming a wetting depth of 90 cm, the maximum available soil water depth (SW) stored in that layer of soil can be calculated by combining Eqs. (1) and (2)

$$SW = \frac{(FC - Y \, 1 \, atm) \times ds \times D}{100},$$

$$SW = \frac{(5.9 - 2.78) \times 1.6 \times 90}{100},$$

$$SW = 4.5 \, cm = 45 \, mm.$$

This amount is indicated at the bottom of the ordinate in Fig. 5A, thereby determining the level of available water in the soil (line a), assuming that on that date – 1 November – the root-zone was saturated. (The initial soil water status has to be assumed.) Along the abscissa, the rainy periods are indicated according to their distribution in time, the vertical columns representing the total effective rainfall that occurred during each period as tabulated in Table 2. The line b represents the evapotranspiration rate as calculated and presented in Table 2, the angles 1–4 of Fig. 5A being respectively equal to the evapotranspiration rates of February (1.5 mm), March (1.7 mm), May (2.8 mm), and June (3.4 mm) as shown in Table 4.

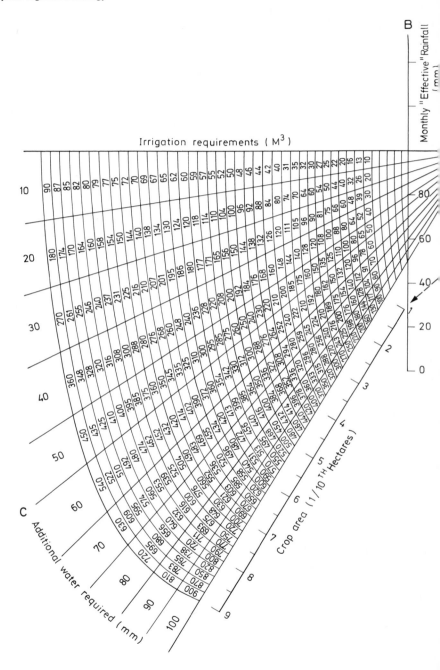

Fig. 13. A graphical procedure of crop water requirements (Method B)

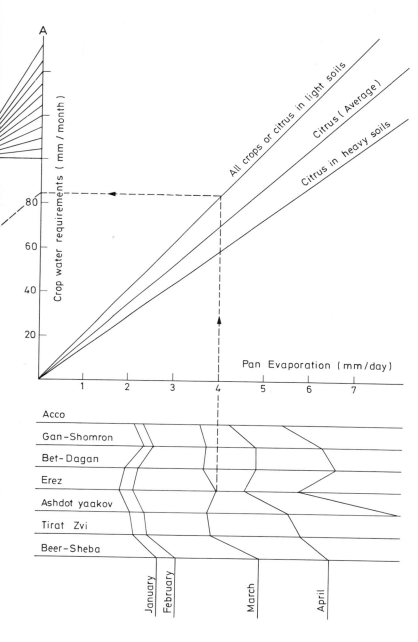

It should be noted that during the rainy periods (periods in which each day received a rainfall amount of at least 1 mm) evapotranspiration is considered to be zero, and the line b is horizontal. The vertical distance between the lines a and b represents the amount of water available to the crop at a given date. When the two lines meet, no water is available and an irrigation is therefore necessary. (The irrigations are indicated by arrows, the length of which represents the amount supplied.)

Conversely, the maximum amount of water that can be stored in a soil of this type being 45 mm, the maximum storage (or vertical distance between a and b) that can be attained by rainfall or irrigation is equal to that amount. The black columns indicate the fraction of the rainfall that is stored at each rainy period in the soil, and the white columns show that fraction of the rain that is lost by deep drainage or runoff.

To increase the speed of the graphic work, a set of transparent quarter-circles indicating the average evapotranspiration rates during the different months of the year (as shown in Fig. 12B) could be made for the different regions to be considered, and the water requirements of the crop could be made quite accurately and rapidly.

This procedure could possibly also be of help to hydrologists for the determination of the amount of water being added to the water table by regional rainfalls.

Method B. This procedure is again based on the known ratio between crop water loss and Class A Pan evaporation (see right-hand side of Fig. 13).

For instance the average daily pan evaporation at Acre in March was 3.6 mm, and at Beer-Sheba during the same month, about 5 mm.

The vertical axis at the centre of the graph indicates the amounts of "effective" rainfall as defined in Sect. 3.2.4. Table 5 indicates the "effective" rainfall expressed as a percentage of daily (or on consecutive days) downpours of up to 50 mm (precipitations that fell on a single day or on consecutive days and exceeded 50 mm were effective at 100%).

The left-hand side of the graph indicates the amounts of irrigation water that have to be applied as a consequence of climatic conditions.

Example: One wants to know the water requirements of a 50-ha citrus orchard on sandy soil at Nahal Oz, in October, when the total rainfall of the month has been 45 mm. From Table 5 one finds that the "effective" rainfall for the station for that month was 61% therefore:

Eff. R. $= 45 \times 0.61 = 27.4$ mm.

The evaporation measuring station that is the closest to Nahal Oz is Erez, which indicates a mean daily Class A Pan evaporation of 4 mm per day. This value crosses the top sloping line (light soils) at a height (on the vertical axis) of 85 mm (see Fig. 13).

Connecting the two values, 85 mm on the requirements axis (A) and 27 mm on the rainfall axis (B) provides us with a straight line. Searching the left-hand side "fan" of curves for a straight line that is parallel to it, we find curve C, that carries the mention "*55 mm additional water required*". The figures within the fan indicate the amounts required for various areas of crop: 5 dunams (1 dunam $= 0.1$ ha) need 250 m^3, 500 dunam (or 50 ha) 25,000 m^3.

Table 5. "Effective" rainfall expressed as percentage of rainy-period precipitations of up to 50 mm in six stations in a semi-arid environment

Month	Station					
	Ein Hachoresh	Rishon Lezion	Merhavia	Gan Shmuel	Dafna	Nahal Oz
October	78	62	64	70	75	61
November	64	68	76	51	76	67
December	82	84	63	57	75	47
January	63	83	72	–	57	61
February	66	80	81	81	79	71
March	70	79	67	67	64	61
April	58	58	60	73	76	71

Method C. A general equation relating soil water deficit under citrus and previous rainfall was developed (Lewin, 1973):

$$SWD = \sqrt{FC} \times D(0.23 - 0.002x_1 - 0.0004x_2), \tag{7}$$

where SWD = soil water deficit (in mm) of the layer considered, FC = field capacity of the soil considered (percent of dry weight), D = depth of layer considered (in cm), x_1 = rainfall during previous 10 days (mm), x_2 = rainfall during the month preceding the 10-day period (mm).

Example: One would like to know what the soil water deficit is in the top 90 cm of soil in a citrus orchard on sandy soil (with field capacity = 5.9) when rainfall during the preceding 10 days was 20 mm, and during the month before that 50 mm. From Eq. (1), one obtains:

$$SWD = \sqrt{5.9} \times 90[0.23 - (0.002 \times 20) - (0.0004 \times 50)]$$
$$= 218.5(0.23 - 0.04 - 0.02)$$
$$= 37.2.$$

The soil water deficit in the top 90 cm is equal to 37 mm.

Equation (1) can be presented in the form of nomograms when FC and D are given (see Figs. 14 and 15).

It should be emphasized that the relation has been developed from data obtained under the climatic conditions of the arid Northern Negev of Israel and the value of the parameters will probably be different under different climatic conditions.

As an illustration, the approach used by us to build a simple soil moisture model for field crops in arid zones will be detailed here:

Available soil water is the amount of water retained in the root zone of a soil between field capacity and the permanent wilting point, and it is assumed to be available for plant growth.

The rate of transpiration by a crop has been shown to be a function of both soil-water availability and atmospheric demand with the soil-water supply decreasing more sharply under high radiation than under moderate atmospheric conditions (see Fig. 1). Research done on wheat by Shimshi et al. (1963) indicated

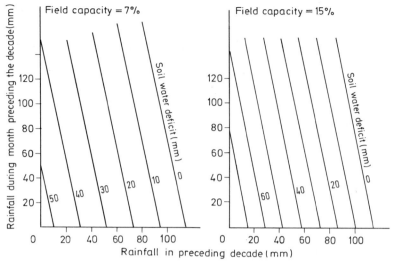

Fig. 14. Water amounts required to bring to field capacity 90-cm thick layers of soils with a field capacity of 7–15 % dry weight, as a function of rainfall

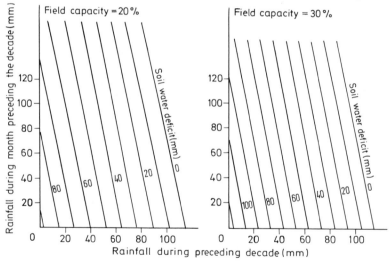

Fig. 15. Water amounts required to bring to field capacity 90-cm thick layers of soils with a field capacity of 20–30 % dry weight, as a function of rainfall

that this process could be approximated by a linear relationship (see dashed lines in Fig. 16).

$$-\frac{dw}{dt} = a + bw,\qquad(8)$$

where $\dfrac{dw}{dt}$ is the water loss per unit time, w is the soil water content at time t ; a and b are parameters of the regression.

The parameters a and b are changing with the changes in the climatological conditions during the year, and with the development of the crop.

Assuming for the moment that both climatic conditions and phenological stage are constant for a given period of time during the year, knowing the value of the parameters of the regression would allow the assessment of rate of crop water loss as a function of soil water content for that given period of time:

From Eq. (1) one can deduce that

$$\frac{dw^2}{dt^2} = -b\frac{dw}{dt}, \quad \text{and} \quad \text{if} \quad \frac{dw}{dt} = X,$$

$$\frac{dx}{dt} = -bX, \quad \text{and} \quad \frac{dx}{X} = -b\,dt. \tag{9}$$

Integrating Eq. (9) from X_0 to X_1 gives

$$\int_{X_0}^{X_1} \frac{dx}{X} = \ln\frac{X_1}{X_0} = -bt \quad \text{and} \quad \ln\left(\frac{a+bw}{a+bw}\right) = bt, \tag{10}$$

where w_0 is the soil water content at time $=0$ and W_1 is the soil water content at time $\delta = t$.

Assuming a to be a constant, (see the left hand group of dashed lines in Fig. 16) one can obtain an accurate approximation of b from soil water measurements (w_0, w_1, and t are therefore known) by the Newton-Raphson iterative technique, which is generally expressed as

$$b_{n+1} = b_n - \frac{F(b)}{F'(b)},$$

where b is the variable for which an approximate solution is sought

F(b) is the function of b

and F'(b) is the first derivative of the function F(b).

This type of situation may arise in extremely arid environments, where the plants cover only a small fraction of the surface, and where soil water loss continues below wilting point.

In the situations that are of more direct and practical interest to us, i.e., crop surfaces that cover the surface almost completely or completely, it can be safely assumed that under all atmospheric conditions crop-water loss ceases completely at or near the wilting point (WP; see right-hand group of dashed lines)

$$-\frac{dw}{dt} = a + bWP = 0$$

and

$$a = -bWP$$

under these conditions Eq. (8) becomes

$$-\frac{dw}{dt} = b(W - WP)$$

and Eq. (10) becomes

$$\ln\left(\frac{W_0 - WP}{W_1 - WP}\right) = bt. \tag{11}$$

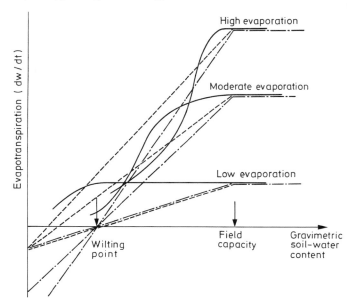

Fig. 16. Linear approximation of crop evapotranspiration as a function of soil water content for different climatic conditions. ——— actual relation (after Denmead and Shaw), – – – linear approximation (assuming parameter a to be constant), —·—·— linear approximation assuming parameter a to be a function of b (see text)

Equation (11) allows for a simple graphic solution for the parameters a and b. One has only to find a ratio $\dfrac{w_0 - WP}{w_1 - WP}$ equal to 2.72 (e) and

$$\ln e = 1, \quad \text{therefore} \quad b = \frac{1}{t}. \tag{12}$$

Soil moisture data under cotton will be used to illustrate the graphic solution. A variety of upland cotton (*Gossypium hirsutum* L.) was sown as the end of April on a clay-type alluvial soil (having a field capacity of 33.2 % by volume and a permanent wilting percentage of 17.0 %) in the central coastal plain of Israel. The crop was given an irrigation of 172 mm in the 11th of August, and one of 69 mm on the 29th of August.

The soil moisture values during the season are shown in Table 6.

Let us consider the situation in May when the initial soil moisture of the 90-cm layer (w_0) is 30 %, and the final soil moisture of the same layer (w_1) is 28 %. Keeping in mind that the wilting point (WP) is 17 %, we find that

$$w_0 - WP = 30 - 17 = 13.$$
$$w_1 - WP = 28 - 17 = 11.$$

If we transpose these values on a semi-logarithmic paper (see Fig. 17), while separating them on the x-axis by the duration of the soil moisture change, and if

Table 6. Soil moisture changes under cotton. (After Marani et al., 1967)

Period	Initial soil moisture (% by volume)	Final soil moisture (% by volume)	Duration of change (in days)	b (see text)	a
May	30	28	41	0.006	−0.0102
June	28	22	30	0.022	−0.347
July	22	19	31	0.033	−0.560
August	29	24	17	0.036	−0.612
September	27	23	13	0.040	−0.680

we connect the two points that are obtained we obtain the straight line $A_1 - B_1$. The intersection of that straight line with two horizontal lines A and B distant by the value of e (2.72) provides us with a value of t, in this case 165 days.

From Eq. (6) we know that $b = \dfrac{1}{t} = \dfrac{1}{165} = 0.006$.

Remembering that $a = -bWP$ gives us

$$a = -(17.0 \times 0.006) = -0.0102.$$

The soil moisture changes in June give us similarly $W_0 - WP = 11$ and $W_1 - WP = 5.5$. Plugging these values on the graph of Fig. 17 gives the straight line $A_2 - B_2$ the intersection with the two horizontal show a distance of 45 days

$$b = \tfrac{1}{45} = 0.0022,$$

$$a = -bWP = -0.374.$$

Similarly the straight line established for the month of July will give a distance on the x-axis of 30 days, and $b = \tfrac{1}{30} = 0.033$. To summarize the procedure, one can extract a series of regression parameters from only one season of soil moisture measurements. But why should we assume that these parameters are still valid during another season, or under a different irrigation treatment?

The answers lies in the climatological particularities of arid regions and the behavior of the crops in that climate. Arid climates are characterized by a very intense radiation often exceeding 200 kcal per cm^2 per year (Budyko, 1968), which is close to twice the amount received in Western Europe, and a concomitant low cloudiness.

As a consequence, characteristic features of an increasing aridity are the enhanced variability in precipitation and the decrease in the variability of evaporation.

Table 7 shows the variability of rainfall and Class A pan evaporation during 10 day periods in January (middle of the rainy season) and April (end of the rainy season) in six stations with an increasing aridity in Israel. The calculations are based on measurements of rainfall over 30 years and evaporation over only 10 years, but they indicate the high variability of precipitation which is increasing with aridity, and the low variability of evaporation. The data show that the coefficient of variation of evaporation is around 20% during the rainy season and decreases to 15% towards the end of the rains. In other words if one accepts these

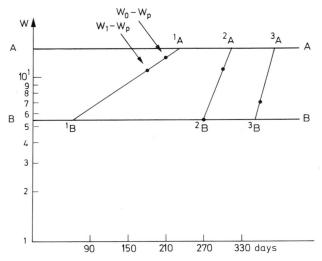

Fig. 17. Graphical method for the determination of parameters a and b of the simulation (see text)

values as acceptable margins of error, one can consider that in arid regions the pan evaporation is a function solely of the time of the year, and that the slopes of the regression, calculated for the different periods of the growth season as shown above, remain valid for other years.

This does not prevent the fact that should the actual evaporation be known for a given season, the model can be corrected by the insertion of the value of that pan evaporation, but our argument here is that it is not indispensable, as the correction will be of the order of the variations in pan evaporation, which have been shown to be small.

It will have been observed that the theoretical analysis has hitherto been based on measurements made in only one layer of soil, referred to variously as the root zone on the 90-cm layer of soil, and this in contrast to most other soil water balance models, who generally take into consideration a succession of horizontal layers of soils. This is not just a simplifying assumption. The basic element of the theory developed here is that the soil water content in the top 90-cm layer of soil is an exponential function of time which can be easily extracted from Eq. (8) as being

$$W = -\frac{a}{b} + \left(\frac{a}{b} + W_0\right) - e^{bt}, \tag{13}$$

where W is the soil-water content of the 90 cm layer at time t, W_0 is the soil-water content of the layer at time 0, a and b are parameters of the regression as shown in Eq. (8).

The underlying assumption is that the soil water content of that layer determines almost to the exclusion of other layers of soil both crop development and crop evapotranspiration. This is derived from the fact that most irrigated crops have shallow root systems, a development which is encouraged by the existing irrigation practices.

Table 7. Variability of rainfall and pan evaporation in arid and semi arid environments

Station	Location	Mean yearly rainfall (mm)	Average 10-day rainfall				Average 10-day evaporation			
			January		April		January		April	
			Amount (mm)	C.V. (%)[a]	Amount (mm)	C.V. (%)[a]	Amount (mm)	C.V. (%)[a]	Amount (mm)	C.V. (%)[a]
GanShomron	35°00'E 32°23'N	576	30.7	97.8	11.1	146.0	23.9	26.2	49.7	13.1
Acre	35°06'E 32°56'N	548	57.0	80.3	10.6	173.0	21.8	21.9	43.4	16.9
Bet Dagan	34°50'E 32°00'N	535	51.2	69.1	8.2	182.9	19.7	17.3	49.7	13.2
Erez	34°34'E 31°34'N	426	45.0	80.1	8.8	196.9	18.5	21.4	45.6	12.0
Beer Sheba	34°47'E 31°14'N	204	20.0	120.0	4.1	187.1	26.8	26.8	65.4	15.1
Eilat	34°57'E	25	3.5	477.0	1.3	255.0	44.2	14.2	97.8	9.4

[a] C.V. = coefficient of variation

In regions where irrigation is necessary, farmers will tend to supply limited applications of water, so as to prevent water loss by seepage below the root zone. Such applications will induce the crop to develop a shallow root system to enjoy maximum exploitation of the wetted soil (see Fig. 6).

It is clear that while frequent irrigations result in an increase in the relative water uptake from the 90-cm layer in the water supply, even under drier conditions and infrequent irrigations, the importance of the layer remains relatively very large.

The effect of shallow soil wetting on water extraction by the crop is even larger than can be inferred from the results shown in Fig. 6. Taylor and Klepper (1971), investigated water uptake by cotton seedlings subjected to an irrigation cycle. They found that water extraction per unit length of root per day was greater in wetter soils and decreased exponentially with the soil water potential, the relation being of the type $y = ae^{bx}$, where y = water uptake in $(cm^3 \cdot cm^{-1} \cdot day^{-1})$, x = soil water potential in bars.

With shallow soil wetting, the increased soil water potential would result in an exponentially decreased water uptake from the lower and drier layers. The principles underlying the assessment of the evapotranspiration have been given in detail. The estimation of the other variables in the water-balance equation can be briefly estimated outlined as follows:

Runoff is considered to be negligible in a well-managed agricultural enterprise situated in the kind of climate considered.

Any amount of water given in excess of the field capacity of the root-zone is considered to be unavailable to the crop and constitutes the deep drainage losses.

One can program the simulation model to print out the total seasonal water loss of the crop, or the summation of the days during which the soil water level was below a given critical value at given periods of the growth season, thereby providing an indication of its expected yield.

Table 8 summarizes the evapotranspiration, lint and seed yields as predicted by the model together with the actual measured values.

Details of the yield prediction procedure are given elsewhere (Lomas and Lewin i. p.). Briefly, the technique is based on establishing a multiple regression relation between y (the yield) and a series of x's, which are the number of "stress days" (days during which the soil water is below a critical level) at different periods of the growth season, i.e.:

$$y = a_0 + a_1 x_1 + a_2 x_2, \ldots, a_n x_n, \tag{14}$$

where y = crop yield, x_1 = number of "stress days" during period 1 (for example the first 10 days of the growth season), x_2 = number of "stress days" during period 2 and so forth, $a_0, a_1, \ldots, a_n$ = regression parameters.

The technique allows the determination of the regression coefficients $a_0, a_1, \ldots, a_n$ which in turn allows the prediction of the yield once irrigation quantity and scheduling are known.

The predicted values in Table 8 are based on all the terms of Eq. (7), even those whose coefficients are not significant; the correlation coefficients, on the other hand, are calculated using only the terms whose coefficients are significant at the 0.05 level.

Table 8. Actual and predicted evapotranspiration, lint and cotton-seed yield for two varieties (Acala and Deltapine) of cotton at Givat 'Brenner (Central coastal plain)

| Treatment | Seasonal evapotranspiration (mm) (0–150 cm) | | Seed cotton yield (kg/ha) | | | | Lint yield (kg/ha) | | | |
| | | | Acala | | Deltapine | | Acala | | Deltapine | |
	Actual	Predicted	Actual	Predicted	Actual	Predicted	Actual	Predicted	Actual	Predicted
1	147	104	2030	2370	2360	2520	960	1000	970	1040
2	253	252	2480	2400	2670	2510	1040	1000	1120	1050
3	287	308	3520	3530	3460	3570	1420	1420	1380	1430
4	180	149	2830	2770	2910	2760	1190	1160	1210	1150
5	312	290	2900	2960	3250	3390	1200	1230	1400	1460
6	309	223	4090	4070	3970	3930	1660	1650	1550	1530
7	218	198	3440	3450	3210	3230	1380	1380	1250	1260
8	347	336	4370	4390	4390	4280	1790	1790	1800	1760
Correlation coefficient	0.919		0.963		0.930		0.955		0.921	

Table 9. Economic gains αij for various
decisions D_i and subsequent weather
developments Yj

		Predictand
		$y_1 \ \cdots \ y_j \ \cdots \ y_m$
	D_1	$\alpha_{11} \ \cdots \ \alpha_{1j} \ \cdots \ \alpha_{1m}$
	$\vdots$	$\vdots \quad \vdots \quad \vdots$
Decisions	D_k	$\alpha_{k1} \ \cdots \ \alpha_{kj} \ \cdots \ \alpha_{km}$
	$\vdots$	$\vdots \quad \vdots \quad \vdots$
	D_a	$\alpha_{a1} \ \cdots \ \alpha_{aj} \ \cdots \ \alpha_{am}$

16.4.2 The Climatic Element in Relation to Operational Decisions

Introduction. We will now consider the question of how to use weather forecasts or climatic information and whatever soil moisture simulation model we have chosen or constructed to make decisions of economic value with regard to water allocation and irrigation planning. The analysis reported here is that of Gringorten (1959) and Gleeson (1959). The procedure was designed by them to maximize gains (or minimize expenses, as the case may be) in an operation or design problem which is affected by weather or climate.

It provides a rational basis for a course of action even when atmospheric developments are uncertain, and long-range probabilities have to be used.

The Method. The Income Matrix. To begin with the operation must be represented symbolically. Our tool is a matrix of numbers aAm, composed of a rows and m columns. Each of the a rows of the matrix corresponds to a course of action or decision. Y, the predictand, is divided into m mutually exclusive and exhaustive classes, each corresponding to a development yj of the weather (the columns).

Each term such as α_{kj} denotes the profit to the operation for course of action D_k and subsequent development in the weather yj. The operational problem is to decide on one course of action D_k, the alternative consequences of which are represented by the figure α_{kj} in the kth row of the matrix.

For each course of action D_k, the expected operational gain

$$g_k = \sum_{j=1}^{m} \alpha_{kj} \times p(y_j),$$

where $p(y_j)$ is the probability of the development y_j in the weather.

The matrix aGl of operational gains then becomes

$$\begin{pmatrix} g_1 \\ \vdots \\ g_a \end{pmatrix} = \begin{pmatrix} \alpha_{11} & \cdots & \alpha_{1m} \\ \vdots & & \\ \alpha_{a1} & \cdots & \alpha_{am} \end{pmatrix} \begin{pmatrix} p_1 \\ \vdots \\ p_m \end{pmatrix}$$

aGl = aAmPl.

The largest g-value in the G-matrix is the maximum expected gain, corresponding to the course of action D_h as follows:

$$g_{max} = \alpha_{h1}p_1 + \ldots + \alpha_{hm}p_m \, .$$

Example 1: One of the classic examples of a problem in cost and loss affected by the weather concerns the pouring of concrete. After the concrete pavement has been poured, two alternative courses of action are open to the operator: to protect the thickening concrete against rain, or not to protect it.

The symbols, C for cost of protective measures and L for loss or damage that can result from rain, may be arranged in matrix form where P is the probability of rain, Eq. (3) for expected gains becomes

$$\begin{pmatrix} g_1 \\ g_2 \end{pmatrix} = \begin{pmatrix} C & C \\ L & O \end{pmatrix} \begin{pmatrix} P \\ P-1 \end{pmatrix} = \begin{pmatrix} C \\ PL \end{pmatrix}.$$

For the indifferent case in which the expected gain resulting from one decision is equal to the expected loss from the other, the probability of rain P_c is given by

$$C = P_c L \quad \text{and} \quad P_c = \frac{C}{L}.$$

P_c, or the "critical probability" is equal to the cost-loss ratio.

For a probability of rainfall greater that $P_c(P > P_c)$, it becomes worthwhile to protect the concrete from rain.

Confidence limits: The method can be improved by considering the confidence limits of the relative frequency of occurrence of a given weather development.

Let p_j'' and p_j' represent the upper and lower confidence limits, respectively, for the relative frequency of class y_j, where

$$1 \ge (p_j'' - p_j') \ge 0.$$

These limits can be determined by any appropriate procedure, but the operator (i.e., the person or team concerned with the economics of the forecast problem) should be quite confident that the true relative frequency is bounded by them.

The method consists of the following two steps:

1) Determine the minimum economic expectation E_i for each decision D_i where

$$E_i = \sum_{j=1}^{m} \alpha_{ij} p_{ij}.$$

Here p_{ij} is a fictitious relative frequency of class y_j, which is a maximum or minimum when α_{ij} is a minimum or maximum respectively, subject to the restriction that

$$p_j'' \ge p_{ij} \ge p_j'.$$

2) Make the decision D_i corresponding to the E_i having the largest value (E_{max}), as determined in step (1).

The example below illustrates how p_{ij}, E_i, and E_{max} are evaluated.

Example 2: Arnold, a farmer, has to choose which crop to plant during the following fall (in a region with fall and winter rainfall). There are four crops to choose from, designated C_1, C_2, C_3, and C_4, in order of increasing moisture requirements.

Table 10. Gains (positive) and losses (negative), minimum expectations (E_i), and upper and lower confidence limits (pj" and pj') in example 2

Decisions	Predictand				
	Y_1	Y_2	Y_3	Y_4	E_i
D_1	5	0	−1	−2	−1.19
D_2	2	3	1	−1	0.62
D_3	−4	1	5	−1	1.38
D_4	−1	0	1	1	0.32
pj"	0.12	0.46	0.77	0.31	
pj'	0	0.12	0.40	0.04	$E_{max} = 1.38$

Table 11. Procedure for computing P_{3j} in example

	Predictand				Sum of relative frequencies	1-Sum
	Y_1	Y_2	Y_3	Y_4		
D	−4	1	5	−1		
pj'	0	0.12	0.40	0.04	0.56	0.44
I	0.12	0.12	0.40	0.31	0.95	0.05
II		+0.05			1.00	0
P_{3j}	0.12	0.17	0.40	0.31		

He has some climatological information about rainfall in his locality during that period. This includes the following relative frequencies: zero for little or no rain (y_1), 27 % for moderate rain (y_2), 60 % for heavy rain (y_3), and 13 % for excessive rain (y_4). (Numerical values for boundaries between rainfall classes are known to him).

Because climatological data are sparse, Arnold chooses confidence limits p''_j and p'_j for these frequencies, as shown at the bottom of Table 2.

The decision D_i to plant crop C_i can lead to a monetary gain or loss depending on subsequent rainfall.

Arnold calculates the various possibilities α_{ij} (we will see below how this can be done) which are given in Table 10, expressed in value units (for example, hundreds of dollars or pounds).

The computation of minimum expectations E_i will be demonstrated for D_3 (decision to plant crop C_3) as a particular illustration. To obtain the *minimum* expectation we will attribute the highest probability possible to the losses and the lowest probability possible to the gains.

Table 11 shows values of gains and losses α_{3j} (corresponding to D_3), the upper and lower limits of relative frequency (respectively p"j and p'j) from Table 2.

It will be noted that the minimum gain (or maximum loss) to be expected is a loss of four units (400 \$) when Y_1 occurs after D_3 is chosen. The first step is therefore to make the relative frequency P_{31} of Y_1 a maximum by equating it to the upper limit of confidence pj". Another loss is to be expected when Y_4 occurs after the decision D_3 has been made, when Y_4 occurs (a loss of 100 $). We will therefore similarly attribute the highest relative frequency possible ($p_4'' = 0.31$) to that case (as indicated in row I).

The other two possible occurrences, Y_2 and Y_3 result in financial gains and the respective lowest possible relative frequencies (p'j) is attributed to them.

All losses from the choice of D_3 have now been considered, but the sum of relative frequencies still is less than one (0.95 in row I). The total is brought up to one by adding 0.05 to the lower frequency limit of the predictand class showing the smallest positive gain (in our case Y_2 where the gain is $= 1$). This is done in row II. The sums at the bottom row are the fictitious frequencies p_{3j}.

The minimum economic return E_3, expected from selection of D_3 is found from Eq. (8) by use of values in Table 3. Thus

$$E_3 = \alpha_{31}p_{31} + \alpha_{32}p_{32} + \alpha_{33}p_{33} + \alpha_{34}p_{34}$$
$$= -4(0.12) + 1(0.17) + 5(0.40) - 1(0.31)$$
$$= 1.38.$$

No doubt exists that 1.38 is indeed the minimum expectation. To decrease E_3 would require that p_1'' and/or p_4'' be increased and/or that p_3' be decreased (but none of these changes are possible) or that p_{32} be decreased (but that would require a compensating increase in p_3' resulting in a net increase of the total expectation.

The last column of Table 10 shows computed values of E_i in this example. The maximum, E_{max}, is 1.38 units, and it lies in row D_3. The decision to plant C_3 should be made because it guarantees the best minimum expectation, 138 $.

The Determination of the Economic Gains α_{ij}. The only practical way of determining the economic gain resulting from a given weather development Y_j and a given decision D_i is to use a computerized simulation program. Two dimentional "production functions" cannot absorb the multiple inputs and provide satisfactory answers.

It will be difficult to obtain precise values for α_{ij} even with a sophisticated program, as the weather developments Y_j are only divided into more or less broad classes, and within each of those there is a range of possible outcomes, resulting each in a different value of α_{ij}.

This can be overcome by choosing either the average value of α_{ij} for each predictand class, or as in Example 2 above, by combining the maximum values of α_{ij} when they are negative (highest losses) with the highest relative frequency and the minimum values of α_{ij} when they are positive (minimum gains) with the lowest possible relative frequency.

16.4.3 Establishing the Complete Simulation System

The United States Department of Agriculture has developed a digital computer model (Anderson and Maass, 1971) that can be used in determining how best to allocate irrigation water resources among crops and among farms when water supply is limiting. The interested reader is referred to it for full details including the complete listing of the formal FORTRAN IV computer instructions.

Our point here will be to indicate how the model can be modified to integrate climatic information or uncertainty, and how in our opinion the assessment of crop response to water supply can be improved.

The same remarks could be made about the computer program developed for irrigation scheduling under arid-climate conditions by Jensen et al. (1969). Their program has been field-tested for several years in Arizona and Idaho. The program calculates evapotranspiration from solar-radiation data and crop-related factors, adds rainfall amounts and calculates the soil moisture depletion. Updated calculations are made twice each week. The irrigator is given a biweekly report which informs him of the approximate number of days before an irrigation is needed, and also suggests the amount of water to apply at that time.

For humid climates a program has been developed by researchers of Auburn University in Alabama (Busch and Rochester, 1975).

Production Function of the Program. The model divides the crop-growing season (April–October) into 14 two-week periods. For each period, estimates of irrigation requirements necessary to produce specified yields of selected crops for a hypothetical area are given (see Table 12).

Estimated yield reductions resulting from missing or posponed irrigations (as specified by Table 12) are shown in Table 13. The information in Tables 12 and 13 forms the "*production function*", to allow the determination of the most efficient allocation of limited irrigation water among competing crops.

While Tables 12 and 13 are based on a number of studies of crop response to soil water stress, the model by its nature labors under considerable limitations, some of which are mentioned by the authors or expressed as assumptions:

The first limiting assumption is that each field is watered with its full requirement or not at all. This assumption is made because the knowledge of crop response to partial water application is not available.

The second limiting assumption is that a field crop is lost if it suffers two successive misses.

The model assumes constant quality of the crops that have suffered water shortage, because of insufficient data on changes in crop quality in response to water supply.

The effect of the changing climatic conditions, (whether an increase or decrease in evaporation, or rainfall occurrence) is not taken into account for the assessment of crop water requirement.

Suggested Modified Structure of the Program. We will give a brief description of the structure of the revised program, modified to allow the integration of a crop-simulation model and the input of climatic information. The main sub-routines will be described, and a flow chart of the program can be seen in Fig. 18.

Executive Program – MAIN. The program is controlled by an executive routine call MAIN. It starts the program for a run and transfers control from one subroutine to the next. At the end of each run, corresponding to a single season, two subroutines are called to summarize production, water use, costs, and returns on farms in the system and to print out the results. Then the MAIN program inquires whether another run is to be made, whereupon it either restarts the cycle or terminates the program.

Table 12. Quantities of water allocated to farms in typical irrigation sequences on selected crops, 14 two-week periods during cropgrowing season[a]. (From Anderson and Maass, 1971)

Crop	2-week irrigation periods														
	April		May		June		July-August					Sept.		Oct.	Total
	1	2	3	4	5	6	7	8	9	10	11	12	13	14[b]	
	Inches														
Alfalfa	–	–	12	–	11	–	–	13	–	–	19	–	–	–	55
Beans	–	–	6	–	7	–	5	5	4	4	–	–	–	–	31
Corn	–	–	6	–	8	–	5	6	8	–	6	–	–	–	39
Small grain	10	–	5	–	8	–	–	–	–	–	–	–	–	–	23
Sorghum	–	–	5	–	6	–	6	6	6	6	6	–	–	–	41
Sugarbeets	–	6	–	7.5	–	5	6.5	–	5	5	8	–	4	–	47
Potatoes	–	6	–	5	4	4	4	4	4	4	4	4	–	–	43

[a] A 50-percent efficiency is assumed from the source of supply to the soil root zone

[b] The above crops are not typically irrigated in period 14 (roughly Oct. 1–15) in the area used for the illustration. The period is included in the simulation model to facilitate its application to situations in which irrigation might be practical

Table 13. Estimated percentage reduction in crop yield when a specified irrigation is not applied to specified crops[a]. (From Anderson and Maass, 1971)

Crop	2-week irrigation periods[b]													
	April		May		June		July-August					Sept.		Oct.
	1	2	3	4	5	6	7	8	9	10	11	12	13	14
	Percent													
Alfalfa	–	–	35	–	30	–	–	30	–	–	20	–	–	–
Beans	–	–	–	–	25	–	30	20	20	15	–	–	–	–
Corn	–	–	20	–	20	–	40	15	20	–	10	–	–	–
Small grain	25	–	25	–	25	–	–	–	–	–	–	–	–	–
Sorghum	–	–	20	–	15	–	20	20	20	20	15	–	–	–
Sugarbeets	–	20	–	20	–	15	20	–	15	15	25	–	10	–
Potatoes	–	20	–	15	15	15	20	20	20	20	15	8	–	–

[a] Assumptions:
1. Each acre during each irrigation period receives either (a) full water requirement or (b) none. Figures represent losses resulting from none during a period
2. Two successive "misses" result in total loss, except alfalfa
3. Percentage reduction is applied to final yield expected at moment – except in the case of the two successive losses
4. A direct relation is assumed between the physical yield of crops and economie values (except potatoes), i.e., water shortage affects yield and not quality. Data on effects of water shortages on quality of crops other than potatoes are not available, but the quality loss is probably not as important for the other crops listed

[b] See Table 1 for irrigation schedules used

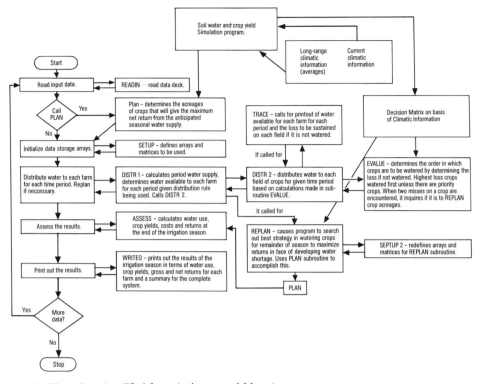

Fig. 18. Flow chart (modified from Anderson and Maass)

Subroutine READIN. At the start of a run, corresponding to an irrigation season, basic data are made inputs to the program by means of subroutine READIN. This information includes the number of farms in the system, the crops raised in the area, and the number of fields for each crop. Also included are the normal yield of each crop in units per ha, gross return from a normal yield, preharvest cost, and harvest cost in dollars per ha. The initial area of each crop and the farmer's share or expected share of water supply are read in, as are cards containing the anticipated water supply for each period and the priority of the crops to be used when the system requires priorities.

Information that was read in here in the original program, such as the irrigation requirements of each crop and the percentage loss in yield (or in quality) associated with increasing water stress is supplied by the crop-simulator subroutine to the PLAN subroutine.

Control is shifted to PLAN, where a check is made to see if the PLAN subroutine is to be called to adjust crop area to maximize income from available water. Because PLAN is a subsidiary routine dependent upon the rest of the program, it is discussed after the regular operating subroutines of the program have been described. If PLAN is not to be used, control is shifted to subroutine SETUP.

Subroutine SETUP. This subroutine initializes various arrays and matrices that are used in the program. These registers are used to store and accumulate information on the status of the system throughout the season.

Subrouting DISTR 1. Once the crop pattern for the farms has been set either as data input to the program or by use of subroutine PLAN, the program begins to run the irrigation season by shifting control to DISTR 1. DISTR 1 calculates the water supply available to the system for the first irrigation period. The period water supply is then allocated to the farms in the system according to the distribution rule in effect. The order of priority may be based on time of settlement or some other standard. This function is similar to the activities of irrigation organization officials who determine how much water the system will have to distribute and of ditch riders who turn water to individual farmers. Once the water supply for single farms has been determined, DISTR 1 passes control to subroutines (below) that determine water use on the farm. After water has been distributed to crops on all farms for the period, DISTR 1 is recalled to distribute the water supply for the next period. This process is continued until the end of the irrigation season. Water supply comes from two sources; direct flow water that is used first and demand water, which is used to make up deficiencies in direct flow supplies.

Subroutine DISTR 2. When the water supply for a particular irrigation period has been allocated to a farm, the next step in the program is to apply the irrigation water to the crops growing on the farm. This subroutine determines how the water is to be used by means of two criteria: (1) A crop priority, or (2) the field of a crop that will suffer the greatest loss in value if not watered.

If crop priorities are in effect, the first-priority crop will be watered without regard to value. (The order of priority is based normally on economic value, but other considerations can be taken into account.) If only one priority is set, the reamining water is used on highest value crops. If several crops have priorities, each will be watered according to priority.

Then the remaining crops will be watered according to value. The program cells subroutine EVALUE (described in the next section) which sorts the crops from highest to lowest priority. Following this the nonpriority crops are sorted with the ranking based on the value lost on each field of each crop if not watered during the period.

With crops arrayed according to priority and/or potential field loss values, DISTR 2 allocates water to priority crops and/or the highest loss fields until the water supply is exhausted or all crops are watered.

The original program allows only for full watering being given to the fields. The program suggested here would allow for only a fraction of the necessary water to be given, as the effect on the crop could be assessed by the simulation subroutine, and a value analysis could be made.

DISTR 2 processes the farms of the irrigation system one at a time, distributing irrigation water to crops according to priority or value and recording the fields of crops watered and those not watered. After the water supply has been allocated to crops on all farms during a period, control is returned to DISTR 1 incremented to another irrigation period. The water supply for the next period is distributed among the farms as before and control is shifted to DISTR 2, which again calls

EVALUE to array crops and then allocates water to the crops. This process is followed for each of the time periods during the irrigation season.

Subroutine EVALUE. The function of EVALUE is to array the fields on the farms according to the highest returns to be gained from the water. EVALUE is called as part of the PLAN subroutine and as part of DISTR 2. Its main use is as part of DISTR 2, where it computes the loss value of all the fields for use in allocating the water supply available during each period.

Climatic data of the current season is fed into the simulation subroutine, which supplies the state of the crop up to the present date. The crop simulation is then run according to the relative frequencies of weather occurrences (long-range data) in the future. These results form the basis of the decision matrix.

Subroutine ASSESS. When the final irrigation period has been completed, subroutine ASSESS is called to analyze the results of the irrigation season. The water applied to each crop, period by period, is recorded, along with total water applied to the crop. Next, the total production of each crop, the costs of production, and the gross and net returns are calculated for each farm and filed. ASSESS also accumulates total water received by each farm and the total water available regionally. The subroutine also accumulates the value of crops produced, total costs, and net value for each farm and for the region.

Subroutine WRITEO. The final step of a run of the program is to print out the data generated during the irrigation season. Subroutine WRITEO prints out data by irrigation periods on individual farms in one section and aggregate data for the whole system in a second section.

In the first section, details on each farm are printed. Each farm is identified and the ha in the farm are specified. Areas of each crop planted are also specified, along with water applications on each crop for each irrigation period. Total water applied on each crop is accumulated at the end of the season, followed by the total yield of each crop. Market values, as calculated in ASSESS, are recorded, and preharvest and harvest costs are subtracted from them, resulting in a net value for each crop. Gross value of all production is accumulated and total costs are subtracted, giving a net value for the farm. This printout allows examination of water use in each period on all crops. It is possible to see where water shortages occurred on each farm during the season. The printout also shows how each farm is affected by the particular water distribution rules being used by the system.

The second section of the WRITEO subroutine creates a summary of crop production for each farm in the region, total water received by each farm, gross value of crops, total costs, and net value, as well as regional total for these factors. With this information, comparisons can be made between different types of water supply situations and water distribution rules used by irrigation systems.

Subroutine PLAN. Subroutine PLAN is designed to allow choice of those crop areas for each farm that will give the greatest net returns to the farm, given the water supply that is forecast for the season and other factors influencing production. PLAN does this by testing to see if the farm's income would be increased or decreased by certain marginal fractions.

The PLAN routine is called by the MAIN program after READIN has entered the data in the computer. PLAN first tests to see if the routine is to be executed. If

PLAN is not to be called, control returns to MAIN program and the run continues with the initial areas. If PLAN is to be executed, the area increments and net return increments specified in READIN are used to compute for each farm the area by which each crop can be increased and the fraction by which net return must increase if PLAN is to continue to add units of area.

All crops are set to their lower limits on each farm and they are arrayed from high to low value. The program increases the area on the highest value crop on each farm by the fraction specified and the resulting net return is determined. If the increase in the net return of a farm meets the fraction that has been specified as significant, the area of the high-value crop on the farm is increased repetitively until there is no further significant increase in the net return, or until the allowable limit for the crop is reached, if at any time the increase in area of a crop does not produce a significant increase in net income, the area is reduced to where it was before the last increment. If there is unused water the program moves to the next highest value crop and increases its area, subject to area limitations. This process of increasing area, then computing and comparing changes in net income continues until there is, within the limits of the area restrictions and water supply, no appreciable increase in net return on any farm. PLAN then returns control to the MAIN program and the program prints out the results obtained with the final crop pattern determined in PLAN.

Subroutine REPLAN. The function of REPLAN is to determine the best use of limited water once the crop season is underway, and it appears the complete loss of some crops cannot be avoided if irrigation continues according to the basic distribution rules. Use of this subroutine is specified on the control card and the subroutine can be called only when PLAN is not requested.

REPLAN is activated whenever, water allocation limitations and weather conditions (i.e., drought) lead the simulation subroutine to conclude to the total loss of the crop. REPLAN stops the watering sequence. By means of SETUP 2 the upper limits of all crops are reset at current productive and lower limits, at zero, or a specified minimum. REPLAN then proceeds, using the PLAN subroutine, to determine the areas of crops that will result in the maximum return for the farm, given the water supply estimated to be available for the remainder of the season. It reruns the remainder of the season repeatedly, incrementing the highest return crop, until the irrigation water available for the remainder of the season is used as efficiently as possible. The subroutine then shifts control back to MAIN, which calls ASSESS and WRITEO to record the results.

Subroutine TRACE. If information is desired on the water available for each farm for each time period, along with an array of dollar losses to be sustained by each crop if it is not watered during the period, a data-retrieval routine named TRACE can be called which lists this information. This information can be used to determine the value of irrigation water at any time in the irrigation season.

At this point, the program inquires if another season is to be run. If so, the data for the next study are given to the READIN subroutine, and another season is run. The program can run as many studies as desired for any system. If no additional runs are required, the program cans END and terminates.

16.5 Brief Review of Other Solution Techniques

Other techniques used to solve the problem of water allocation in irrigation systems have been reviewed in a recent M. Sc. thesis (de la Fuente, 1974):

Dudley et al. (1971a) published results on a dynamic programming model on how to allocate a given irrigation water amount stored in a reservoir over a season when the decision on what crop area is going to be irrigated has already be made. The model considers rainfall and plant water usage as stochastic variables. They also studied the problem of deciding what area of crop to plant at the beginning of the season (Dudley et al., 1971b) considering the water demand of the plant and the water inflows to the reservoir from which irrigations are made, to be both stochastic variables and they obtained a solution to the long-term problem of how to determine the best size of the area to be prepared for an irrigation season (Dudley et al., 1972).

Earlier studies were made by Flinn and Musgrave (1967) and Flinn (1968) on the same short-term problem of allocating a finite quantity of irrigation water over a season.

Hall and Butcher (1968) analyzed a deterministic dynamic programming approach to the same problem. Their model, as Aron pointed out in his review (Aron, 1969) gives an optimal policy which is not affected by irrigation costs.

Dudley et al. (1971a) in their solution to the short-term problem, show that costs do affect the optimal irrigation policy over a season.

For the sake of completeness, the work of the Agrometeorology Section, Plant Research Institute, Canadian Dept. of Agriculture should be mentioned, in particular the computer program of Baier and Russelo (1968) and the work of Coligado et al. (1969), Wilcox (1967), and Wilcox and Sly (1974).

16.5.1 Stochastic Inventory Models

An example of the application of Inventory Theory to irrigation allocation (for the details on the theory, see de la Fuente, 1974), will be given. The problem is to determine the optimum soil water level to be maintained over the crop growth season (i.e., the soil water content that will maximize the net revenue from the crop).

The following definitions will be used:

Set-up cost (C_M): preliminary labor cost of establishing the irrigation system.

Holding cost (C_H): The yield or quality reduction of the crop when the soil water content was higher than needed, and the irrigation water loss from deep drainage and evapotranspiration.

Shortage cost (C_S): value of the reduction in yield due to water shortage and proportional to the severity of the shortage.

Demand: crop water requirements.

The demand Z_1, defined as the potential evapotranspiration rate of the plant, is a random variable. In areas where rainfall occurs during the growing season, the evapotranspiration depends on the rainfall amount (Z_2), another random variable. Thus Z_1 and Z_2 are two dependent random variables with a given joint density function

$$f(z_1, z_2) = f(z_1/z_2) f_2(z_2),$$

where $f(z_1/z_2)$ is the conditional probability density function of Z_1 (the demand) given that $Z_2 = z_2$ (i.e., given that rainfall has a value z_2), and $f_2(z_2)$ is the marginal probability density function of Z_2.

B_c: unit benefit of a crop from satisfying the plant's water requirements.

B_s: salvage benefit per unit of water left in the soil reservoir at the termination of the inventory model.

C_w: cost per unit of irrigation water.

The practical application of inventory models will be illustrated, using the first and simplest model developed by de la Fuente, the single-period model with instantaneous demand (de la Fuente, 1974):

16.5.2 The Single-Period Inventory Model

Theory. As a first approach it is assumed that the plant water usage is instantaneous. This might not be too close to reality, but will permit understanding of the basis procedure.

Let x = the amount of water in the soil (inventory on hand) before an irrigation is placed, y = the amount of water available in the soil after an irrigation of $(y-x)$ units, $y \, x$.

It must be noted that x includes the amount of water in the soil due to rainfall occurrences previous to the inventory period. Depending on the amount of water.

Demanded (instantaneously) z_1, the inventory position right after the demand occurs may be either positive (overirrigation) or negative (underirrigation or shortage). Figure 18 is an illustration of the inventory during the single period in both cases.

Thus, given x and assuming y is a continuous variable, the holding and shortage cost functions, for each given value of $Z_2 = z_2$, will be

$$H(y) = C_h(y + z_2 - z_1) - B_s(y + z_2 - z_1) - B_c(z_1), \qquad 0 < z_1 \leq y + z_2$$
$$G(y) = C_s(z_1 - y - z_2) - B_c(y + z_2), \qquad y + z_2 < z_1 < \infty$$

respectively.

The expected holding cost is given by

$$\sum_{z_1, z_2} H(y) = \int_0^\infty \int_0^{y+z_2} H(y) f(z_1, z_2) dz_1 \, dz_2 .$$

The expected shortage cost is given by

$$\sum_{z_1, z_2} G(y) = \int_0^\infty \int_{y+z_2}^\infty G(y) f(z_1, z_2) dz_1 \, dz_2 .$$

Finally, the expected total inventory cost $L(y)$ for the period is the sum of the expectations of all partial costs and is given by

$$L(y) = C_m + C_w(y - x)$$

$$+ \int_0^\infty \int_0^{y+z_2} H(y) f(z_1, z_2) dz_1 \, dz_2$$

$$+ \int_0^\infty \int_{y+z_2}^\infty G(y) f(z_1, z_2) dz_1 \, dz_2$$

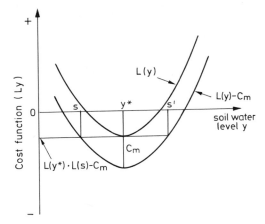

Fig. 19. Convex cost function. F(y) includes the setup cost. $L(y)-C_m$ is the same cost function when there is no setup cost (after de la Fuente)

substituting we have

$$L(y) = C_m + C_w(y-x)$$

$$+ \int_0^\infty \int_0^{y+z_2} [(C_h - B_s)(y + z_2 - z_1) - B_c z_1] f(z_1, z_2) dz_1 dz_2$$

$$+ \int_0^\infty \int_{y+z_2}^\infty [C_s(z_1 - y - z_2) - B_c(y + z_2)] f(z_1, z_2) dz_1 dz_2. \tag{1}$$

The optimum value of y is y* and it is found by equating the first derivative of $L(y)$ to zero. Thus

$$\frac{\partial L(y)}{\partial y} = C + \int_0^\infty \int_0^{y+z_2} (C_h - B_s) f(z_1, z_2) dz_1 dz_2$$

$$- \int_0^\infty \int_{y+z_2}^\infty (C_s + B_c) f(z_1, z_2) dz_1 dz_2$$

$$= C_w + (C_h - B_s + C_s + B_c) \int_0^\infty \int_0^{y+z_2} f(z_1, z_2) dz_1 dz_2$$

$$- (C_s + B_c)$$

$$= 0.$$

This equation becomes

$$\int_0^\infty \int_0^{y^*+z_2} f(z_1, z_2) dz_1 dz_2 = \frac{C_s + B_c - C_w}{C_h - B_s + C_s + B_c} = \frac{P}{Q}. \tag{2}$$

It is clear that given z_2 the value of y* that satisfied this equation is unique. Taking the second derivative

$$\frac{\partial^2 L(y)}{\partial^2 y} = Q \int_0^\infty f(y^* + z_2, z_2) dz_2$$

wee see that the second factor of the right hand term is positive [since $f(y^* + z_2, z_2)$ is a joint density function] and $Q = C_h - B_s + C_s + B_c$ must be positive, otherwise it would mean that $B_s > C_h$, or that overirrigation would always give a higher profit. Thus, the second derivative of $L(y)$ is always positive which is the same as saying that $L(y)$ is a convex function (a graphic illustration is given in Fig. 19). Hence, since y^* is unique, it gives a global minimum of $L(y)$ which is what we are looking for.

The inclusion of the setup cost C_m in the model leads to the following discussion: Since C_m is a constant, y^* is the minimizing value for both $L(y)$ and $L(y) - C_m$. If we let s be such that

$$L(y^*) = L(s) - C_m$$

we see from Fig. 19 (the other value s' satisfying this can be disregarded) that given x

$$L(y^*) < L(x) - C_m \quad \text{for} \quad x < s,$$

and

$$L(y^*) \geq L(x) - C_m \quad \text{for} \quad x \geq s$$

therefore it pays not to irrigate when $x \geq s$.

The optimal policy, given x, will then be given by

Irrigate $(y^* - x)$ units,   if $x < s$

Do not irrigate,   if $x \geq s$.

Example

It is not possible, in the models developed in the last section, to give a closed form solution for y^* unless $f(z_1, z_2) = f(z_1 | z_2) f_2(z_2)$ is specified.

For the arid conditions of the southwestern United States the conditional density function for an index of potential evapotranspiration (demand) was obtained from two histograms on pan evaporation rates which was assumed to be identically distributed as evapotranspiration. It was found that pan evaporation data could be grouped into two classes corresponding to days in which rain occurred and days when no rain occurred (see Fig. 20). No significant difference was observed in pan evaporation for varying amounts of rainfall.

The mean values of pan evaporation on rainy and clear days shown in Fig. 20 are surprisingly similar, and may be wrong, but this should not reflect on the validity of the model as such.)

To simplify the problem, a triangular distribution was tested and a good fit was obtained at the 0.05 level of confidence. The triangular density function obtained will be used throughout these examples, and since the dependence of the pan evaporation rates was reduced to either rain or no rain, the random variable for the rainfall (Z_2) will be limited to take on only the values zero when no rain occurs and some expected value k given in mm/day when rain occurs. Thus, the marginal

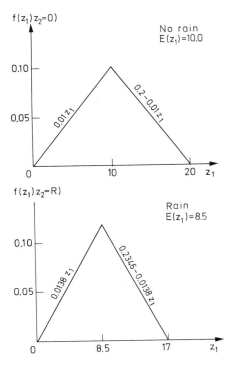

Fig. 20. Histograms on pan evaporation rates for wet and dry days during the summer in the south-western United States (after de la Fuente)

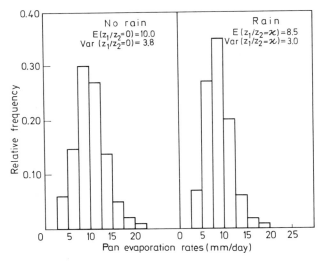

Fig. 21. Graph of the conditional probability density of the demand z_1 given that there is no rain and that rain occurs (after de la Fuente)

density for the rainfall is given by

$$f_2(z_2) = \begin{cases} \alpha, & \text{if } z_2 = 0 \\ 1-\alpha, & \text{if } z_2 = k. \end{cases}$$

The conditional density for the demand was found to be (see Fig. 21)

$$f(z_1|z_2) = \begin{cases} f(z_1|0) = \begin{cases} 0.01z_1, & 0 \leq z_1 \leq 10 \\ 0.2-0.01z_1, & 10 \leq z_1 \leq 20 \end{cases} \\ f(z_1|k) = \begin{cases} 0.0138z_1, & 0 \leq z_1 \leq 8.5 \\ 0.2346-0.0138z_1, & 8.5 \leq z_1 \leq 17. \end{cases} \end{cases}$$

According to Eq. (2) the optimum value y^* has to satisfy

$$\int_0^\infty \int_0^{y^*+z_2} f(z_1|z_2)f_2(z_2)dz_1\,dz_2 = \frac{P}{Q}.$$

In this case since $f_2(z_2)$ is discrete, this relationship becomes

$$\sum_{z_2}\left[\int_0^{y^*+z_2} f(z_1|z_2)dz_1\right]f_2(z_2) = \frac{P}{Q}.$$

Since

$$\frac{P}{Q} = \frac{C_s+B_c-C_w}{C_h-B_s+C_s+B_c},$$

this equation is only valid when $C_s+B_c>C_w$ (the opposite would call for discarding the inventory system completely); analyzing the denominator it is clear that B_s, the salvage value of water for the next period, is a small value compared with the rest, therefore $C_s+B_c-C_w<C_h-B_s+C_s+B_c$ and $0<P/Q<1$ as it was expected since this quantity is a probability. A good estimate for this value is

$$\frac{P}{Q} = 0.8.$$

Roughly speaking, for the southwestern United States area during the summer season, the probability of a wet day is $\alpha=0.2$ (and hence $1-\alpha=0.8$) and the expected amount of rainfall in a wet day is $k=6\,\text{mm/day}$.

Furthermore, the triangular distribution is not continuous, thus in order to perform the integration and find y^* we have to consider two cases, namely, $y^* \leq E(z_1|z_2)$ and $y^*>E(z_1|z_2)$.

Consider the case $y^* \leq E(z_1|z_2)$, then the value y^* of y must satisfy

$$\frac{P}{Q} = \alpha \int_0^{y+0} f(z_1|0)dz_1 + (1-\alpha)\int_0^{y+k} f(z_1|k)dz_1$$

which becomes

$$0.8 = 0.8\int_0^{y+0} 0.01z_1\,dz_1 + 0.2\int_0^{y+6} 0.0138z_1\,dz_1$$

or

$$0.00538\,y^2 + 0.01656\,y - 0.75032 = 0$$

the solutions for this equation are

$$y = \begin{cases} \text{neagive (unfeasible)} \\ 10.37 > E(z_1 | z_2), \quad \text{which contradicts the assumption } y^* \leqq E(z_1 z_2). \end{cases}$$

Consider $y^* > E(z_1 | z_2)$, then y is such that it satisfies

$$0.8 = 0.8 \left[\int_0^{10} 0.01\,z_1\,dz_1 + \int_{10}^{y} (0.2 - 0.01\,z_1)dz_1 \right]$$

$$+ 0.2 \left[\int_0^{8.5} 0.0138\,z_1\,dz_1 + \int_{8.5}^{y+6} (0.2346 - 0.0138\,z_1)dz_1 \right]$$

$$= -0.00538\,y^2 + 0.19036\,y - 0.767275$$

or

$$0.00538\,y^2 - 0.19036\,y + 1.567275 = 0.$$

The solutions are

$$y = \begin{cases} 22.35 \text{ (unfeasible)} \\ 13.04 \text{ optimum value}. \end{cases}$$

Therefore $y^* = 13.04$, this is, 13.04 mm/day are needed to maintain an optimum economic water soil content.

This value appears to be excessive, which may be due to an incorrect assessment of pan evaporation (see above). The approach, however, is undeciably worthy of further interest.

Literature

Agroclimatological methods, p. 392. Proc. Reading Symp. UNESCO, Paris 1968

Anderson, R.L., Maass, A.: A simulation of irrigation systems (the effect of water supply and operating rules on production and income, on irrigated farms). Tech. Bull. 1431, USDA, ERS, Washington D.C. (1971)

Aron, G.: Optimal timing of irrigation: discussion. J. Irrig. Drain. Div., ASCE, 95 (IR 1) 254–257 (1969)

Austin-Bourke, P.M.: The aims of agrometeorology. Proc. Reading Symp. UNESCO, Paris 1968

Baier, W.: Crop-weather analysis model: review and model development. J. Appl. Meteorol. 12 (6), 937–947 (1973)

Baier, W., Robertson, G.W.: A new versatile soil moisture budjet. Can. J. Plant Sci. 46, 299–315 (1966)

Baier, W., Robertson, G.W.: Estimating yield components of wheat from calculated soil moisture. Can. J. Plant Sci. 47, 617–630 (1967)

Baier, W., Robertson, G.W.: The performance of soil moisture estimates as compared with the direct use of climatological data for estimating crop yields. Agric. Meteorol. 5, 17–31 (1968)

Baier, W., Russelo, D.A.: A computer program for estimating risks of irrigation requirements from climatic data. Tech. Bull. 59. Agrometeorological Section, Plant Research Institute. Can. Dep. Agric. Ottawa 3 (1968)

Banin, A., Amiel, A.: A correlative study of the chemical and physical properties of a group of natural soils of Israel. Geoderma 3 (1969/1970), 185–198 (1970)

Bielorai, H.: The irrigation of field crops – cotton. In: Irrigation in arid zones. Volcani Inst. Agric. Res., Bet Dagan Israel (draft edit.) (1969)

Bielorai, H.: The effect of irrigation on water consumption, yield and salt accumulation in a grapefruit orchard. Progress Report 1968/69, Volcani Inst. Agric. Res. Rehovot, Israel (In Hebrew) (1969)

Bielorai, H., Levi, J.: The effect of irrigation on water consumption, yield and salt accumulation in a grapefruit orchard. Annual Report 70/02, Volcani Inst. of Agric. Res., Rehovot, Israel (In Hebrew with English summary) (1970)

Bielorai, H., Levi, J.: Irrigation regimes in a semi-arid area and their effects on grapefruit yield, water use and soil salinity. Isr. J. Agric. Res. *21* (1), 3–12 (1971)

Bierhuizen, J.F., de Vos, N.M.: The effect of soil moisture on the growth and yield of vegetable crops, pp. 83–92. Rep. Conf. Suppl. Irrigation. Comm. VI Int. Soc. Soil Sci. Copenhagen 1958

Bresler, E., Shmueli, E., Goell, A., Kalmar, D., Angelchik, M.: Auxiliary experiments in citrus irrigation (at Ein Hamifrats and Saar). Progress report 1963–1965. Ministry of Agriculture, Tel Aviv (In Hebrew) (1965a)

Bresler, E., Shmueli, E., Goell, A., Kurtz, H.: Auxiliary experiments in citrus irrigation (at Kfar Mordechai). Progress report 1963–1965. Ministry of Agriculture, Tel Aviv (In Hebrew) (1965b)

Budyko, M.I.: Proc. Reading Symp., UNESCO 7 (1968)

Busch, C.D., Rochester, E.W.: Scheduling and application rates of irrigation in a humid climate. Bull. 470, Agric. Exp. Sta., Auburn University, Auburn, Alabama 1975

Coligado, M.C., Baier, W., Sly, W.K.: Risk analyses of weekly climatic data for agricultural and irrigation planning, p. 32. Summerland, British Columbia. Tech. Bull. *57*, Agrometeorology Section, Plant Res. Inst., Can. Dep. Agr., Ottawa 1969

Dale, R.F., Hartley, M.: Computer program of estimating soil moisture under corn. Appendix to Final Rpt. CWB-10554, Iowa State University of Sc. and Tech. (stencil) 1963

Dudley, N.J., Howell, D.T., Musgrave, W.F.: Optimal intraseasonal irrigation water allocation. Water Resour. Res. 7 (2), 770–788 (1971a)

Dudley, N.J., Howell, D.T., Musgrave, W.F.: Irrigation planning 2: choosing optimal acreages within a season. Water Resour. Res. 7 (5), 1051–1063 (1971b)

Dudley, N.J., Musgrave, W.F., Howell, D.T.: Irrigation planning 3: the best size of irrigation area for a reservoir. Water Resour. Res. 8 (1), 7–17 (1972)

Fitzpatrick, E.A., Nix, H.A.: A model for simulating soil water regime in alternating fallow-crops systems. Agric. Meteorol. *6*, 303–319 (1969)

Fliin, J.C.: Allocation of water resources: A derived demand function for irrigation water, and efficiency of allocation. Farm. Manage. Bull. 1. Univ. of New England, Armidale, Australia 1968

Flinn, J.C., Musgrave, W.F.: Development and analysis of input-output relations for irrigation water. Aust. J. Agric. Econ. *11* (1), 1–19 (1967)

Gleeson, T.: A prediction and decision method for applied meteorology and climatology, based partly on the theory of games. J. Meteorol. *17*, 116–121 (1959)

Gringorten, I.: Probability estimates of the weather in relation to operational decisions. J. Meteorol. *16*, 663–671 (1959)

Halkias, N.A.: Evaluation of the proper water supply systems as a basis for planning complete programmes of collective irrigation. J. Agric. Eng. *8* (1), 74–77 (1963)

Hall, W.A., Butcher, W.S.: Optimal timing of irrigation: J. Irrig. Drain. Div. ASCE *94* (IR 2), 267–275 (1968)

Heller, J., Shalhevet, J., Goell, A.: Response of a citrus orchard to soil moisture and soil salinity. In: Ecological studies analysis and synthesis. Hadas, A., et al. (eds.), Vol. 4, pp. 410–419. Berlin, Heidelberg, New York: Springer 1973

Jeffers, J.N.R.: The challenge of modern mathematics to the ecologist. In: Mathematical models in ecology. Jeffers, J.N.R. (ed.), pp. 1–11. Blackwell Scientif. Publ., Oxford: 1972

Jensen, M.D., Middleton, J.E.: Scheduling irrigation from pan evaporation. Circ. No. 527, Wash. Agr. Exp. Sta. Washington State U. Washington: Pullman 1970

Jensen, M.E., Robb, D.C., Franzoy, C.E.: Scheduling irrigation using climate-crop-soil data. Paper presented at the National Conference on Water Resources Engineering of A.S.C.E., New Orleans, Louisiana 1969

Kalma, J.D.: Some aspects of the water balance of an irrigated orange plantation. Ph.D. Thesis, Hebrew University of Jerusalem, Israel 1970

Kalma, J.D.: Some aspects of the water balance of an irrigated orange plantation. Volcani Inst. Agric. Res., Bet Dagan, Israel, Spec. Publ. *70* (1970)

Kalmar, D., Goell, A., Colomb, A., Heller, J.: The response of a Shamouti orange grove on S.O. stock planted on heavy clay soil to different irrigation regimes in spring and summer. Ministry of Agriculture, Acre, Israel 1973

Kautilya, 321–296 B.C. *Asthasastra*, as quoted by G. Stanhill 1973. Plant Response to Climatic Factors. Proc. Uppsala Symp. (Ecology and Conservation, 5) UNESCO 1970

Lewin, J.: A simple soil moisture simulation model for assessing the irrigation requirements of citrus. Isr. J. Agric. Res. *23* (1), 3–12 (1973)

Lomas, J., Lewin, J.: Yield prediction of a cotton crop by computer simulation (in preparation)

Mantell, A.: The irrigation of field crops – peanuts. In: Irrigation in arid zones. Volcani Inst. Agric. Res., Bet Dagan Israel (draft edition) (1969)

Mantell, A., Goell, A.: The response of Shamouti orange trees to different regimes, Preliminary Report, No. 744, Volcani Inst. Agric. Res., Bet Dagan, Israel (In Hebrew) (1973)

Marani, A., Shimshi, D., Amirov, A.: The effect of time and duration of soil moisture stress on flowering, boll shedding, seed and lint development, yield and seed quality of cotton. First Ann. Rep. 1965–1966, Project No. A 10-CR-64, Univ. Jerusalem, Faculty Agric., Rehovot, Israel (1967)

Marcia A. Gonzales de la Fuente: Inventory theory application to the optimum control of irrigation water. Arizona Univ., Tucson Dep. Syst. Ind. Eng. (1974)

Mathematical models in ecology. Proc. 12th Symp. Brit. Ecol. Soc. Jeffers, J.N.R. (ed.), p. 398. Oxford: Blackwell Scientif. Publ. 1972

Nix, H.A., Fitzpatrick, E.A.: An index of crop water stress related to wheat and grain sorghum yields. Agric. Meteorol. *6*, 321–337 (1969)

Penman, H.L.: Natural evaporation from open water, bare soil and grass. Proc. R. Soc. London A *193*, 120–145 (1948)

Plant response to climatic factors, p. 504. Proc. Uppsala Symp. UNESCO Paris 1972

Prediction and measurement of photosynthetic productivity, p. 632. Proc. IBP/PP Tech. Meet. Trebon Pudoc, Wageningen 1970

Proc. regional training seminar on agrometeorology, p. 419. WMO, Wageningen 1968

Proc. seminar on agricultural meteorology, p. 357. WMO 310, Geneva 1972

Rawitz, E.: The influence of a number of environmental factors on the availability of soil moisture to plants. Ph.D. thesis, Hebrew University of Jerusalem, Israel (Hebrew with English summary) (1965)

Russelo, D., Edey, S., Godfrey, J.: Selected tables and conversion used in agrometeorology and related fields. Can. Dept. Agric., Publ. *1522* (1974)

Shimshi, D., Gairon, S., Rubin, I.: Irrigation experiments with winter cereals in the Northern Negev. Report No. 419, Natl. Univ. Inst. Agric. Res., Rehovot, Israel (in Hebrew) (1963)

Shimshi, D.: The irrigation of field crops – Wheat. In: Irrigation in arid zones. Volcani Inst. Agric. Res., Bet Dagan Israel (draft edition) (1969)

Shmueli, E., Bielorai, H., Heller, J., Mantell, A.: Citrus water requirements conducted in Israel during the 1960's. In: Ecological studies. Analysis and synthesis. Hadas, A., et al. (ed.), Vol. 4, pp. 340–350. Berlin, Heidelberg, New York: Springer 1973

Skellam, J.G.: Some philosophical aspects of mathematical modelling in empirical science with special reference to ecology. In: Mathematical models in ecology. Jeffers, J.N.R. (ed.), pp. 13–28. Oxford: Blackwell Scientif. Publ. 1972

Slayter, R.O., McIllroy, I.C.: Practical micrometeorology. UNESCO (1961)

Stanhill, G.: The control of field irrigation practice from measurements of evaporation. Isr. J. Agric. Res. *12* (2), 51–62 (1962)

Stanhill, G., Vaadia, Y.: Factors affecting plant responses to soil water. In: Irrigation of agricultural lands. Hagan, R.M. (ed.). Agronomy 11, pp. 446–457. Am. Soc. Agronomy. Wisc.: Madison 1967

Taylor, H.M., Kleppner, B.: Water uptake by cotton roots during an irrigation cycle. Aust. J. Biol. Sci. *24*, 5, 853–859 (1971)

Training in agrometeorology and research on photosynthesis of crops in relation to productivity, p. 343. Guelph Project Final Report, University of Guelph, Guelph, Canada 1974

Wilcox, J.C.: Credit to give for rain when scheduling irrigation. Can. Agric. Eng. *9*, 103–107 (1967)

Wilcox, J.C., Sly, W.K.: A weather-based irrigation scheduling procedure. Tech. Bull. 83, p. 23. Agrometeorology Res. Serv., Chem. Biol. Res. Inst. Can. Dep. Agric., Ottawa 1974

17 Special Forecasts

Y. I. CHIRKOV

In recent decades, special forecasts – specifically, forecasts of weather phenomena potentially damaging to crops (frost, hail, dust storms, etc.), of soil moisture reserves, the distribution and extent of the destruction of crops by pests and disease, and of onset times for basic plant development stages as well as of maturation schedules – have found, along with weather forecasts, increasingly wider application in agriculture. Yield forecasts have also acquired significant value in recent years.

All these forecasts supply data on anticipated crop growth conditions, and can also be used for the timely application of measures aimed at controlling adverse conditions and for making arrangements related to harvest management.

In the sections following, we will examine frost forecasts (Chap. 19), forecasts of the incidence of pests and plant diseases in connection with developed and anticipated weather conditions (Chap. 20), and forecasts of crop water needs (Chap. 21). Chapter 18 discusses agrometeorological forecast systems, among which are forecasts of plant development and maturation onset times, of soil productive-moisture reserves, and of yield structure of major crops as affected by agrometeorological factors.

18 Agrometeorological Forecast Systems

Y. I. Chirkov

Correlation of perennial agrometeorological observations made by a network of weather stations and of field-experiment data form the basis of agrometeorological forecast systems.

Agrometeorological observations include regular instrumental analyses of the status of key meteorological factors and field (and orchard) observations of crop condition. A program of field observations includes the following types of observation and analysis.

1) Phenological observations and onset of plant development stages. These observations are made regularly, usually every other day (on even-numbered dates). Specially selected (tagged) plants at four locations on a control plot are inspected. The stage of development of each plant is noted, the number of plants in a given stage is tabulated, and the percentage of plants within each stage is determined. If a given stage is noted in 10% of the plants, stage onset (a) has begun; if a stage is noted in 50% or more, major onset (b) has started.

The developmental stages being differentiated are linked to the biological features of the plants. In cereal crops, for example, the following are observed: sprouting, seedlings, third-leaf stage, tillering, stem extension (shooting), appearance of lower stem node above the soil surface, heading (ear emergence), blooming, milky ripeness, gold ripeness, complete ripeness (in winter crops, growth resumption also occurs). In sugar beets and other root crops, sprouts come up, first-third leaf pairs appear, the vegetable begins to develop, and the drills are covered by the growing leaves. Phenological observations are made in fruit orchards on permanently tagged trees. The following stages eventuate: swelling of buds, blooming of buds, opening of the first leaves, flower clusters, blossoming, ripening of the fruit, autumnal coloring of the leaves, and exfoliation. Tables 1–3 list the phenological stages of major farm crops as recorded by a network of meteorological stations in the Soviet Union.

2) Plant spacing. This is performed four times for field crops. In drilling, the number of plants in a row is calculated (one meter apart in close drill sowing and 5 m apart in adjacent lengths in wide row sowing). Analysis times are fixed according to crop features. In spring wheat, oats, and barley, for example, are analyzed four times: in the third-leaf stage, after the appearance of straw above the surface, during heading, and in the milky ripeness stage. Spacing is analyzed twice in the sunflower: after thinning and during blossoming.

3) Plant height variation. In field crops, the height of each of 40 plants is measured in centimeters on the day of major stage-onset and on the last day of den-day periods. These measurements are not taken for root crops.

4) Weediness of the planting. This analysis is carried out while making phenological observations on the control plot. The degree of weediness is

Table 1. Phenological stages of major grain, bean, and perennial cereal crops, recorded by a network of meteorological stations

Stages	Rye	Wheat	Barley	Oats	Millet	Rice	Cereal grasses	Corn	Peas	Legumes	Kidney beans
Sprouting	+	+	+	+	–	–	–	+	–	–	–
Seedling	+	+	+	+	+	+	–	+	+	+	+
Third leaf	+	+	+	+	+	+	+	+	+	+	+
Tillering	+	+	+	+	+	+	+	–	–	–	–
Stem extension (shooting)	+	+	+	+	+	+	+	–	–	–	–
Appearance of lower stem node above the surface	+	+	+	+	–	–	+	–	–	–	–
Flower clusters (heading, tasseling)	+	+	+	+	+	+	+	+	+	+	+
Blooming	+	+	–	–	–	–	+	+	+	+	+
Cob blooming	–	–	–	–	–	–	–	+	–	–	–
Milky ripeness	+	+	+	+	–	+	–	+	–	–	–
Gold ripeness	+	+	+	+	–	–	–	+	–	–	–
Complete ripeness (maturation)	+	+	+	+	+	+	+	+	+	+	+

Table 2. Phenological stages of major industrial and melon corps, recorded by a network of meterological stations

Stage	Cotton	Sunflower	Hemp	Cucumber	Musk melon, Pumpkin, Watermelon
Seed germination	+	−	−	−	−
Seedling	+	+	+	+	+
First leaf	+	−	−	+	+
Third leaf	+	−	−	+	+
Fifth leaf	+	−	−	−	−
Second pair of leaves	−	+	+	−	−
Flower cluster (budding)	+	+	+	+	+
Blooming	+	+	+	+	+
Opening of first boll	+	−	−	−	−
Maturation	−	+	+	−	−
Termination of growth	+	−	−	−	−

Table 3. Phenological stages of potato and major vegetable crops, recorded by a network of meterological stations

Stage	Potato	Beets	Carrot, Turnip, etc.	Tomato	Eggplant, Pepper plants, etc.
Seedling	+	+	+	+	+
First leaf	−	−	−	+	+
First pair of leaves	−	+	−	−	−
Third leaf	−	+	+	+	+
Fifth leaf	−	+	+	−	−
Side shoots	+	−	−	+	+
Hypocotyl thickening	−	+	+	−	−
Flower clusters	+	−	−	+	+
Blooming	+	−	−	+	+
End of blooming	+	−	−	−	−
Top wilt	+	−	−	−	−
Yellowing of outer leaves	−	+	+	−	−
Maturation	−	−	−	+	−

determined visually on a scale of zero to four. Zero signifies "no weeds", four means "numerous weeds".

5) Crop damage due to adverse meteorological phenomena. The nature and extent of damage caused to the plants by frosts, droughts, dry wind, hail, showers, dust storms, and strong wind are recorded.

6) Crop damage due to pests and disease. The nature and extent of the damage is noted.

7) Plant bow, moisture content of the straw and the grain, and grain intergrowth under adverse harvest conditions.

8) Quantitative evaluation of the condition of the seedlings of major crops according to special indices.

9) Formation of productivity elements and crop yield structure.

10) Observations of field operations and an evaluation of their quality and of the effect of weather conditions on them.

11) Cattle grazing conditions in terms of pasture state.

12) Condition of hibernating field crops and fruit trees.

13) Temperature, depth of freezing and thaw of the soil, and snow cover above winter crop fields and in orchards.

14) Moisture content of crop soil. Instrumental observations are made regularly during the warm season on the eighth day of ten-day cycles from the beginning of field operations to the harvest of spring crops, and, for winter crops, from harvest to late autumn and from growth resumption to harvest.

Many systems of agrometeorological forecasting (forecasts of onset times of developmental stages of corn, of the yield of winter and spring wheat and corn, as well as of the hibernation of winter crops) are based on morphophysiological analysis and a method of biological inspection developed by professor F. M. Kuperman of the Soviet Union. This method is indispensable for diagnosing the state of crops as it enables one to determine precise boundaries between the 12 stages of plant development. Inasmuch as plant response to meteorological conditions varies according to the stages of organogenesis, it is obviously important to consider these stages when establishing quantitative agrome-teorological relationships (especially the first seven stages, where developmental phases, in defining only the external changes in a plant, do not allow one to determine such relationships sufficiently accurately).

This method of biological inspection not only augments the phenological observations significantly, but it also enables one to determine more objectively the condition of the plants, and to detect damage caused by intrastem agents, dry wind, frosts, etc. After the plants are prepared for the inspection, the apical cones are then examined under a microscope, establishing the stage, viability, and anticipated productivity of a given plant from specific criteria. Early stages of organogenesis and the condition of the cone are thus identified.

Correlating and statistically processing perennial agrometeorological obser-vation data have made the calculation of prognostic equations expressing the quantitative relationship between forecasted factor and key predictors possible in the Soviet Union.

18.1 Soil Moisture Reserve Forecasts

Soil moisture reserve forecasts are of paramount importance in an arid region, not only for evaluating crop growth conditions, but also for determining irrigation standards.

A method has been developed for forecasting the spring soil moisture reserves in regions with a moderate climate where the ground freezes in the winter. It is based on the date of the stable transition of average diurnal atmospheric temperature through $+5\,^\circ$C which permits growth resumption of vegetation and the commencement of spring field operations.

Since it is necessary to know beforehand what the moisture reserves prior to sowing will be in the top meter of soil, in order to decide upon optimal dates for planting spring crops, methods of presowing soil treatment, and crop assortment, this forecast is very valuable in regions with insufficient and fluctuating humidification. (If, for example, the soil is considerably dehydrated in the spring, it is advisable to sow drought-resistant crops like millet rather than spring wheat.)

The method was developed by L. A. Razumova of the Soviet Union. The forecast is usually compiled by 1 March, i.e., 30–50 days before the planting of early spring crops in the steppe zone. The base data for the forecast are: (1) productive moisture reserves (W_1) in autumn, before the ground freezes; (2) the productive moisture deficit (d) in autumn (the difference between the minimum water capacity of the top meter of soil and the autumn productive moisture reserves in that layer); (3) amount of precipitation (m) occurring during the period from the previous soil determination to 1 March (observation data) and from 1 March to the transition of the mean diurnal air temperature through 5 °C, which defines the commencement of spring field operations (these data are calculated from a long-term weather forecast).

The relationship between the parameters affecting productive-moisture reserve variation (ΔW) from the time the ground freezes to the transition of the mean diurnal air temperature through $+5\,°C$ in the spring has been established by means of observation data. This relationship is expressed by means of the equations:

$$1/\Delta W = 0.115\,m + 0.56\,d - 20 \text{ (in regions with unbroken winter)}$$

$$2/\Delta W = 0.21\,m + 0.62\,d - 33 \text{ (in regions with winter with frequent thaws)}.$$

If fall productive moisture reserves and their winter variation are known, one can estimate the moisture reserves that will be present in the soil at the start of spring field works.

Let us analyze the structure of a sample forecast of spring productive moisture reserves for a fall-plowed field.

Base data:

1) Date of the previous soil-humidity determination is 10 November.
2) Fall productive-moisture reserves (W_1) in a layer 0–100 cm thick is 50 mm.
3) Minimum water capacity for the given layer of soil (W_0) is 180 mm.
4) Total precipitation from 10 November to 1 March is 80 mm.

Weather forecast data:

5) Date of the transition of the average diurnal atmospheric temperature through 5 °C will be 18 April.
6) Total precipitation expected from 1 March to 18 April is 38 mm.

Calculated data:

7) Fall soil-humidity deficit (d) is $W_0 - W_1$, which equals 130 mm.
8) Total precipitation for the whole period (m) is $80 + 38$, which equals 118 mm.
9) Change in productive-moisture reserves between 10 November and 18 April (calculated with the following equation for regions with a stable winter) is

$$\Delta W = 0.115\,m + 0.56\,d - 20 = 0.115 \cdot 118 + 0.56 \cdot 130 - 20 = 6.5\,mm.$$

10) Spring moisture reserves (W_2) are: $W_2 = W_1 + \Delta W = 50 + 65 = 115$ mm.
11) W_2 as a percentage of the minimum water capacity is:

$$\frac{W_2}{W_0} \cdot 100 = \frac{115}{180} \cdot 100 = 64\%.$$

12) The average perennial spring moisture reserves (W_{cp}) are equal to 120 millimeters; hence, the moisture reserves, in this case, expressed as a percentage of the perennial average value, are:

$$\frac{W_2}{W_{cp}} \cdot 100 = \frac{115}{120} \cdot 100 = 96\%.$$

Using this method, one can, therefore, calculate the spring productive moisture reserves by correlating their magnitudes with the minimum water capacity and with the perennial average value. This makes the characterization of moisture assurance conditions for spring cereal crops possible.

Equations based on the relationship between the change in moisture reserves during a ten-day period and initial reserves, precipitation, and average atmospheric temperature for that period have been formulated for the purpose of computing anticipated crop moisture reserves.

The general form of these equations is:

$$I = a^x + b^y - cz + m,$$

where I is the anticipated productive moisture reserve (in mm) in a given soil layer, x is the initial productive-moisture reserve (in mm) on the date of the forecast compilation, y is the total precipitation (in·mm) for the forecasting period (ten days), and z is the average diurnal atmospheric temperature for the period. The coefficients of the equation (a, b, and c) are a function of the physical properties of the soil, the heterosis of the soil bed, the species of the plant, and the developmental stage.

In practical application, many authors have adapted these equations to different crops in various soil-climatic zones based on stage of plant development. Yu. I. Chirkov and G. V. Belukhina, for example, in forecasting moisture reserves for corn, have derived over 30 equations for various stages (seedling-tenth leaf, tenth leaf-tasseling, tasseling-young maturity) in different soil beds of a corn belt stretching from northern Caucasus across the Ukraine to the nonchernozem zone.

18.2 Systems for Forecasting Onset Times of Major Crop-Development Stages

Plant growth rate depends primarily on environmental temperature and it increases with limited elevation of temperature. Hence, the onset of a developmental stage will occur earlier in warm weather than in cool weather. This function is expressed by the equation $n = \dfrac{A}{t - B}$, where n is the length of the period; A is the total effective temperature (the total effective temperature is the sum of average diurnal temperatures above the minimum temperature required for plant development) necessary for the onset of a given stage; B is the minimum temperature required for plant development in a given stage; and t is the mean temperature of a given period.

The function referred to above forms the basis of the existing systems for forecasting the onset times of developmental stages. A. A. Shigolev first developed the procedure for this forecast for a number of grain and fruit crops. He established the total effective temperature (above 5 °C) for the principal developmental stages of many strains of grain and fruit. According to Shigolev, the total effective temperature (above 5 °C) necessary for the period between stem extension and heading for the majority of spring wheat strains is between 305°–375 °C; while it is 490°–540 °C for the period from heading to yellow ripeness. In many strains (Lutescens 62, Diamant, Tetcher, Melyanopus 69, and others), these aggregates are 330 °C in the first period and 490 °C in the second. The majority of oat strains require 378 °C in the first period and 428 °C in the second; barley requires 330° and 388 °C; winter wheat, 330° and 490 °C. For most apple strains, the total effective temperature from the start of the growing season to the beginning of blooming is 185 °C (plus or minus 10°), whereas to the end of blooming it is 310 °C (plus or minus 25°). Shigolev assumed the minimum temperature for most plants in temperate climates to be 5 °C.

The forecast of a given stage is calculated with the equation:

$$D = D_1 + \frac{A}{t-5},$$

where D is the anticipated onset time for a given stage; D_1 is the date of onset of the preceding stage; A is the total effective temperature necessary for the onset of the anticipated stage; and t is the anticipated mean air temperature for the interstage period.

As an example, we shall calculate the onset time of the yellow ripeness stage in spring wheat (Lutescens 62) from the heading stage. The date of the onset of the heading stage (D_1) is 20 June. The total effective temperature (A) is 490 °C and the mean temperature of the period (t), determined by the weather forecast or from climatic data, is 20 °C. We insert these data into the equation:

$$D = 20/VI + \frac{490}{20-5} = 20/VI + 32.6 = 23 \text{ July}$$

It should be noted that a soil moisture deficiency, which retards germination, affects the growth rate considerably in the period between sowing and seedlings. Equations allowing for the effect of moisture reserves on germination rate have been calculated for winter wheat by Ye. S. Ulanova and for corn by Yu. I. Chirkov.

Further studies have demonstrated that the growth rate does not increase at high air temperatures. Consequently, the effective-temperature aggregates loose their constancy and are thus a function of the temperature level. Their relative stability is observed primarily within the range of mean diurnal temperatures from 5° to 20 °C, i.e., in northern temperate zones.

In the subtropics and tropics, therefore, it is necessary to include ballast temperatures in the calculation of equations involving the relationship between plant growth rate and atmospheric temperature. For example, Yu. I. Chirkov developed a method of forecasting tasseling onset times for corn which considers the high temperatures of the steppe zone as well as number of leaves as the index governing the early maturation of a strain (hybrid). This method is expressed in the equation:

$$D = D_1 + \frac{(0.101x^2 - 0.5x + 27.4) \cdot (N - 2)}{x},$$

where D is the anticipated tasseling time, D_1 is the date of the onset of the third-leaf stage, x is the average, effective air temperature (higher than 10 °C) for the period, and N is the number of leaves characteristic to the given strain.

Calculations using this equation demonstrate that, with an increase in the mean diurnal air temperature above 20 °C, the growth rate does not increase and the onset of ear emergence does not occur earlier. This is empirically confirmed. A reduction in moisture reserves increases the length of the period between tasseling and cob blooming. The maturation rate for corn, as for other crops, is directly related to the magnitude of the plan diurnal temperatures and the atmospheric humidity deficit.

18.3 Yield Forecast Methods for Major Farm Crops

The relationships uncovered by agrometeorologists between developmental growth and productivity formation of agricultural plants, on the one hand, and meteorological factors, soil moisture reserve dynamics, and features of agrotechnics, on the other, form the scientific basis for agrometeorological yield forecast methods.

By and large, light, heat, moisture, and soil fertility determine yield formation processes. Plant response to meteorological factors varies in relation to the stage of development of the plant. Using a method of biological control developed by F. M. Kuperman, it has been found that productive moisture reserves and air temperature affect the productivity formation of cereal crops as early as stages III and IV of organogenesis (the stages are taken from Professor Kuperman's classification); while the number of grains in spikes (ears) with an equal number of spikelets is a function of moisture provision in stages V and VI. In constructing a matrix of correlation coefficients for selecting forecast predictors, it is therefore necessary to consider the periodicity of the influence of meteorological factors, examining their effects on yield not only as a whole during the growing season, but also with respect to specific interphase periods or stages of organogenesis.

Forecasting Winter Wheat Yield. A forecasting method based on crop strain composition has been developed by Ye. S. Ulanova and has been supplemented in recent years with prognostic equations for the promising new strains, Bezostaya 1 and Mironovskaya 308. Ulanova formulated multifactor prognostic equations for the various developmental stages of winter wheat, beginning with growth resumption in the spring.

Correlation analysis has shown that, upon growth resumption, the closest tie between wheat yield and the number of stalks is in the stem extension stage and that the closest tie between yield and the number of spike-bearing stalks is in the heading stage. An intimate connection also exists with the productive moisture reserves in a soil layer 0–100 cm deep and plant height during heading. The inclusion of these factors, which express the condition of the crop and its potentialities, affords great accuracy to the equations (over 80%).

Of the many multifactor equations derived on the computer, we shall examine only the simplest one.

During the spring (within a ten-day spring inspection period), one can calculate the anticipated productivity (y) of winter wheat per ha, given the data on moisture reserves in millimeters in a soil layer 1 m deep (x_1) and knowing the number of stalks per m^2 (x_2), with the equation:

$$y = 0.059x_1 + 0.024x_2 - 2.97.$$

The multiple correlation coefficient R equals 0.82. If we assume that x_1 is 120 mm and x_2 is 1500 stalks, then

$$y = 0.059 \times 120 + 0.024 \times 1500 - 2.97 = 40.1 \, y/ha.$$

It should be noted that the average forecast compiled with this equation for a regional (district) harvest has a term of about three months, and is highly accurate.

During the stem extension stage, one can calculate the anticipated yield (y) by using the average productive moisture reserves (x_1) present in a soil layer 0–100 cm thick during the period from growth resumption to stem extension, the mean air temperature for that same period (x_2), and the number of stalks per m^2 (x_3).

The equation takes the form:

$$y = -12.8 + 0.29x_1 - 10^{-3}x_1^2 + 0.4x_2 - 10^{-5}x_2^2 - 0.72x_3 + 0.03x_3^2.$$

Its accuracy is 82%.

In Czechoslovakia, the relationship of winter-wheat yield (y) to the mean October air temperature (x) was established and expressed (Chermak, 1967) with the equation:

$$y = -29.92 + 8.09x - 0.5398x^2.$$

Dr. M. Gangapadkhiaia of India formulated an equation defining the dependence of wheat yield (y) on the amount of precipitation falling during specific interstage developmental periods $(a^1, b^1, c^1, etc.)$. The equation, computed with data from the Dzhalgan bureau (state of Makhapashtra), for example, has the form:

$$y = 36.399a^1 + 181.989b^1 + 1356.441c^1 + 573.452d^1 - 1481.055e^1 - 1051.496f^1.$$

The precipitation pattern during the presowing period and throughout the growing season is a factor which determines the size of the wheat yield in the majority of regions of India. Precipitation exceeding the norm during the month before seedlings and during seed germination increases the yield significantly by raising the soil moisture reserve, which ensures the formation of spike-producing elements during stages IV through VI of organogenesis.

Forecasting Spring Wheat Yield. Humidification indices are the controlling factors for spring wheat, which grows primarily in regions with insufficient and fluctuating humidity levels. A. V. Protserov and K. V. Kirilicheva have proposed several of the equations being used for yield forecasting.

The equation for the Skala strain, by way of example, common to Eastern Siberia and parts of Western Siberia, has the form:

$$y = 0.090v_2 + 1.840x_5 + 0.012x_2 - 16.824 \times R = 0.78$$

where v_2 is moisture provision (in percent) from stem extension to heading, x_5 is the number of developed spikelets on a spike during the heading stage, and x_2 is the number of stalks per m^2 during the stem extension stage.

By moisture provision (v_2) is meant the ratio of moisture resources to evaporativity, as indicated in the equation:

$$v_2 = \frac{W_1 - W_2 + r}{0.45 \sum D},$$

where W_1 is the productive moisture reserves in a soil layer 0–50 cm deep during the stem extension stage, W_2 is the moisture reserves during heading, r is the total precipitation for the period between the two stages, and D is the sum of the average diurnal atmospheric humidity deficits for the same period.

The gain-corn yield forecast, developed by Yu. I. Chirkov, is based on the relationship of yield to soil productive moisture reserves and to the leaf surface area of the young crop. It expresses the photosynthetic potential, i.e., the yield potential of a crop as a function of solar radiation. Here, the effect of meteorological conditions on productivity formation of the crop in stages IV through VII of cob organogenesis is taken into account.

The grain-corn yield forecast is built upon equations which include magnitude of leaf surface area (S) in thousands of m^2 per ha, productive moisture reserves in millimeters in a soil layer 0–50 cm deep (W), the air temperature for the period of formation of cob-producing elements in stages IV and V of organogenesis (t_1), and the air temperature during the month following tasseling (t_2), when blooming, cob fertilization, and grain forming occur.

In practice, a system of equations based on the different magnitudes of leaf surface area for the crop during tasseling is used in forecasting corn yield.

The general equation for this has the form:

$$Y = \frac{(-aW^2 + bW - c)Kt_2}{Kt_1 W_1}$$

where Y is the yield per ha with the given leaf-surface area (S), W is the productive-moisture reserves in mm in a soil layer 0–50 cm thick, and the coefficients a, b, and c are a function of (S) given in thousands of m^2 per ha. For example,

if $S = 30$, then $a = -0.0071$, $b = +1.41$, $c = -3.2$;

$\quad S = 20$, $\qquad a = -0.006$, $\; b = +1.1$, $\; c = -4.2$;

$\quad S = 10$, $\qquad a = -0.0029$, $b = +0.53$, $c = -1.5$.

A graphic expression of these equations is given below for application purposes. The term Kt_2 is obtained from Table 4.

When the mean air temperature is above 20 °C and the soil humidity is lower than 50 mm, the term $Kt_1 W_1$ is calculated with the equation:

$$Kt_1 W_1 = 0.065t_1 - 0.016W_1 + 0.46,$$

where t_1 is the mean air temperature in stages IV through VI of cob organogenesis and W_1 is the average moisture reserves for the same period.

Table 4. Correction coefficient at prescribed temperatures
in the course of the month

Storage of productive humidity (mm)	Mean air temperature (°C)				
	16	18	20	22	24
100	0.68	0.86	0.97	1.00	0.96
80	0.72	0.88	0.99	0.98	0.90
60	0.78	0.90	1.00	0.93	0.80
40	0.84	0.93	0.97	0.86	0.55
20	0.90	0.92	0.90	0.80	0.50

Thermal conditions are used in the prognostic equation above as correction factors. However, the reduction in yield can be substantial with high temperatures against a background of insufficient humidification. The atmospheric temperature, allowed for by the term Kt_1W_1, is the observed value. The temperature of the period following tasseling (t_2) is the value predicted in the monthly weather forecast. Let us assume that S equals $30,000\,m^2$ per ha, W is 60 mm, Kt_2 equals 1.00, t_1 is 22°C, and W_1 is 40 mm. Kt_1W_1 is taken as $0.065 \cdot 22 - 0.016 \cdot 40 + 0.46 = 1.43 - 0.64 + 0.46 = 1.25$. We then substitute the obtained values into the general equation:

$$Y_{30} = \frac{-0.0071 \cdot 3600 + 1.41 \cdot 60 - 3.2}{1.25} = \frac{-25.56 + 84.6 - 3.2}{1.25}$$

$$= \frac{55.84}{1.25} = 44.67\,\text{per ha}.$$

The sunflower seed productivity forecast is formulated with a method developed by Yu. S. Melnik and based on the relationship of seed yield (Y) to a moisture-provision factor (K). The prognostic equation has the form $Y = 23.44(K = 0.46)^{0.8}$.

The yield forecast for rice, which is cultivated by flooding, rests chiefly upon data of atmospheric temperature and solar radiation ingress. In Japan, Khaniu (1965) established a connection between rice yield in kilograms per 10 acres (Y) and both the number of hours of sunlight over the 40-day period following heading (S) and the mean air temperature for the same period (t). The equation acquires the form $Y = S[4.14 - 0.13(21.4 - t^2)]$.

Murata (1968) proposed another form of the yield forecast equation:

$$Y = St_1(21.95 - 0.72t_2),$$

where Y is the yield from 10 acres in kilograms, S is the mean diurnal influx of sunlight (cal/cm^2 per day) during August and September, t_1 is the mean air temperature during May and June, t_2 is the mean air temperature during August and September. There are a number of other equations for rice which include data on relative atmospheric humidity, the humidity deficit, etc., along with the above-mentioned parameters.

The prognostic equations discussed (developed in a number of countries for wheat, corn, rice, the sunflower, and other crops) illustrate the various methods of agrometeorological yield forecasting. The inclusion of limiting factors in the equations is characteristic of the existing methods. These factors vary in relation to crop, cultivation technique, soil and climate conditions. For example, equations for arid regions include moisture provision indices (productive moisture reserves in the soil, precipitation, etc.); whereas for rice (cultivated by flooding), atmospheric temperature and solar radiation values serve as the parameter. Data on crop conditions (number of stalks, leaf-surface area, plant heights) are used in an array of methods. The majority of existing theoretical and applied yield forecast methods are based on statistical analysis of agrometeorological observation data and on correlation and regression analyses. The equations derived in these instances, however, refer only to specific regions and cannot be used in others.

Biometeorological models using cybernetics have been developed in recent years in a number of countries (Holland, Soviet Union, United States, Federal Republic of Germany Poland, and others). This trend is promising. Many mathematical models, however, in attempting to represent the complex process of yield formation by allowing for a great many factors (including physiological processes, the stereometry of a crop, subtleties of gas exchange in a grass stand, energetics of photosynthesis, and microflora activity in the soil), cannot be used at the present time to forecast yields in production conditions involving millions of ha. The primary reason for this is the infeasibility of organizing observations of the above-mentioned complex processes and characteristics in a massive network of meteorological stations. The second reason is the efficiency required for synthesizing a forecast (presuming, wherever possible, the use of the simplest, least laborious form of calculation, one which permits the rapid retrieval of vast amounts of information even with a limited number of predictors).

Consequently, the existing methods, based on the statistical analysis of a large stock of interdependent observations, are currently being employed successfully on a practical basis due to the unfailing biological and physical validity of the predictor sample.

Verification of the accuracy of the yield forecasts is accomplished by means of comparing the predicted values with the actual size of the yield recorded by statistical agencies. In the Soviet Union, for example, the average accuracy of forecasts over a ten-year period is 90–92 %.

Further refinement of the existing yield forecast methods requires considerable elaboration of reference point and, primarily, allowance for the patterns of the spatial distribution of the factors comprising the prognostic equations. Increasing the term of the yield forecasts is of prime importance. For that, factors and processes more constant than changeable weather conditions (e.g., soil moisture reserves, developed photosynthetic potential of young crops, etc.) are being employed. However, the wider utilization of climatological patterns with allowance made for periodical climatic fluctuations has promise. The extent of damage caused by phytopathological and entomological organisms, which in itself is related to weather conditions, should be included as a correction factor. International collaboration among agrometeorologists will promote the most rapid refinement of existing yield forecast methods and the development of new methods for many other farm crops. This will be the contribution of agrometeorologists to the rise in world agricultural productivity.

19 Forecasting the Minimum Temperature

B. PRIMAULT

19.1 Need for Such Forecasts

Most human activities are limited by temperatures below certain thresholds. Thus, in construction work, it is impossible to pour concrete below the minimum temperature, which depends on the composition of the concrete. It is possible to lower these limiting thresholds by adding to the mixture chemical products which facilitate the setting of hydraulic cements. In industry, low temperatures also prevent some work, particularly in the autogenous or electric welding of metals. Other industries, however, in order to develop, must have very low temperatures. Thus, in the production of special cardboards, and particularly of printing pads, a much greater uniformity in the alveoli is obtained by freezing the paper paste very rapidly (at temperature of $-10\,°C$ or more). Certainly, those specialized industries could use refrigerated rooms in order to counter the unsuitable climate; but their profitability then becomes questionable.

With respect to housing, low temperatures require a more consistent heating, which means that an estimate of the duration and frequency of such low temperatures, this time on the basis of climatological data, is of primary importance in making possible the selection of heating plants of reasonable size.

It is, however, especially in agriculture that low temperatures are harmful, because very often they cause the death of all or part of the plant tissues. A very sound knowledge of climatic conditions constitutes the basis for the introduction of new crops or the preservation of damaged crops.

In this field, it is necessary to determine the frequency with which the critical thresholds are crossed, and the dates on which such meteorological accidents take place more often.

Forecasting low temperatures, however, involves two aspects which must be dealt with separately: the overall risk evaluation, which is a climatological aspect; and the forecasting properly so-called, which is a synoptic aspect.

The former is a frequency analysis over periods as long as possible of the occurrence of the accident that is feared. Thus, it is a statistical calculation.

The latter is a forecast, which means that it is no longer a case of studying within which limits the accident has already occurred in the past, and therefore may occur again under certain conditions, but of attempting to explore the future: to forecast tomorrow's minimum temperature. It thus constitutes an essential part of the forecasts specially prepared for agriculture, particularly during the seasons of the year particularly critical in this respect.

19.2 Definitions

19.2.1 Cold by Advection

When one attempts to imagine the causes of a decrease in temperature, one thinks most often of the arrival of a mass of cold air invading little by little an entire region. That air, of polar or arctic origin, is driven by the gradient winds prevailing on a continental scale. Generally, it follows a cold front. The decrease in temperature can be gradual or sudden. The final temperature, however, can be estimated in a relatively simple manner by considering the temperature differences between the two masses present.

19.2.2 Cold by Radiation

Any body – solid, liquid or gaseous – whose temperature is above the absolute zero ($-273.25\,°C$ or $0\,K$), constantly emits heat radiation proportional to that temperature. As a result, this temperature gradually decreases as long as this emitted radiation is not compensated for by heat from another source (usually in the form of received radiation). There is therefore a constant exchange of radiated heat among bodies. However, in the absence of counter-radiation (sunshine, for example), bodies, and thus plants, tend to cool off. The rate of cooling depends on the temperature of the body, on its mass, on the nature and color of its surface. Consequently, during clear nights, the air, the ground, and the plant cover will cool very considerably, because their own radiation will be dispersed into the sky, without being compensated for fully or in part by a counter-radiation produced, for example, by a cloud cover.

19.2.3 Cold by Evaporation

In order to lower the temperature of an overheated body, evaporation heat is used very often in industry. Approximately 590 calories are necessary to evaporate 1 g of water. Thus, in the cooling towers, the heat from water evaporation returns the coolant water of thermal plants to an appropriate temperature.

A similar lowering of temperature, also due to the evaporation of water, is, however, also found in nature. After rainfall, the wet bodies dry out. This phenomenon is always accompanied by a considerable decrease in the surface temperature of the body itself. Rainfall or water sprinkling may therefore have very serious consequences if the lowering in temperature caused by evaporation goes beyond the critical threshold. It may even cause the death of the tissues.

19.3 Orographic Effects

In the great plains, the masses of cold air driven by the continental winds advance without obstacles, and that air gradually invades farm areas, depending on wind velocity.

If the decrease in temperature is due to radiation emitted and not compensated for, it is practically uniform over very large areas and the resulting cold air (cooling of the air in contact with the cooled solid bodies) is uniform. The thin layer which results from it is usually very low and does not exceed a few meters.

In mountain areas, the situation changes completely. On the one hand, the invasion of the cold masses is deflected by the mountain ranges, and the latter may produce protected pockets within which the mass only penetrates with difficulty. Only if the cold air mass has a considerable thickness, higher than the passes, or even the summits of mountain ranges, does the inflow reach all parts of a region.

In the case of cooling by radiation, things are quite different. Actually, as a result of the lower temperature of the solid bodies, the air which is in contact with those bodies is cooled. Consequently, its density increases. The colder air therefore tends to flow toward the lower parts of the terrain (ravines, valleys, etc.). Thus, the formation of the so-called cold air pockets takes place.

In its original place, which means around the plants and on the ground located above the upper limit of the accumulated cold air, the latter is replaced by milder air, so that the process is repeated. Thus, the warmer air originating from the nearby environment reheats the solid bodies, which in turn continue to emit radiated heat. Therefore, there is a perpetual motion of heat exchange between the surrounding air and the solid bodies, and the cold air produced by the heating of the plants flows toward the lower ground, while the warm air replacing it is in turn cooled. On the low ground this supply of heat does not exist, although the solid bodies, like the others, by radiating heat continue to become cooler, and this loss is not compensated for at all. This results in a more marked cooling, which means a greater risk of damage. Thus, ravines and valleys are often called "frostbitten areas".

Terrain features are often not necessary to create a cold air pocket. Buildings, the edge of a forest, a live hedge, or even a metal lattice are sufficient to stop, at least in part or temporarily, the flow of cold air toward the plain, and to cause behind the artificial obstacle a small cold air pocket, which may be of considerable economic importance.

19.4 Forecasting the Minimum Temperature

19.4.1 Cold by Advection

By a study of the movements of air masses from day to day, and of the displacement of anticyclonic and low-pressure centers, it is relatively easy to predict the position of weather fronts. Since a cold air mass generally follows a cold front, it is relatively easy to predict when it will reach a certain point. The study of the first decameters of aerological soundings makes it possible to determine what the temperature of the air mass will be when it reaches the critical point. Furthermore, a comparative study of the aerological soundings shows the thickness of the cold mass and, consequently, the mountain ranges which it will directly overcome. Thus, a study of the synoptic charts and of the aerological soundings and the interpretation of the information provided by them make it possible to forecast the minimum temperature in case of an inflow of cold air, which means of cold by advection.

19.4.2 Cold by Radiation

In the stable situations which follow an inflow of cold (and no longer cold) air, the decrease in temperature due to radiation is no longer directly and only a synoptic function. The simple study of the meteorological charts and aerological soundings and their interpretation are no longer sufficient. The former only shows us the initial temperature and the probability of the absence of wind. The fall in temperature is then the result of the cloud cover, or, more precisely, of the absence of cloud cover. A forecast of the minimum temperature is much more difficult in the latter case that in the former. It requires great experience on the part of the forecaster to calculate in which area the cloud cover will disappear during periods of great losses of heat energy (by night). In mountain areas, the forecast will be much more local in nature, and therefore more accurate, because the cloud cover is a factor which varies enormously from place to place. It is affected by the relief and by local conditions, even though the general synoptic conditions may be identical everywhere. Lower temperatures of this kind generally occur in areas affected by anticyclones. Consequently, the study of the velocity of secondary air movements here plays a predominant role. Furthermore, as we have seen above (see Sect. 19.2.2), the formation of cold air pockets adds a further difficulty in determining an exact forecast, applicable to an entire region. Actually, in the case of frost by radiation, very considerable differences in temperature are found over very short distances.

19.5 Work in Progress

The importance of forecasting the minimum temperature has long been recognized in the fields of farming, industry, heating, and ventilation. For this reason, various WMO technical commissions, their working groups, or the scientists reporting to them, have studied these problems. The latest measure in this direction is that taken by the Commission on Agricultural Meteorology (CAgM), which has ordered a massive compilation of the procedures presently used in the various regions of the world by the national meteorological services in preparing such forecasts.

   The correlations used between cloud cover and the decrease in temperature by night are generally empirical. They are based on measurements taken at a very great number of places particularly exposed to the cold or, instead, protected by this natural meteorological factor. By statistical procedures, it is then possible to calculate in advance the most probable minimum temperature on the basis of given meteorological conditions. Those conditions include the synoptic situation on a continental scale, the relative position of mountain ranges in relation to great air currents, the distribution of cloud cover in the afternoon and evening preceding the critical night, and finally the temperature prevailing at that time in the areas to be protected.

   These procedures have so far produced excellent results and have made it possible to save many crops by their practical application. Their main drawback, however, is that they require long series of observations before it is possible to determine the firecasing pattern. For this reason, physicists are looking for a more general method, based this time only on physical axioms. It should make it possible to prepare calculations – by computer and not by man, as in the case of the empirical methods – of the nightly minimum temperatures in figures, and this irrespective of the previous observed data, in other terms of the presence or absence of long series of observations. Such work is in progress, but will still require many years of research and testing.

19.6 Measuring Methods

So far, we have always talked about minimum temperatures, without specifying their nature. However, depending on the application, the subject to which the temperature is related becomes predominant. Thus, we can think of the air temperature in a shelter, of that of solid bodies, or of the temperature measured at different heights above ground level. This distinction is of primary importance in the definition and identification of cold air pockets.

19.6.1 Temperature in a Shelter

The WMO technical regulations provide that temperatures are measured in a shelter 2 m above the ground. Consequently, all the calculations performed on temperatures transmitted by the synoptic or climatological networks refer to those particular conditions. It must be emphasized that measuring the air temperature in a shelter excludes most of the effects of radiation. Consequently, the reported temperatures of a cold air mass will be those of that air, but 2 m above the ground. It is known, on the other hand, that by night as one gets closer to the surface of the ground the temperature decreases. The difference increases as a function of the radiation produced. Therefore, it is inversely proportional to the cloud cover. The difference is practically nonexistent with an overcast sky. Under a clear sky, on the other hand, it may be considerable.

19.6.2 The "Grass-Minimum"

In order to avoid this drawback, a second minimum temperature has been introduced at the synoptic stations. It is the temperature shown by a thermometer placed 5 cm above a lawn, which is called the "grass-minimum". Then the instrument is not protected from radiation by a screen. This second minimum is regularly reported once a day and gives an idea of the importance of radiation by comparing it with the minimum read in the shelter. This second temperature may contribute a new factor in forecasting nightly minimum temperatures, the more so because the thermometer itself radiates. Therefore, there is a double meteorological component contained in the results of the measurement.

19.6.3 VAH Temperature

In central Europe, where spring frosts practically every year cause very great damage, an attempt was made to depart from air temperature measurements in a shelter, because they do not correspond to the temperature of the plants in case of night radiation. An instrument was studied which might simulate what happens to the plant.

Numerous studies have made it possible to demonstrate the similarity in the evolution of the temperature of a shoot of fruit tree or vine, on the one hand, and of the alcohol in a freely radiating minimum temperature thermometer, on the other. The radiating properties of the fluid used in measuring the temperature play a predominant role in this area. On the market, most minimum temperature thermometers are filled not with alcohol but with toluol. Toluol, however, does not have the same radiating properties as alcohol. Consequently, it is not possible to compare directly the "grassminimum" mentioned above with the VAH temperature obtained by this new method (V = Vergleichskörpertemperatur comparature body temperature, A = actinothermic index, and H = Höhe or height). It has been necessary to introduce the concept of height above ground in this method because, in measuring temperatures for purposes of protection against frost, it plays a not inconsiderable role. Thus, to each VAH temperature, a figure will be added which indicates, in cm, the height above ground at which that temperature was measured.

The use of alcohol thermometers in measuring temperatures presents, however, a very important drawback: under the action of sunshine, the alcohol is distilled and often interruptions appear in the column of liquid. Consequently, when alcohol thermometers are used in measuring the minimum night temperature (VAH temperature), it is essential either to remove the instrument in daytime, or to cover it with a projective screen, in order to prevent the formation of vapor bubbles. (It is because of this drawback that alcohol has been replaced by toluol in the production of the common minimum temperature thermometers).

19.6.4 Standard Ground

We have seen under Sect. 19.6.2 that one of the proposed methods to determine the minimum temperature which actually occurred in a given place was that of the "grass-minimum". This method, however, required the presence of a lawn at the observation site. Furthermore, the term "lawn" does not guarantee a sufficiently uniform environment to permit comparisons between one place and another. However, such comparisons should be useful as a basis for the establishment of the physical model referred to under Sect. 19.5.2 above.

Depending on the region, the terrain, and the climate, the composition of a lawn and its rate of growth, and therefore the number of mowings required during the year, are quite varied.

After a long dry spell, the lawn has radiating properties, and therefore minimum night temperatures, totally different from those which it shows after a long period of rain. Should it then be required of the observers that they regularly water their lawn in order to keep it permanently green? This would be an unacceptable interference, because the temperatures measured on that artificial surface would not correspond at all to those actually prevailing in that region. Desert areas present a particular case: should one bring soil there, which will be watered to keep a lawn green? And in polar regions: should the ground be heated to enable the grass to grow?

With respect to the lawn itself, which should be its botanical composition? It is not conceivable that it would be possible to maintain permanent lawns of the same botanical composition in the tropics or in the vicinity of the polar regions. There are so many difficulties, which are practically insurmountable if one wants to obtain comparable results in different places. For this reason, a slow start has been made toward the study of totally artificial surfaces, be they made of concrete, of bitumen, of wood, or of other plastic materials, in order to create everywhere an identical environment, even if it must differ considerably from the natural environmental conditions. It would then be possible to compare the results of measurements taken on this standard (*totally artificial*, we emphasize this point) surface, and to proceed from there to the natural conditions typical of each place. The procedures for passing from the artificial environment to the natural environment might then be differentiated from region to region, and even – a very important advantage – from season to season in the same place. It would then be possible to determine a physical formula corresponding to the artificial environment and to that environment alone, and to insert in it a correction varying from place to place and from season to season. The synoptic services would thus determine a forecast of the minimum temperature of the artificial environment, and the practitioner would apply the appropriate correction on the basis of local experience and the specific properties of his crops.

20 Epidemiology of Insects and Other Diseases

B. PRIMAULT

20.1 Statement of the Problem

All living beings are affected by meteorological factors during their growth. Those which live outside houses or shelters specially built for them feel the effects of the climate. Thus, not only plants, but also insects and some domestic animals are subject to the direct effects of the various meteorological parameters, as they are measured at the observing stations. However, depending on the case, this influence is quite varied. It follows necessarily that the spreading of an epiphytic or epizootic disease will have to be considered differently, depending not only on the type of pathogenic agent, but also, or even more, on the type of host.

20.2 Effect of Weather on Pathogenic Agents

20.2.1 Insects

Since the most ancient times, farmers have feared attacks on their crops by predatory insects. The Old Testament already mentions one of the most harmful parasites in the Mediterranean basin and Africa as one of the greatest plagues: the locust or migrating grasshopper. These insects are doubly affected by the prevailing meteorological conditions. First of all, in their breeding places, that is, where their eggs have been laid, they require particular temperatures and humidity levels for the eggs to hatch and for the larvae to be able to perform their various transformations. The fully grown insect (imago), once ready for its great migration, will have to wait for favorable winds, because its wings, while they enable it to rise in the air, do not however enable it to cover great distances. This only occurs if the swarms of locusts are carried by the wind. Therefore, it is not surprising that the WMO has been dealing with this problem for many years. Special observations have been carried out and many studies of this subject have been published. In this particular case, the study of development alone is not sufficient. It is necessary to add to it considerations concerning the possibility of propagation or dissemination.

Locusts are, however, not the only insects which prey on plants. Therefore, many studies have been carried out over the entire world to determine the most important meteorological conditions affecting their development. Let us mention, among a very great number of examples, the carpocapse of apples and pears, the cherry fly, the cotton pink worm, the corn pyralid, or the rice rhyncote.

One thing is surprising in reading different reports by researchers, and it is that, with a few exceptions, the only meteorological factor affecting the development of insects is the temperature. Actually, each insect requires a fixed amount of

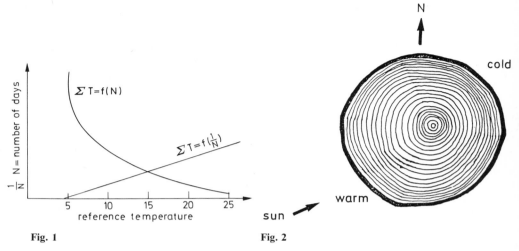

Fig. 1   **Fig. 2**

Fig. 1. Determination of zero growth point (zero vegetation point)

Fig. 2. Temperature differences on a tree trunk (note that the lower the sun is on the horizon, the maro important the difference)

temperature above a given threshold in order to pass from one phenological state to the next: the egg from laying to hatching, the larva from one molting to the next, the nymph from the cocoon to the imago.

In order to obtain these "biometeorological indexes", one can work either in situ, which requires many years, or in the laboratory. In the latter case, the work will be more rapid, but it is more complicated to determine the zero growth level (minimum temperature required for development).

If insects are placed in incubation chambers kept at constant temperature, but different for each group, they require different periods of time (number of days or hours) to reach the same stage of development. By recording that time as a function of the test temperatures, a curve is obtained, which rises asymptotically toward the lowest temperatures. The zero growth level will then be the limiting point, that is, the abscissa of the point whose ordinate is infinite. Often it is very difficult to determine this point with accuracy. Thus, by taking the inverse of the time, a straight line is obtained, which crosses the x axis $\left(\frac{1}{N}=0\right)$ at a point which is then the zero growth point (Fig. 1).

The practical application of these laboratory studies obviously varies from insect to insect. Those which undergo a certain stage of their development in the ground are easier to follow than those which hibernate in the bark of tree trunks or branches. Let us compare, for example, the nymphal conditions of the carpocapste and those of the cherry fly. In the case of the latter, this stage of development takes place in the ground, which is a relatively homogeneous environment where the temperature can be easily monitored, and where measurements at given points are also valid for large areas. It will then be easy to determine, by studying the successive temperatures, the time of appearance of the perfect insects. In the former case, instead, this nymphal stage takes place in the cavities in the bark of trees. The larvae which have woven their cocoon on the south side of the trunks will develop much more rapidly than those whose cocoons are located on the north side. Actually, the radiation from the sun heats the south side while leaving in the shade, and therefore relatively cold, the north side of the trunks and branches (Fig. 2). For this reason, the flight of the insects is spread over a much longer period in the case of the carpocapse than in that of the cherry fly.

With respect to locusts (development stage on the ground), conditions similar to those of the cherry fly are observed.

20.2.2 Fungi

An important group of enemies of our crops is formed by fungi. In contrast with the case of insects, the temperature is no longer the only factor in the development of fungi. The water found on the plant organs and/or the moisture in the air here play a very important, and even predominant, role. Consequently, in order to monitor the development and spreading of a fungus disease in plants, it will no longer be sufficient, as in the preceding case, to measure the temperature and to determine its sum total above one or more appropriate thresholds. It will also be necessary to examine, together with the temperature, the presence of water on the plants (precipitation, dew, etc.), as well as air humidity. Thus, the temperatures will be added up, as in the case of insects, above a certain threshold, but only if a second condition is fulfilled: the air humidity exceeds a certain level and/or the sensitive organs are covered by a film of water.

For many fungi, the germination of the spores is actually determined not only by a certain temperature, but also by the presence of water in liquid form on the plant. Consequently, the study of the conditions for germination will involve the measurement or observation of a particular condition of the plant: the presence of water on its organs (wet leaves).

Depending on its development cycle, the fungus can spread in different particular forms. Each one of them will be the subject of a special study.

20.2.3 Bacteria and Bacilli

Certain diseases, particularly of cattle, are caused by bacteria or bacilli which develop inside the organism of the host. These pathogenic agents are absorbed with the grass and eliminated through the feces or the urine (in some cases, through the saliva or local exudation).

Upon leaving the organism of the host, these pathogenic agents assume a particular form which enables them to survive under external conditions which are harmful to them, particularly outside a watery environment. In this temporary form, they remain in a latent condition, before being absorbed by new hosts, which they infect.

In this particular state, however, which gives them a great resistance to a hostile environment, the meteorological conditions may have an unfavorable effect. Generally this happens with dry spells – since certain bacilli or bacteria need water to survive, such as, for example, those of cholera – heat or cold. Thus the prevailing weather may act as a disinfectant for prairies and pastures.

In the case of bacterial plant diseases, the pathogenic agent is carried from one host to the next by insects. We have then reported under item 20.2.1 with respect to the development of an epiphytic disease.

20.2.4 Viruses

As in the case of bacteria and bacilli, one must distinguish two aspects of the infection, depending on the nature of the host: plant or animal. A plant is generally infected by viruses when it provides sustenance to insect vectors. In the case of animals, the viruses pass from one host to the next either through the food, or through the respiratory system. Therefore, there is a direct infection.

As in the case of bacteria, however, the virus vegetates in a latent state outside the host. It may be destroyed by particular meteorological conditions, especially ultraviolet rays. Thus, generally, a prolonged period of good weather ends an epizootic of foot-and-mouth disease. The meteorological conditions at the time may, however, also be favorable to the spreading of the disease, either by increasing the virulence of the pathogenic agent, or by decreasing the natural immunity of the host.

20.2.5 Other Pathogenic Agents

Among the pathogenic agents different in nature from those examined so far, and which are most affected by the meteorological conditions, we shall particularly mention worms.

A close correlation has often been found between the development of an epizootic due to parasite worms and particular meteorological conditions. We shall mention here as examples the liver fluke and the ascarides.

20.3 Forecasting Development

As research develops on both the development of the pathogenic agents and the host reactions, the meteorological services are increasingly contributing to the preparation of special forecasts concerning the spreading of the disease. The forecasts generally concern the timing of the appearance of the parasite (qualitative forecasts), in order to make possible more specific treatment, and therefore more effective, but their main purpose is to decrease the number of chemical treatments. Those forecasts may also concern the size of the parasite population (quantitative forecasts). Both qualitative and quantitative forecasts are essential in an integrated campaign (see Sect. 20.4).

20.3.1 Insects

The struggle against a predatory insect generally is limited over time, because the parasite only reacts to the chemical products at a well-defined stage in its development. Thus eggs, larvae, and perfect insects will be treated with different products and appropriate concentrations. Consequently, in order to be able to fight in a sensible, which means economical, way, it is necessary to know when the form particularly sensitive to a product is actually present on the farm. For this purpose, meteorological measurements and their interpretation are much more effective than the observance of a strict schedule. Actually, in view of the fluctuations in time from one year to the next, the most appropriate time for a treatment often varies within a very long period.

20.3.2 Fungi

As stated above, the development of fungi is determined by given temperatures and humidity levels, typical of each species. Since the humidity is generally reached at night (most often, relative humidity levels are involved, which then depend, among other factors, on the temperature of the surrounding air), the climatological observations performed in daytime are, in this case, of only very limited usefulness. It will then be necessary to rely either on much more frequent observations, or on recordings, which involve complex and delicate equipment. The use of automated observing stations and of small programmable computers (microprocessors) may provide a not negligible help in the preparation of such forecasts.

Here, as in the case of insects, the time of treatment will depend above all on meteorological conditions favorable to the development of the parasite.

20.3.3 Bacteria and Bacilli

In Sect. 20.2.3, it was shown that bacteria and bacilli are little affected by meteorological conditions at the time of their action on the plant or animal. The spreading of an epizootic or epiphytic disease will largely depend on the potential development of vectors. In most cases, they are insects (see then Sect. 20.2.1).

Since, however, particular meteorological conditions may destroy the pathogenic agent, the national meteorological services may prepare development forecasts. The forecasts will however be only of a negative nature, which means that they will be limited to an evaluation of the losses suffered by the population. They will not be able, instead, to show how many individuals have survived, but only to estimate the percentage of individuals which have been destroyed.

20.3.4 Viruses

In the case of viruses, the disease can be transmitted either by insect vectors or by the wind. In the case of plants, the former phenomenon generally applies; in the case of animals, the latter.

The virulence of the viruses themselves can, however, be affected by certain particular meteorological phenomena (atmospherics, among others). The rapidity of the spreading of the disease after the appearance of the first symptoms – which spreading generally takes place in successive waves – may be estimated by monitoring very closely the evolution of particular meteorological phenomena (number of atmospherics, pressure fluctuations, etc.) and of the wind.

Until the last few years, the idea that viruses might be carried by the wind in the form of aerosols, that is, as isolated viruses, had always been rejected. Recent experiments, however, have demonstrated that very often pathogenic viruses are present in the air in an endemic state. Furthermore, we have pointed out above that ultraviolet rays have the property of destroying the protein molecules forming the viruses. Consequently, the meteorological forecasts prepared to predict the spreading of virus epizootics (direction, area involved, time) will have to take into account not only the wind, pressure, and atmospherics, but also, or even more, the cloud cover.

20.3.5 Other Pathogenic Agents

The other pathogenic agents are also affected by meteorological conditions. Consequently, again in this case, special forecasts can be prepared to provide the herds with immunity, protecting them from massive attacks.

20.3.6 Reliability of Forecasts

However, in both cases, the "forecasts" are not based on an evaluation or estimate of the future evolution of the weather, but only on observations covering a period prior to the broadcasting of each forecast. Therefore, they are "forecasts" quite different in nature from the ordinary meteorological forecasts, although they are also "meteorological". To prepare them, it is necessary to have a dense network of observing stations in the regions where the parasites or the pathogenic agents develop (they are rarely regions where airports are located). The observation posts will immediately report their figures to the collecting center, which will process them as appropriate. Certainly synoptic messages – intended primarily for the preparation of ordinary and aviation meteorological forecasts – can be used for this purpose. Their accuracy, however, particularly with respect to temperature and humidity, is generally insufficient (temperature reports in whole degrees and humidity reports in the form of dew point, again in whole degrees).

The processing of the information to determine the development of the pathogenic agents or parasites is generally sufficiently simple not to require highly qualified personnel. Class 3 technicians or meteorologists are perfectly capable of doing this. Since the forecasts have a limited space coverage, because they are only applicable to relatively small areas, it is necessary to establish a large number of centers located in the vicinity of farming areas, on the one hand, and of the meteorological observing stations, on the other. Thus, the reliability of the information in the opinion of the user will be enhanced: he may personally know the person who processes it. In his mind, it will not be an elucubration coming from a scientist "up in the clouds", and therefore totally unrelated to the real world.

20.4 The Integrated Campaign

All predators and all their enemies are present endemically in the natural environment, but they remain balanced. Only the breaking of this balance is dangerous. The breaking point may be reached either by the explosion of one of the populations (the predator), or by the reduction or even the disappearance of another (their enemy), which in the end is the same thing. Therefore, the struggle against parasites must aim not at the elimination of harmful parasites, but at maintaining them at an economically tolerable level.

With the development of modern chemistry, the farmer is tempted to use increasingly specific, but also increasingly toxic products, to fight parasites harmful to his plants or to his animals. Those toxic chemical products (at least the last mentioned) destroy the undesirable parasites, but also other insects, which are instead useful in the natural biological community. An imbalance thus appears, which imbalance requires the use of products increasingly dangerous to man and to his environment. A modern trend aiming at a greater respect for the dynamics of ecosystems advocates a "return to nature" in the fight against parasites as well. This basic idea has given rise to the so-called integrated campaign, which means the use of natural products or beings in order to keep the parasite populations at an economically tolerable level.

The integrated campaign is presently used profitably especially against insects. Some studies, however, lead to envisage integrated campaign systems also to fight fungi or worms.

20.4.1 Insect Against Insect

In order to prevent a proliferation of one species of insects harmful to all, nature acts in two different ways: self-destruction and the proliferation of enemies.

The first method consists in the destruction of the plants on which they feed by the insect population itself. As the number of individuals increases, the possibility of feeding each one of them consequently decreases. A cyclic pattern is then established. The population gradually becomes so numerous that it destroys its entire food reserve before the individuals have reached their sexual maturity, and thus before they can reproduce. A very large part of the population thus dies of starvation, and the few individuals which manage to survive (the strongest) are called upon to preserve the existence of the species. Since the predator population then is very limited, the plants can recover and proliferate before a new wave of predators comes again to destroy the entire harvest.

In the second case, the cyclic pattern of the population of parasite insects originates from fluctuations in the population of the natural enemies of the predator insect itself. Actually, at a particular stage in their development, other varieties of insects feed on the insect which we would like to see disappear in order to protect our farms. By artificially supplying a large population of insects which are enemies of those who prey on our crops, it is possible to activate this cyclic pattern and to maintain the latter at an economically acceptable level. This method, however, will never permit the eradication of the harmful insects, but will only keep tham at a tolerable level.

A further method – this one not a natural one – to reduce the population of so-called harmful insects is to release previously sterilized males shortly before the mating season. They "fecundate" the females, which then lay eggs which will never hatch.

In both cases, a thorough knowledge of the physiology of insects is essential. It must, however, be complemented by studies concerning the action of meteorological factors on their development. Such development is then closely monitored in two parallel ways. In the laboratory, both the enemies of the predatory insects and the males of the parasites themselves, which will be sterilized as they reach sexual maturity, will be bred. This breeding must be carried out very carefully, so that the counter-measure (the insects) is available at the desired stage of development at the precise time when, in nature, the parasite is most vulnerable to its action. By alternating periods of warm and cool environmental conditions, is is possible to control the development of large amounts of insects. It is also possible to stop that development just before the critical stage. The insects are then kept in reserve for several days, or even weeks, in order to bring them to the desired stage of development only precisely when, in nature, the predator to be fought is also ready.

The results thus obtained are surprising and make it possible to state that, in a short time, it will be feasible to maintain predator populations at an endemic state sufficiently low to be economically tolerable.

20.4.2 Bacteria of Bacilli Against Insects

Insects are not the only natural enemis of other insects. Just like the higher animals, insects are subject to attacks by bacteria or bacilli. The latter are transmitted either directly from individual to individual within the same species, or through the intermediary of other species which act as vectors, as we have seen above in the case of plant diseases.

Then those bacteria will be bred in the laboratory and will be caused to infest either the insect vectors, or insect of the same species intended to contaminate their companions. At the selected time, the disease-carrying insects will be released into the environment and will spread the destructive agent.

20.4.3 Viruses Against Insects

Just as in the case of bacteria or bacilli, viruses can be used in an integrated campaign in order to keep populations of predatory insects under control. The operating methods will be the same as those described above (Sect. 20.4.2). In order to multiply the viruses, it will be necessary to keep in the laboratory populations of the insect to be fought, which will be contaminated and will be used to multiply the viruses.

20.5 Effect of the Weather on the Hosts

So far, we have only considered the populations to be fought and the natural means that can be used for that purpose. However, when a parasite insect or a pathogenic agent, whether it be bacteria, bacilli, or viruses, is present in a region in an endemic state, there it generally attacks weakened hosts. Thus, in a forest, xylophagous or phytophagous insects will first attack trees suffering from diseases or drought or grown in the shade of their neighbors, and therefore weak. The same applies to bacteria in the case of domestic animals or game. They first attack wounded, old, or weakened animals.

However, the weakening of an individual, which causes its susceptibility to a pathogenic agent (bacteria or viruses) may be induced by particular meteorological conditions. In the case of man, for example, a cold (attack by a virus) will follow exposure to cold weather, which is a precise meteorological parameter in a certain direction: low temperature.

The same applies to plants and animals. If the environmental meteorological conditions are not the optimum for their development, their metabolism is altered and their natural immunity to pathogenic agents is thereby impaired. Therefore, they will constitute a particularly suitable prey.

Therefore, the special meteorological forecasts designed for the struggle against epiphytic and epizootic diseases must not be limited to the harmful agents, such as insects, bacteria, bacilli, viruses, worms, etc., but must also take into account the host: the plant or the animal. In both cases, a careful examination of past meteorological conditions makes possible to predict whether a host population is particularly exposed or not to the attack by a specific parasite.

21 Water Requirements of Plants

J. Seemann

No concrete answer can be given in respect of water requirements of plants. The amount of water required by plants for their development depends on the most varying conditions. Among these are, first of all, the plant itself, according to its type and stage of development, the soil with its differing properties and, not least, the meteorological conditions in its location.

The plant consumes water, essentially for two processes, first for photosynthesis (6 molecules H_2O per molcule $C_6H_{12}O_3$), and secondly for respiration. Water take-up takes place almost exclusively through the roots and primarily through the root hairs. These aspire it osmotically in the same manner as every other plant cell. The root hairs will take moisture from the ground until they are either saturated or for as long as their absorptive strength suffices to extract the water that is held by the soil. The latter presents no difficulty as long as there is enough moisture in the soil. This water that is available to the use of the plants is that amount that lies between field capacity and wilting point. This portion is called usable capacity (UC). The field capacity (FC) is known to be that maximum amount of water that is capable of being held by the soil in opposition to gravity. The wilting point (WP), in contrast, corresponds to that amount of water that opposes the absorptive strength of the plant. The amount of usable capacity differentiates itself in different soils, it is the smallest in sandy soils, the highest in loam and loess (see Table 1).

Most plants obtain by far the highest percentage of their water requirements from the upper soil layers. Deep-reaching roots are generally to be considered only as "life insurance". The primary supply zone, i.e., the soil layer from which the plants obtain up to 90 % of their water, can be established for grass to be the upper 25 cm, for potatoes and beets, the upper 60 cm and for grains the upper 50 cm. The deep-reaching roots of grains and beets can, however, penetrate to as far as 1.5 or 2 m into the ground. The useful capacity (UC) in the primary root zone, is called the root space capacity (RC). Table 2 presents the date on root space capacities for several cultures on a variety of soils. One can conclude from many years of

Table 1. Field capacity (FC), wilting point (WP), and useful capacity (UC) for a 10-cm thick soil layer (in mm)

	FC	WP	UC
Sand	13	3	10
Loamy sand	22	4	18
Sandy loam	27	5	22
Loam, loess	36	10	26
Clayey loam	34	14	20

Table 2. Root space capacity (RC) of some cultures on different soils (in mm)

Soil type \ Culture	Grass and flat-growing vegetables	Potatoes, beets	Grains
Sand	25	60	50
Loamy sand	45	100	90
Sandy loam	55	130	110
Loam, loess	65	150	130
Clayey loam	50	120	100

Table 3. Optimal water supply of various soils (80–50 % RC) for sume cultures in mm ($=1\cdot m^{-2}$)

Soil type \ Culture	Grass and flat-growing vegetables	Potatoes, beets	Grains
Sand	20–12.5	48–30	40–25
Loamy sand	36–22.5	80–50	72–45
Sandy loam	44–27.5	104–65	88–50.5
Loam, loess	52–32.5	120–75	104–65
Clayey loam	40–25	96–60	80–50

experience, especially in field sprinkling technology, that the optimal water supply of agricultural plants is attained when the root space capacity (RC), is between 50 and 80%. In the presence of soil water contents in a range above 80% RC, a detrimental effect on the growth can be expected because of lacking air in the soil. Down to a RC of about 50%, enough water is available to the plant for potential transpiration. At a RC below 40%, yield-limiting damage can already occur. Table 3 compiles the values for optimal water supply for various soil types and some agricultural plants.

If the optimal water supply remains maintained in the soil by natural precipitation or by means of artificial irrigation, the water requirement of the plants is assured. This, however, says nothing about the actual water requirement of the plants. It is, as we consider the already described water household conditions of the soil as being a fact, determined primarily by evapotranspiration. The primary factors that exert an influence on evapotranspiration are the meteorological conditions (temperature, radiation, humidity, and wind), and the behavior of the plants in regard to the control of transpiration. Control of transpiration of the plant takes place, on the one hand, by the opening and closing of stomata and, on the other, by processes in the plant cells that have a detrimental effect on water transport in the plant. These processes are described more closely in the chapter on the greenhouse climate (Chap. 10; evaporation and water consumption). Thus, for example, a high heat conversion on the leaves causes the plant to enter into a stress situation in respect to its water household. These

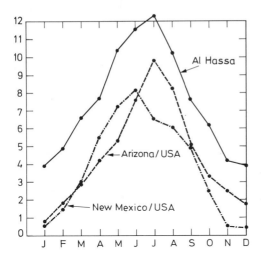

Fig. 1. Water consumption of alfalfa between January and December in various climatic areas. (From Ohlmeyer and v. Hoyningen-Huene, 1975)

considerably influence the growth processes, especially inasmuch as photosynthesis is closely coupled to transpiration.

How differently great the water consumption of plants can be in different climatic regions, can be seen from an example on alfalfa in Fig. 1. In the arid climate of the Oasis Al Hassa (Saudi Arabia), the actual evapotranspiration rises in July to 12 mm/day. In the agricultural areas of Arizona, the maximum values also is in July, but it only amounts to 10 mm/day. Only 8 mm/day can be determined as maximal water consumption in New Mexico, where this highest value for the year occurs in June.

In agriculture, the water requirements of plants are often expressed by the "transpiration coefficient". Because of the strong coupling of photosynthesis to transpiration, it is to be expected that certain relationships must exist between the water quantities consumed (T) during the course of a vegetative period and the dry substance (A) produced by the plant during the same time. The transpiration coefficient of a plant type $(C = T/A)$, indicates the number of liters of water that the plant requires in order to produce 1 kg of dry substance. Such determinations have been carried out on a broad basis in agriculture. The values determined for a specific agricultural plant are, however, characteristic within a specific climatic area. Table 4 shows transpiration coefficients for various agricultural plants in the continental, moderate climatic zone.

In general, the transpiration coefficient of agricultural plants is, on an average, between 400 and 600. In lighter soils in dry climates and in poor nutrition supply, these coefficients will be higher in one and the same plant than on heavier soils, cool-moist climate, and well-fertilized soils. Larger fluctuations of the values of the transpiration coefficients can naturally also be determined in the same location and with the same plant-type from year to year, as a function of the corresponding course of the weather.

Table 4. Transpiration coefficients of agricultural plants

Rye	685	Potatoes	636
Wheat	435	Beets	397
Corn	368	Alfalfa	844
Millet	293	Cotton	462

Table 5. Increase of water consumption as a function of harvest yields of alfalfa

Yield t/h	% of maximum yield	Increase in (%) water consumption
1.12	40	0.0
1.40	50	16.0
1.68	60	17.2
1.96	70	23.5
2.24	80	31.0
2.52	90	45.4
2.80	100	125.0

It becomes clear with a more precise study of the transpiration coefficient, that the transpiration of the plant must be considered to be unproductive to a large part. The transport of mineral salts from the roots to the leaves takes place, as is known, by way of the transpiration stream. For this, however, such a strong water consumption would not be required. There does not exist any direct relationship between salt intake and transpiration. Often, the usefulness of transpiration is overly accentuated, even in textbooks. This often produces a false picture of the actual relationships. In respect of the production of the plant, it is not of such great importance that it converts a great deal of water by transpiration, but the fact that, based upon meteorological conditions, the plant is often forced into high transpiration and, therefore places heavy requirements on the water supply in the soil, something that then has a detrimental effect in situations of insufficient water re-supply by rain or irrigation. Too small a water consumption is not detrimental to plant production, but instead, too small an availability of water. The actual water requirements of the plants – differing as they are, dependent on the course of the weather – can be covered by a sufficiently large water supply. Water requirements are always higher in an arid climate than in a humid one. The fact that the amounts of the yields are dependent, largely, on water availability, can be shown by means of the values in Table 5. These represent planting experiments and their results on alfalfa in the Oasis Al Hassa in Saudi Arabia (Ohlmeyer and v. Hoyningen-Huene, 1975).

Up to an increase of the yield to about 70% of the maximum possible water consumption increases within sensible limits. For an increase of the yield from 90 to 100%, however, so much more water is required that the requirements for artificial irrigation, by means of which in an arid climate an appreciable yield increase can only be attained, are surely uneconomical.

Literature

Ohlmeyer, Hoyningen-Huene, J. v.: The problem of the prognosis of water consumption of a planting, presented on the example of the extremely aric climatic conditions of the Oasis Al Hassa/Saudi Arabia. Mitt. Leichtweiß-Inst. Braunschweig *40* (1975)

22 The Climate of Agricultural Produce in Transport and Storage

J. SEEMANN

22.1 Problems in Transport

Special importance must be assigned to ship-transport, along with railway and truck transportation. Ship loads are under way commonly for extended periods of time and pass through various climates. All goods that are being transported carry their temperature and moisture into the carge hold. Thus, temperature and humidity of the cargo space atmosphere are already, indirectly, determined by growth, harvest transport to loading and the conditions during possible prior storage. It is important to consider the fact that the goods to be transported generally have a high heat capacity.

In the cargo hold, therefore, the climatic conditions are primarily dependent on the difference between temperature and water content of the load and the dew point of the surrounding air. Since the loads always take in a large part of the cargo space, they will largely determine the storage climate. Hygroscopic goods always strive for an equilibrium condition in their moisture content. Intense and rapid changes of the temperature and the air humidity disturb this equilibrium during the trip. The higher the speed of travel through various climatic areas, the greater also is the danger of damage to the cargo. A change in weather with strong temperature changes during extended berthing periods of ships in harbors also endangers the load.

The most dangerous situation with which transport climatology must concern itself is the formation of condensation. During a transport of goods from cold into warm areas, condensation causes moisture deposits on the cold goods in the load. This occurs when the dewpoint of the surrounding air becomes greater that the dewpoint of the temperature in the cargo hold. This results in corrosion on metallic goods and caking and sticking in products, such as fluor and sugar. The temperature differential becomes especially pronounced by passage through cold drift waters along the western coasts of subtropical continents, whenever ships, coming from cold ocean areas enter tropical waters. Höller (1966a) has measured, according to his reports concerning the main shipping route around West Africa, in an extreme case as much as 12 °C temperature change within 2 h of travel. The change is, however, considerably smaller on the average and amounts to 0.8–2.5 °C per day. Ventilation of the cargo hold is to be recommended in order to avoid condensation, however, only for as long as the dew point of the outer air remains below the temperature of the cargo. If ventilation does not take place over extended periods under the given conditions, the danger upon opening the holds in the tropics becomes especially high, primarily when off-loading consumes an extended period of time.

Differing radiation conditions can also promote undesirable condensation processes. Appropriate knowledge is available from Höller's (1966b) investigations. Thus, for example, on a trip from Europe into the tropics, one side of the ship was continuously exposed to solar radiation. Increased temperature on the sunny side led to evaporation from the relatively moist cargo. The water vapor then condensed on the correspondingly colder cargo on the side that was turned away from the sun.

In voyages from the tropics into moderate zones, a different type of condensation can occur, ship's condensation. This danger is especially great with vegetative cargo with high moisture content and correspondingly low dew point difference. This ship's plate cools off rapidly be heat conduction with reducing temperatures and, in doing so, drops below the dew point of the air in the hold. Condensation is produced along the plate of the ship. The cargo, continuing to evaporate water in order to establish an equilibrium, promotes the condensation on the walls. This, in a manner of speaking, pruduces a drying out of the cargo.

Great temperature differences between harbors and the open sea must also be counted as danger points during voyages from the tropics into moderate zones. Höller was able to determine this (Höller, 1966c), especially on trips through the North American Lakes. Ship's condensation can also occur during radiative nights, when the ship's hull cools off more rapidly.

The removal of this effect by ventilation, perhaps during the night, is possible only with larger dewpoint differences. During transports into different climatic regions, the temperature of the cargo should, as much as possible, be adjusted to the temperature expected at the end of the voyage. This, however will be difficult to attain in bulk loads, since, in any case, it will only be the upper layers that are heated or cooled. If the formation of ship's condensation has already started, and one wishes to eliminate this by way of intensive ventilation, the temperature discontinuity layer, and, hence, the condensation zone, is shifted from the ship's plate along with the cool ventilation air, into the surface of the cargo. This, in turn, results in cargo condensation. The magnitude of the damage that will occur will depend on the temperature difference between cargo and the cool outside air.

Added to the simple loading into the hull of the ship, is, in recent years, shipment be means of containers. Here also, there are dangers from condensation. The container is a miniature cargo hold. In the case of air-tight containers, a container-specific climate is formed, whereby the container hull protects the content against processes within the hold. Containers loaded on the deck, however, are fully exposed to the weather. Because of the lacking ventilation, the dangers are, therefore, even greater than in the hold. Added to this is the fact that the proportion of remaining air is considerably smaller and that hence greater limitations are set for equilibrium conditions of temperature and humidity.

Vegetable cargoes are damaged not only by condensation. A humidity of 80 % can already support the development of mold and bacteria. During the formation of mold and the spoilage of the goods by microorganisms in these ranges, further moisture enrichment takes place which in turn supports further biological degradation. Temperature rises that result from this lead to selfheating. The heated cargo again becomes the culture medium for further microorganisms, and burning of the cargo is at the end of the line of development. Even small amounts of moist cargo, that in their total are not very consequential, can initiate this process. Even the destruction of the temperature-humidity equilibrium can reduce the quality of fruits without the formation of water or mold (Höller, 1967). The climate of the hold should, therefore, remain as constant as possible during the voyage.

In order to eliminate damage, one utilizes modern air conditioning equipment along with ventilation. These see to the appropriate temperature, humidity, and ventilation, depending on the environmental climate. In this connection, the CO_2-content of the air must also be observed. It should not exceed certain limits. $\leqq 2\%$ has found to be suitable for apples and $\leqq 10\%$ for bananas.

22.2 Storage Climate

Agricultural products can spoil for physical and chemical, as well as biological reasons. The processes that lead to destruction are essentially controlled by temperature and humidity. It can be stated that, generally, products spoil more readily at higher temperatures than at lower. Certain laws must be observed in respect of humidity. The stored goods attempt to establish a moisture equilibrium with the surrounding air.

Two types of storage can, in principle, be differentiated:

a) Dry storage, when a drying process is introduced between harvest and storage, such as, for example, hay, grains, cotton, and dried fruit.

b) Moist storage when agricultural products are stored in their fresh condition, such as fruits, vegetables, potatoes, and flowers.

22.3 Storage of Grain

For the storage of grain, that is to be treated more closely in this context, it is necessary that the grain moisture be reduced to 10–14% by means of a prior drying process. This inhibits germination on the one hand and it also stops the respiration processes. Respiration leads to the consumption of assimilate materials stored in the grain, and produces heat in the store, which, in turn and in combination with moisture, supports the development of bacteria and fungi. The protection of the dried grain directs its primary attention against damage by insects and against heat nests. Insect development is reduced through a water vapor content below 9% and temperatures below 5 °C. It is especially supported by moistures above 13% and temperatures around 20 °C. Heat nests of about 30° or 40 °C are especially dangerous because these cause convection currents to start within the stored goods. If in this manner moist outside air reaches into the bulk of the grain, this can cause condensation, which, in turn, forms the precondition for the storage damage that has already been mentioned. Condensation streams (flows) can also be initiated by temperature fluctuations. Daily temperature fluctuations reach about 15 cm into the depth of the grain, annual fluctuations as far as 3 m. Strong heating of the stored material produces upward movement along the walls of the store and a down-flow of air in the center. Forced ventilation during periods with advantageous outside temperatures and humidity counteract store-damage. In hot climates, the air must generally be pre-cooled. A detailed presentation on the storage climate for grains is available from Smith (1969). The storage of other dry products generally takes place in a manner similar to grains. For this, Villiers has compiled a summarizing presentation (1968).

22.4 Storage of Fruits and Vegetables

It is the aim of extended storage to limit, as far as possible, the respiratory processes and ripening processes of the fruits, in order to avoid age-related decay occurrences. The requirements that must be made upon the storage climate for the fruits in respect of temperature, humidity, and air composition, vary considerably according to type and kind.

Many investigations are available concerning the optimal conditions of the parameters of the storage climate. Herregods (1963) and Lyons and Rappaport (1959) give suggestions for temperature conditions in vegetable storage. Fischenich et al. (1959), as well as Thiel (1961) and also Jenny (1959), have carried out temperature investigations on potatoes. The investigative results of a working group on the construction and equipment of storage spaces for fruits and vegetables have been published by Henze (1972). Here, one can find, in addition to suggestions on technical problems of storage, also extensive tabular data on temperatures, humidity, composition of the atmosphere and storage duration. Table 1 presents data on the storage conditions for some fruits and vegetable types. They have been derived from the general literature.

Temperatures that are too high, or too slow in cooling, do not sufficiently inhibit the ripening processes. The consequences of this are decay, peel- and meat-browning, fruit rot and susceptibility to parasite infestation. The more "short-lived" a fruit or vegetable sort, the more rapid must be the cooling-off, especially for vegetables. High humidity (generally between 80 and 95%) is to prevent, primarily, shrinkage, and wilting.

Temperatures, humidity, air supply and its composition can be controlled in storage spaces. According to the technical possibilities, one generally distinguishes three methods of climatic control of storage areas:

  fresh-air cooling,
  machine cooling, and
  machine cooling with controlled atmosphere.

In the case of fresh-air cooling, colder outside air is merely supplied to the store. Seemann (1954) has developed a special thermostat for this purpose. It starts up the fans only when the outside temperature is lower than that of the storage space and stops the addition of air when the desired storage temperature has been attained. The application of fresh-air cooling can, of course, only be recommended in climatic areas where at the start of storage temperatures occur frequently enough that are near to the storage temperature or even below.

In machine-cooled storage, the course of temperature is dependent upon the cooling output of the machine, the heat penetration of the walls, and the respiratory value of the stored material.

Some problems will, however, develop even with machine-cooling. Even though fruits or vegetables can be brought to an optimal storage temperature

Table 1. Storage conditions

Stored goods	Machine cooling			Storage with controlled atmosphere			
	Temperature (°C)	Relative humidity (%)	Storage duration (W = Weeks, M = Months)	Temperature (°C)	CO$_2$ (%)	O$_2$ (%)	Storage duration (W = Weeks, M = Months)
Apples	1 to 4	90–95	3 to 6 M	1 to 4	2–4	3	6 to 8 M
Bears	− 1 to 0	90–95	2 to 5 M	− 1 to 0	2–4	3	6 to 7 M
Stone fruit	About 0	90–95	2 to 5 W	0 to 2	3–5	3	2 to 6 W
grapes	− 1 to 0	95	1 to 6 M	−1	3	2	to 6 M
Vegetables (general)	About 0	92–95	1 to 4 W				
However							
Egg plants	8 to 10	92–95	to 2 W				
Green beans	7 to 8	92–95	to 1 W				
Cucumbers	7 to 10	92–95	1 to 2 W				
Paprika (green peppers)	8 to 9	90–93	2 to 3 W				
Tomatoes (half-ripe)	12 to 14	85–90	to 3W				
Potatoes	4 to 5	92–95	8 M				
Oranges	6 to 8	85–90	3 to 5M				
Tangerines	6 to 7	85–90	to 6 W				
Lemons	3 to 4	85–90	3 to 4 W				
Pineapple (hard ripe)	over 10	90–95	to 5 W				
Bananas	13 to 15	95	to 2W				
Avocado	10 to 12	90	to 4 W				
Grapefruit	8 to 15	85–90	to 3 W				
Watermelon	5	85–90	3 to 4 W				

Table 2. Storage conditions for cut-flowers

	Temp. (°C)	Storage duration	Remarks
Asparagus	1 to 2	2 Weeks	
Chrysanthemum	0 to 2	3 Weeks	
Dahlias	5	3–5 Days	Store in full bloom
Lilac	2	10 Days	
Gladiolus	7	8 Days	
Narcissus	About 0	2 Weeks	Store as green buds
Carnations	2	3 Weeks	Back in foil
Roses	About 0	10 Days	in bud-stage
Tulips	About 0	4 Weeks	In bud-stage

(1–2 days), at the same time it is usually difficult to maintain the storage climatic conditions on an equal basis throughout the storage area. This is caused by a generally insufficient turn-over of the air. In addition, condensation will occur frequently on the cooling surfaces. In this manner, in the closed storage area, water vapor is removed from the air, which, in turn, causes the stored material to react with increased transpiration. In the machine-cooled storage area, there must, therefore, be equipment for humidity control. In addition, such storage must be provided, within certain periods, with fresh air in order to avoid an undesirable accumulation of respiration products.

In machine cooling with controlled atmosphere, the effect of ripening in-hibition is increased by an increase of carbon dioxide contents in the air, and a reduction of the oxygen content. In this manner, the respiratory process of the stored goods is inhibited. Frideghelli and Gorni (1966) as well as Marcellin et al. (1965) have published appropriate experimental results in connection with this problem. Within a storage room with controlled atmosphere, the CO_2-content of the air should, generally, be between 2 and 4 %, and that of oxygen approximately 3 %. Too high a content of carbon dioxide can lead to discoloration of the fruit-meat and to tears or hollow spaces within the tissue. Too low a content of oxygen influences primarily the flavor of the stored goods.

In addition to the storage methods that have been described, there is also the practice of so-called short-period storage (a few days). In the case of fruit and vegetables, this form of storage can, essentially, be equated to the climatic transport conditions. In this context, generally similar conditions are aimed at for storage over extended periods. In respect of temperature, however, the spreads are selected to be somewhat higher, where especially the upper values are set higher than in the case of usual storage.

Special importance, however, is to be assigned to short-term storage of flowers. Here, it is intended that the course of blooming be retarded by several days. In this manner, for example, a larger available amount of cut flowers, that cannot be disposed of in the market at one time, can be distributed over a somewhat extended period. Table 2 presents suitable storage temperatures for a selection of cut flowers and data concerning the duration of possible storage.

Literature

Fideghelli, C., Gorni, F.: Frutticoltura 28 (1966)

Fischnich, O., Thielebein, M.: Useful potato storage. Kartoffelbau *10* (1959)

Henze, J.: Construction and installations for storage rooms for fruits and vegetables. Kurator Tech. Bauw. Landwirtsch. KTBL-Schrift *154* (1972)

Herregods, M.: Storage of tomatoes. Tuinbow. Meded. 27, 11 (1963)

Höller, E.: Some further remarks on the shipping lane to West Africa. Wetterlotse *18* (1966a)

Höller, E.: Boxes from fresh lumber, causes for dorrosion damage. Wetterlotse *18* (1966b)

Höller, E.: Area roadstead and harbor weather. Wetterlotse *18* (1966c)

Höller, E.: Fruit transport from the Canary Islands to Bremen. Seewart *28* (1967)

Jenny, J.: Some fundamental concepts on the ventilation of stored potatoes. Le Tracteur at la Machine agricole 7 (1959)

Lyons, J.M., Rappaport, L.: Effect of temperature on respiration and quality of Brussels Sprouts during storage. Proc. Am. Soc. Matic. Soi. *73* (1959)

Macellin, P., Lebliondet, C., Leteinturisr, J.: Les méthodes de conservation des fruits en atmosphère controllée 96. Cong. Soc. Pomol. Paris (1965)

Seemann, J.: On the question of natural cooling of fruit storage cellars in the Rhineland. Rhein. Monatsschr. *6* (1954)

Smith, C.V.: Meteorology and grain storage. WMO Techn. Note 101, Geneva, 1969

Thiel, R.: Temperature control in potato stores. Landbauforsch. Völkenrode *11*, 2 (1961)

Thiel, R.: Temperature control in potato stores. Landbauforsch. Völkenrode *11*, 2 (1961)

Villiers, G.D.B.: Climatic and microclimatic aspects of the drying storage and transport of agricultural products. Ag. Meteorol. Proc. WMO-Seminar, Vol. II, Melbourne, 1968

23 Usefulness of Agroclimatology in Planning

B. PRIMAULT

23.1 Presentation of Climatological Data

For centuries – Louis XIV founded the Paris Observatory as early as 1667 – meteorological data have been collected in various parts of the world. The data are generally used by geographers to define the climatic zones, and to draw from them conclusions concerning their suitability for agriculture, tourism, or industry. For this purpose, they generally use averages regularly published by the appropriate national services. It might then be assumed that the immediately available data, namely those contained in the geography manuals or in the records periodically published by the national climatological services, would also be sufficient for agricultural purposes.

However, upon reflection, it appears that the available averages are only aids of very limited use in planning agricultural production. The latter must be based on consistent yields and the profitability of the enterprise. In calculating profits, however, it is necessary to include a certain risk of accidents.

23.1.1 The Averages

As stated above, a great number of documents are available concerning the averages of the various meteorological parameters. Those averages are generally based on many years for time periods ranging from one day to the entire year. An average, however, is of little significance for agricultural planning purposes. Actually, in agriculture, it is important to know not a hypothetical, so-called *normal* weather, which never occurs, and which is based on the simultaneous appearance of the averages of all the parameters under consideration, but rather the possible variations of each element and the combination of some of them.

Since an average is the quotient of the sum total of a certain number of observations divided by that number, it provides no indications concerning the possible fluctuations and the frequency of such fluctuations; it may consist of clusters of compact points, but quite far apart from one another (example: see Table 1A), of a uniform distribution of such points (see Table 1B) or of a frequency distribution more or less corresponding to the standard distribution or Gauss curve (see Table 1C). Consequently, the average, and even the standard variation accompanying that average, would not fulfill the requirements of agriculture.

Therefore, as one begins to deal with planning problems or profit calculations, it becomes necessary to compile again the original data, in order to present them in an adequate form, which is rather different from that established by classic climatology.

Thus, the three series shown in Table 1, although they present perfectly identical averages, have a totally different significance in agrometeorology. Even

Table 1. Comparison of three sets of figures

Example	A	B	C					
	Absolute maximum							
	30.0	30.0	30.0					
	29.9	29.0	27.3	} 10%				
	29.8	28.0	24.2					
	Effective maximum							
	29.7	27.0	22.1					
	29.6	26.0	21.0					
	29.5	25.0	20.0		25%			
	29.4	24.0	19.1					
	29.3	23.0	18.3					
	First quartile							
	29.2	22.0	17.6					
	29.1	21.0	17.0					
	29.0	20.0	16.5					
	28.9	19.0	16.1					
	28.8	18.0	15.8					
	28.7	17.0	15.6					
	28.6	16.0	15.5					
	Median or second quartile			} 80%	} 50%			
	2.4	15.0	15.5					
	2.3	14.0	15.4					
	2.2	13.0	15.2					
	2.1	12.0	14.9					
	2.0	11.0	14.5					
	1.9	10.0	14.0 .					
	1.8	9.0	13.4					
	Third quartile							
	1.7	8.0	12.7					
	1.6	7.0	11.9					
	1.5	6.0	11.0					
	1.4	5.0	10.0					
	1.3	4.0	8.9		} 25%			
	Effective minimum							
	1.2	3.0	6.8					
	1.1	2.0	3.7	} 10%				
	1.0	1.0	1.0					
	Absolute minimum							
Average	15.5	15.5	15.5					

Normal range   Effective range   Total range

though those examples are entirely hypothetical and are never found in nature, they are nonetheless quite significant for our development.

23.1.2 Ranges

As we have already stated, the extreme points within which a meteorological factor may vary at a given place already provide much more useful information for our purposes. Thus, the three series in Table 1 are all contained between 1 and 30.

It can therefore be stated that, from the viewpoint of the total range (interval separating the absolute minimum from the absolute maximum), the three series have the same significance. The total range, however, which is the interval within which an element may vary, is not the only important factor in agriculture. Two more highly relevant ranges are distinguished: if one leaves out 10% of the number of data on both sides of the series, the effective minimum and maximum, respectively, are obtained. In our case, out of 30 data, 3 high and 3 low ones will be left out, retaining the limits of the median 80% in the series. Thus, one obtains 29.7, 27.0, and 22.1 as the upper limit, and 1.3, 4.0, and 8.9 as the lower limit. The interval contained between these limits is the effective range. It shows that production is still profitable 8 years out of 10 (one year of deficit by excess and one by default), which is quite permissible.

As we pointed out above, in agriculture one must always expect meteorological accidents which may cause insufficient harvests. It is still possible to achieve greater accuracy in calculations and planning: it is possible to know the yields for one out of every two years, which leads us to the "normal range", knowing the median 50% in the series. This is obtained by leaving out 25% of the number of data on both sides of the series (first and third quartiles). In our example, the normal range (not to be confused with the "normal" of climatologists) is the range limited by 29.2, 22.0, and 17.6 at the top, and 1.8, 9.0, and 13.4 at the bottom, respectively.

Thus, the more one moves toward the center of the series (median), the greater become the differences among our three examples. This shows that their ultimate significance will be very different (see Table 1).

23.1.3 Frequencies

By dividing the total range into six classes of the same size and distributing among them the various data, the frequency distribution is obtained. It is noted then that, in example A, the four central classes (Nos. 2–5) do not contain any data. In example B, each class contains an equal number of data, while in example C there is a considerably greater number of data in the two central classes (see Fig. 1).

Such a frequency analysis provides very substantial additional information in calculating the profitability or in the agrometeorological planning of the enterprise.

23.1.4 Relative Values

So far, we have been talking primarily about absolute values, either those read directly from the instruments or those obtained by visual observation (for example, fog). In some particular cases, however, the actual value is of only limited interest. Actually, it is often very difficult to pass from one place to another, because of the scarce density of climatological networks. It is then necessary to extrapolate the data obtained at one or more relevant climatological stations in order to obtain information valid elsewhere. Actually, very rarely is a climatological station where observations have been performed for many years (at least 30) located in the immediate vicinity of the agricultural enterprise under consideration. It is then necessary, when one wants to determine its suitability, to

No. of cases

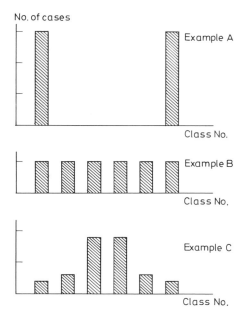

Fig. 1. Frequency histogram

extrapolate the data on the basis of data from stations located under similar conditions.

Thus, for example, with respect to the duration of sunshine, the natural horizon of the site – especially in a mountain area – plays a very important role in direct observation. To move from one place to another, it is desirable to have a horizon profile, that is, the duration of maximum possible sunshine for each day of the year (measurement of the daily angles).

Consequently, the national services will be required to add, in their publications concerning this parameter, the relative value, which is the ratio existing in each period between the value actually measured and the maximum possible sunshine (see Table 2).

23.1.5 Combined Indexes

In order to solve a great number of agricultural problems, the simple compilation of even numerous series of isolated meteorological parameters is not sufficient. Such a study only gives the fluctuations in each parameter, without taking the others into account. On the other hand, the development of plants and – to a lesser extent, it is true – of domestic animals is not determined by a single meteorological factor. More often, it is the combined action of several of them. Consequently, if one wants to obtain information concerning certain particular phenomena, it is necessary to know the frequencies of simultaneous events (for example, crossing of thresholds) of two or more meteorological parameters. For example: temperature/humidity, temperature/sunshine, temperature/wind, humidity/precipitation, etc. Let us here point out specifically the importance of the duration of the crossing of the temperature and humidity thresholds in the fight against harmful fungi.

Table 2. Duration of sunshine at the Neufchâtel station

Period 1931–1960		Jan.	Feb.	March	April	May	June	July	August	Sept.	Oct.	Nov.	Dec.	Year
Maximum possible duration	Hours	256	269	345	379	431	434	441	411	347	319	262	248	4142
Absolute maximum	Hours	70	171	252	249	298	319	342	296	227	157	82	79	2056
	%	27.3	63.6	73.0	65.7	69.2	73.5	77.6	72.0	65.5	49.2	31.3	31.9	49.6
Effective maximum	Hours	58	119	182	231	250	286	317	275	195	134	64	45	1857
	%	22.7	44.2	52.8	61.0	58.0	66.0	71.9	66.9	56.2	42.0	24.4	18.1	44.8
1st quartile	Hours	47	82	164	200	240	272	282	250	183	120	56	36	1700
	%	18.4	30.5	47.5	52.8	56.7	62.1	64.0	60.9	52.8	36.6	27.4	14.5	41.0
Median	Hours	38	70	148	179	215	230	241	227	160	101	42	29	1676
	%	14.8	26.0	42.9	47.2	49.9	53.0	54.1	55.2	46.1	37.7	16.0	11.7	40.5
3rd quartile	Hours	28	54	115	151	182	196	212	197	138	78	32	19	1603
	%	10.9	20.1	33.3	39.8	42.3	45.2	48.1	47.9	39.8	24.5	12.1	7.7	38.7
Effective minimum	Hours	18	48	96	137	162	179	194	188	126	74	23	14	1563
	%	7.0	17.8	27.8	36.2	37.6	41.3	44.0	45.8	36.3	23.2	8.8	5.0	37.7
Absolute minimum	Hours	14	26	60	101	122	158	162	161	120	34	15	6	1346
	%	5.5	9.7	17.4	26.7	28.3	26.4	36.7	39.2	34.6	10.7	5.7	2.4	32.5
Average	Hours	39	78	148	179	210	232	251	226	162	101	44	29	1699
	%	15.2	29.0	42.9	47.2	48.8	53.5	57.0	55.0	46.7	31.7	16.8	11.7	41.0

[a] Normal range. [b] Effective range. [c] Total range.

23.1.6 The Risk of Meteorological Accidents

Until now, we have only talked about the frequency analysis of each of the traditional meteorological factors, taken alone or in groups. In the case of agriculture, however, that analysis only makes possible a first approximation in the determination of the suitability of a place for farming. While doing so, one should also, however, take into account certain precise meteorological facts which are of paramount importance in determining the growth of plants or the development of parasites.

We have mentioned in the preceding paragraph the combination of temperature and humidity in the fight against cryptogams.

The frequency with which the temperature crosses certain thermal thresholds may, depending on the crops, play a controlling role in planning. Let us think, for example, of crossing below the $0\,°C$ threshold, which means danger of frost.

The possibilities of farming are not only affected by low temperatures: high temperatures may also be very harmful to the development of certain plants. Let us mention here a single example: the scorching threshold ($30\,°C$), beyond which wheat no longer develops normally, but follows a process of early ripening (decrease in the weight per 1000 grains and lower baking quality as a result of the disappearance of gluten).

Other meteorological accidents may be serious enough to cause the complete loss of a harvest. We shall mention hail, hurricanes, prolonged droughts, floods, etc. Knowing the frequency of such phenomena in advance means to a great extent being capable of evaluating the risk of damage.

The frequencies may be taken from series of exisiting observations. Only their recompilation and presentation in appropriate formats, however, will make it possible to detect the risks faced by farmers.

23.2 Determination of Suitable Crops for a Certain Place

It would certainly be presumptuous to decide to set up an exhaustive list of the meteorological parameters and their determinant thresholds making it possible to calculate the suitability of a place for certain crops. Depending on the nature of the soil, the geographical location, and the economic conditions, each region has its own standards. Therefore, we shall limit ourselves to give here only an example and not a complete list.

23.2.1 The Climatological Method

As we have stated above, the national climatological networks generally are not very dense. Such an arrangement is amply sufficient on flat ground, where climatological changes operate gradually. In the mountains, on the other hand, very considerable differences are found over short distances. In order to meet the requirements, it would be necessary to increase the density of stations to such an extent that the resources required would be prohibitive. One is then forced to proceed to numerous extrapolations to move from the conditions at an observation post to those of a given agricultural enterprise. For this reason, first of all, a regional study will be carried out, which will include, among others, relatively large-scale charts (1:200,000, for example), showing the evolution of the various meteorological parameters. They will contain not only the averages, but especially the ranges and frequencies of those factors, as described under Sects. 23.1.2 and 23.1.3, respectively. After making those calculations, it is possible to prepare regional charts showing which are the crops that can be cultivated in one place or another. For this purpose, precise definitions are necessary. Of course, this procedure implies that the "place" is no longer a given farm, but rather an *area*, that is, a group of farms of the same kind. On flat ground, this first approximation is quite sufficient, because the climatic changes to which the various meteorological parameters are subject, as we stated above, are small over distances between 5 and 20 km. In mountain areas, on the other hand, this method only permits a first approximation to the farming conditions, because the very considerable differences over short distances necessarily affect the suitability of crops. Thus a change in exposure may, by itself, involve radical changes in the radiation received, in the radiation emitted, in the wind, in the duration of sunshine, or even in temperature and humidity. Furthermore, radiation and the duration of sunshine are of primary importance in affecting the evolution of the climate of the first centimeters of soil, particularly their temperature. Consequently, minor changes, due to the local orography, may cause major changes in the conditions for plant development. A climatological analysis alone is incapable of dealing with details of this kind; it cannot provide sufficient information to determine the crops suitable for a place, which here means a field, a small vineyard, or an orchard.

23.2.2 The Ecological Method

The climatological method is thus inappropriate to determine the crops suitable for small plots within a nonhomogeneous climatic zone, determined by the configuration of the terrain. The data available to us are too few for this purpose.

It has been necessary to find another way of differentiating among the various plots: the best is certainly the ecological method. It takes into account, on the one hand, the information from the climatological analysis, as described under Sect. 23.2.1, and, on the other hand, local observations.

It is actually preferable to refer to figures which are relatively inaccurate, but reflect the evolution of the weather over a very long period, in order to estimate the possible fluctuations of combinations of meteorological parameters or the risk of specific accidents. If one wanted only to use local information, it would be essential to proceed to perform observations over long periods, which would postpone the availability of the result to an unjustifiable extent, in economic terms.

The local observations are performed over a short period of time. They concern exposure, the presence of natural or artificial obstacles (rock walls, edges of forests, road or railroad embankments, groups of buildings, etc.), but above all the nature of the soil and such local features as springs, swampy areas, sand or gravel banks, etc. Furthermore, they are complemented by the phenological observations, which show the effect on the climate of the soil of total radiation and other meteorological phenomena, such as duration of snow cover, wind, sunshine, water supply, etc. In certain cases, in order to define this groups of variables, the term "temperature zones" has been used. It is, however, erroneous, because plant development does not depend on the air temperature alone, but also, or even more, on other meteorological parameters, such as those mentioned above. Consequently, it would be preferable in this case to use the term "zones of similar phenological development".

The study of the soil and of the orographic features, complemented by accurate phenological data, will lead to the preparation of detailed charts, which, added to the regional climatological suitability charts, will make it possible to determine exactly the crops susceptible of being cultivated on a given plot of land.

It follows then that, while the climatological method makes it possible to determine the farming suitability of areas of a few km^2, the ecological method goes into details which may affect on order of magnitude of one hectare, or even a few ares.

23.3 Use of Biometeorological Indexes

The simultaneous study of climatic conditions and of plant development provides the necessary basis for statistical comparisons. The so-called "biometeorological indexes" or "models" can be deduced from them (for example, the sum total of the temperatures required by a given crop to pass from one phenological state to the next). After being specified, those indexes may be used as the basis for calculating the economic viability of the enterprise. Actually the climatological data, calculated from those indexes expressed in the form of formulas, equations, etc., will show how many times the crop would have been successful at a given station in the observation network. The calculations may tend to express the development of the crop from its quantitative or qualitative aspects, or to show the deadlines for planting or harvesting.

23.3.1 Profit Calculations

The examination of the climatological data in conjunction with the bio-meteorological indexes for a given product in quantitative and qualitative terms will show the probable amount produced, at least each year, 9 out of 10 years, 8 out of 10 years, etc. The use of profitability thresholds (production costs, sale price, amount of product required to balance them, taking into account the desired profit, etc.) will show in advance if a certain crop is suitable. In expressing such profitability, one will always take into account a certain permissible risk of meteorological accidents, foreseeable or not. Therefore, it will never be 100% guaranteed.

Such profitability calculations should precede the introduction of any new crop. Actually, the suitability of a crop, that is, the possibility, in view of the climate, of making the plant grow, does not necessarily assure that its cultivation is profitable for the producer.

23.3.2 Crop Selection

The informed manager of an agricultural enterprise who wants to develop it will first of all perform a climatological analysis following the two methods described under Sect. 23.2. Subsequently, by means of the profitability calculations as suggested under Sect. 23.3.1, he will know the crops which can be introduced on his land. At the beginning, the choice will then already be rather limited. It will be even more so after the examination of the machinery required by those crops (see Sect. 23.4) and of the need for antiparasitic measures (Sect. 23.3.3). From that time on, only questions of ability and personal preferences will guide the producer in his selection.

23.3.3 Need for Antiparasitic Measures

We have seen above that it is possible, from simultaneous phenological and meteorological observations, to determine the biometeorological indexes affecting the growth of crops. The same does not apply to parasites, whether they be insects or fungi. The application of those indexes to sets of climatological observations will make possible to estimate the number of interventions required to keep a population below the level of economic danger.

This method of analysis of the need for antiparasitic measures, however, does not apply only to chemical means. It can also be usefully employed in an integrated campaign, by showing the conditions encountered by the parasite's predators and, thereby, the respective development of the two opposing organisms. It will then be possible to see whether laboratory breeding is required to remedy the deficiencies of the local climate acting on the development of the former.

23.3.4 Calculating the Potential Production or Harvest Forecasts

We have stated in Sect. 23.3.1 that the application of the biometeorological indexes to certain climatological thresholds would make possible to estimate the amount of the harvest. Those figures will be the essential basis for determining the size of the storage areas in which the harvest will be placed, whether they be grain elevators, barns for fodder or any other warehouse.

An analogic analysis of those meteorological elements, however, this time performed during the year and not retrospectively, makes possible the preparation of both quantitative and qualitative harvest forecasts and the sale of the harvest with a good knowledge of the fluctuations of supply and demand.

23.4 Possibilities of Working in the Fields

While briefly describing the ecological method (see Sect. 23.2.2), we have seen the importance of the reciprocal action of the meteorological elements, on the one hand, of the soil and orographic conditions, on the other, in determining the suitability of crops for a plot. This reciprocal action is even more marked, however, in determining the possibility of working in the fields, that is, in calculating in advance the number of days available for farm work (tilling, seeding, antiparasitic treatment, harvesting, etc.) during the different periods of the year.

23.4.1 Soil Plasticity

The texture of the soil changes according to its nature, the frequency and amount of precipitation, and the intensity of evaporation and transpiration. A clay soil, for example, subjected to abundant and frequent precipitation, at low temperatures and high humidity, will fill with water: it will become sticky and it will be practically impossible to work it without causing irreversible damage (compacting in depth, for example). Tilling cannot take place and the seeds risk rotting. The opposite extreme occurs with a sandy soil, located under hot and dry climatic conditions. Strong radiation and little precipitation will make it dusty. In that case again tilling will be very difficult, because the turf will not turn over and will fall back as dust onto the plowshare. On the other hand, it will be easy to move on that soil, without damaging it either on the surface or in depth, because it has considerable mechanical firmness.

Consequently, the preliminary calculation of soil fluctuations and plasticity by means of climatological data and of the analysis of its physical properties will make it possible to know the number of farming operations feasible in that field in each season.

23.4.2 Adaptation of Farming Methods

The climatological and physical analysis of soil plasticity, together with a similar study of the water balance, will make it possible to supply the basic information required to determine the farming methods most appropriate to that place.

23.4.3 Selection of Machinery

After performing these preliminary analyses, the planner knows the evolution over time of the various vegetative stages in which farming operations are required. Therefore, he was able to select the most appropriate crops and farming methods. He can then proceed to the selection of the machinery.

The nature of the machinery is determined on the one hand by the crops grown, and on the other hand by the farming methods. The freedom of choice of the planner is thus limited by the suitability of his land for certain crops.

When the type of machinery is fixed, however, its weight and type of construction are still left to the free choice of the planner. At this point, the results of the soil plasticity analysis again intervene. The maximum permissible pressure per square centimeter, be it that of a pneumatic tire, of a track, or of any other support, depends on them. Thus, by knowing the maximum pressure tolerated by the soil, it is possible to vary the number of wheels, the width of the tires or the size of the tracks, so as not to exceed that limit. In the calculations, not only the gross weight of the machinery itself will be considered, but also, if applicable, that of the equipment carried (plowshares, seed dispenser, manure spreader, etc.).

23.5 Summary

If, in addition to the foregoing, one takes into account what was mentioned concerning the ideal conditions for cattle growth (Chap. 11) and the possibility or adapting the construction of shelters (Chap. 13). it appears that any agricultural planning involves an appropriate analysis of the local climate. Agriculture can only be developed under the express condition of having available exact figures, reliable and sensibly obtained.

The present demographic explosion and the general rise in the standard of living require a greater consumption of food products of higher quality. Only a rationally planned agriculture can meet this need. This concept applies everywhere, even in countries with a high standard of living and intensive agricultural production.

In agriculture as well as in industry, the evolution of the modern world requires managers who often reconsider their working methods, if they want constantly to improve the quality of the increasingly diversified products which they supply, and increase their quantity. Those products, however, must remain fully suitable to the external conditions of the enterprise, among which climate is one of the most basic, the other being the soil.

Literature

Bergeiro, J.M.: La Meteorologìa aplicada al estudio de la vegetación. Revista meteorologica *I*. 62–86. Montevideo (1942)

Boughton, W.C.: Effects of land management on quality and quantity of available water. Univ. New South Wales. Water Res. Lab. Rep. *120*, 330 (1970)

Burgos, J.: El termoperiodismo como factor biclimàtico en el desarollo de los vegetales. Meteoros *2*, 215–242 (1952)

Capman, L.J., Brown, D.M.: The climates of canada for agriculture. The Canada Land Inventory. Report *3*, 24 (1966)

Davitaja, F.F., Kulick, M.S.: Agrometeorological problems, p. 91. Hydrometeorological Publishing House (Moscow Department 1958)

Derevyanko, A.N.: Calculation of date for the beginning of field works and sowing of early spring cereal crops in the non-Chernozem zone of the European territories of the USSR. Meteorologia i Hidrologija. 1. Mockba 74–77 (1969)

Golzow, M.M., Maximow, S.A., Jaroschewski, W.A.: Praktische Agrarmeteorologie. Dtsch. Bauernverlag. Berlin *2*, 310 (1955)

Griffiths, J.F.: Applied climatology. An introduction, p. 118. London: Oxford University Press 1966

Hills, G.A.: A ready reference to the description of the land of Ontario and its productivity, p. 10. Ontario Dep. Land and Forest. Research Branch 1959

Lelouchier, P.: Contribution à l'étude écologique des versants de vallée. Bull. Soc. Roy. Bot. Belg. *92*, 39–76 (1960)

McQuigg, J.D.: A review of problems, progress and opportunities in the use of weather information in agricultural management, p. 38 UNESCO/AVS/NR/200 (1966)

Müller, W.: Zur Bestimmung der effektiv möglichen Sonnenscheindauer in stark kupiertem Gelände. Geofisica e Meteorologia *XI*, 279–280 (1963)

Quellet, C.E.: The usefulness of climatic normals for plant zonation in Canada. Naturaliste Canada *96*, 507–521 (1969)

Pascale, A.J., Damario, A.E.: Agroclimatologia del cultivo de trigo en la Repùblica Argentina. Revista de la Facultad de Argonomia y veterinaria *XV*, 115 (1961)

Polentika, W. von: Agrarklimatische Synthese der russischen Landwirtschaft. Berichte des Deutschen Wetterdienstes in der U.S.-Zone *38*, 249–254 (1952)

Primault, B.: Nouvelle conception de la présentation, pour l'agriculture, des séries climatologiques. Publications de l'OEPP. Série A *57*, 41–53 (1970)

Primault, B.: Etude méso-climatique du Canton de Vaud. Cahiers de l'Aménagement régional. *14*, 186 (1972)

Primault, B.: De la présentation de documents climatologiques pour les praticiens. Atti del Congresso Internazionale di Climatologia Lacustre. Como, Italia. Lecco 1972, 100–104 (1971)

Russel, J.S., Moore, A.W.: Detection of homoclimates by numerical analysis with reference to the brigalow region (eastern Australia). Agr. Meteorol. 7, 455–479 (1970)

Schreiber, K.-F.: Les conditions thermiques du canton de Vaud et leur graduation. Cahiers de l'Aménagement régional 5, 31 (1968)

Smith, L.P.: Farming weather, p. 208. London: Thomas Nelson and Sons Ltd. 1958

Utaaker, K.: The local climate of Nes, Hedmark. Univ. Bergen Skrifter 28, 117 and XXVI (1963)

Wagooner, P.E., et al.: Agricultural meteorology. American Meteorological Monographs 6, 188 (1965)

Williams, G.D.V.: Aplying estimated temperature normals to the zonation of the canadian great plains for wheat. Can. J. Soil Sci. 49, 263–276 (1969)

Yamamete, S., Asano, Y., Kakiuchi, G.H.: Areal functional organisation in agriculture: an example from Japan. Sci. Rep. Tokyo Kyoku Daigaku. Section C 10, 165–209 (1969)

24 Use of Agroclimatology in Crop Distribution

Y. I. Chirkov

An agroclimatic basis is becoming ever more common in the placement of specific crop strains and hybrids, as well as in the planning of the specialization of branches of agriculture, the introduction of valuable plants outside their traditional areas, and the systematization of measures designed to increase the productivity of the agricultural industry. In connection with this, the evaluation of climatic resources as applied to agriculture is assuming an increasingly practical value.

The agricultural climate evaluation methods currently used in various countries are aimed primarily at giving a quantitative appraisal of thermal and water resources, based on the recurrence period of their levels in specific years.

The total temperature method is used extensively in the appraisal of thermal resources. The works of M. I. Budyko of the Soviet Union have established on a global scale the intimate correlation between the average perennial values of total active temperatures and the radiation balance value.

The total active temperature method has been employed extensively in the Soviet Union for general agroclimatic zoning (Selyaninov et al.), while the total effective temperatures have been used for zoning specific crops. In the latter case, heat provision during the growing season to strains and hybrids differing in early development has been determined by comparing thermal resources with the heat demands of the crops (which is expressed in the total effective temperature). Total active, effective, and bioclimatic temperatures have been established for many crops. These methods of expressing crop heat requirements are widely used for agricultural climate evaluation in the Soviet Union, Bulgaria, Poland, Rumania, and a number of other countries.

The degree day method, which expresses numerically the relationship of plant development and growth to atmospheric temperature, was developed in the United States in the first half of the twentieth century. In the degree day method, the mean diurnal air temperatures above the minimum plant heat requirements are totaled (in essence, this is analogous to the total effective temperature method). This method was used in connection with the determination of crop maturation times.

Significant research has been done in many countries on temperatures critical to plants, and this, along with the aggregate evaluation of thermal resources, has made possible a substantially more accurate determination of climatic heat provision to crops.

Moisture resources have been studied in connection with crop needs, using various methods whose underlying bases are the amount of precipitation during the growing season, as well as humidification indices expressing the correlation between precipitation and temperature aggregates (Selyaninov hydrothermal coefficient, Morton index, etc.) or the correlation between precipitation and evaporativity (the Thorntwaite, Ivanov, Shashko, Koloskov, and Buchet methods, etc.), or the ratio of the radiation balance to annual precipitation multiplied by the latent heat of vaporization coefficient (Budyko, Soviet Union).

The use of soil productive moisture reserves as an index of climatic moisture provision has found application in agroclimatology comparatively recently, primarily in the Soviet Union, where a large stock of observations on the dynamics of productive moisture reserves for crops in various climatic zones is accumulated.

Solar radiation has become a target of study for agroclimatologists in recent decades. In a number of countries (Soviet Union, United States, Japan, and others), numerical expressions of the effects of length of day and total ingress of solar radiation on farm crops have been formulated and their agroclimatic interpretation given.

It should be pointed out that mathematical statistics methods coupled with computer technology and mathematical modeling of the relation of crop development and productivity growth to climatic factors are assuming increasingly greater significance in agroclimatology, by enabling the formulation of multifactor agroclimatic indices.

Average perennial figures, as well as an annual modulus of climatic elements essential to plants, are used in the agricultural evaluation of climatic resources.

Total temperature methods (active and effective) express in the first approximation the total thermal energy requisites essential for plants to pass through specific developmental stages as well as the growing season as a whole. However, this method must be supplemented by allowances for ballast temperatures, temperature amplitude, and elevation above sea level in order to give the temperature aggregates a more complete ecological content.

Methods defining the significance of solar radiation to plant growth rate and productivity are extremely promising. They are chiefly expressed with equations denoting the relationship between plant productivity and radiation balance, radiation influx, and the number of hours of sunlight.

The dependences of crop productivity on productive-moisture reserves have been determined primarily in the Soviet Union, where perennial observations of the productive moisture dynamics of crops are available from thousands of meteorological stations. Methods for the numerical expression of plant moisture needs through the use of various humidification and water balance indices are being developed in a number of countries. These approaches should be compared under similar conditions.

The development of techniques establishing the complex relationships between plant needs and climatic resources is of great importance. Such techniques have considerable value in the evaluation of climatic resources.

It is necessary to study the light, heat, and moisture requirements of plants during specific stages of organogenesis, particularly during critical periods, in order to determine the biological validity of the effects of climatic factors on plant developmental growth and productivity formation.

Data on agricultural climate evaluation are usually presented in terms of agroclimatic zoning.

Agroclimatic zoning is the division of a territory according to similar and dissimilar features of climate conditions essential to agricultural objectives.

General agroclimatic zoning defines the degree of favorableness of climatic conditions in different sections of a territory to agriculture as a whole. Here the quantitative characteristics of climatic factors, principally of heat and moisture, are used.

Essentially, the following indices are used for characterizing thermal resources:

1) Total active temperatures (the sum of the mean diurnal air temperatures for a period with temperatures above 10 °C).

2) Total effective temperatures (the sum of mean diurnal temperatures above biological zero, i.e., the temperature at which plant growth and development begins). This index is used primarily for zoning specific crops as is the degree day index, developed in the United States.

3) The average of the absolute annual air temperature minima, which characterizes the severity of the winter.

4) The mean temperatures of the warmest and the coldest months.

The humidification of an area is defined with the following indices:

1) Annual precipitation aggregates and aggregates for the warm and cold seasons as well as for the growing season.

2) Selyaninov hydrothermal coefficient (HTC), which expresses the ratio of precipitation totals for a given period to the aggregate of temperatures above 10 °C for that same period. A modified HTC allows for winter precipitation measured by various factors.

3) The ratio of precipitation to evaporativity (evapotranspiration), calculated with various methods (Thorntwaite, Buchet, Penman, Pandakis, Tyurk, and Shashko formulas and others).

4) Budyko formula – an index of dryness – which expresses the ratio of the radiation balance to annual precipitation multiplied by the latent heat of vaporization.

5) Humidification balance, where evaporativity is calculated with the Ivanov formula or with the formulas of the authors mentioned under point 3.

It should be noted that the different methods for assessing the humidification of an area are very heterodynamic. Their application in a number of instances is restricted by climatic conditions and the scale of the area involved. For example, the Tyurk formula is more suitable for determining the humidification of large regions, while the Buchet formula is better for individual farms.

Until recently, sunlight resources were not taken into account in agroclimatic zoning. Only in the past decade have agroclimatic charts been prepared for the ingress of photosynthetically active radiation for the growing season and for the period with temperatures above 10 °C (Yu. K. Ross and M. I. Budyko of the Soviet Union). Sunlight resources are now beginning to be incorporated in the zoning of specific crops (corn, rice, soy beans, and others).

Specific (specialized) agroclimatic zoning is conducted for the purpose of singling out regions with varying measures of climatic factors favorable to the growth of specific crops, their strains and hybrids, as well as for the purpose of specializing farm production, soil management techniques, and farming practices.

In agroclimatic zoning (particularly in specific zoning), it is necessary to first formulate numerical expressions for crop climate requirement, which are identified as the agroclimatic indices. Second, one must ascertain the geographic distribution of the obtained indices by region.

Under the established methodology in the Soviet Union, the following indices essential to agroclimatic zoning of specific crops:

1) Minimum temperature required for plant growth and development during the principal stages (phases) of growth.

2) The temperature aggregate (active and effective) essential for the onset of the principal stages of development, and for the growing season as a whole.

3) Temperature levels (high and low) causing damage to plants during their growing season and during the dormant period (for winter crops and perennial plants).

4) Magnitudes characterizing the effects of solar radiation (length of day, hours of sunlight, influx of radiation, etc.) in specific periods of plant development.

5) Levels defining plant moisture requirements during the growing season as a whole and in critical periods of development.

6) The resistance of plants to drought and other adverse climatic factors (dry winds, strong winds, heavy showers, etc.)

This chart is augmented by still other indices, which are used for the purpose of especial elaboration of climatic conditions.

Data from long-range observations of meteorological stations are used for establishing the geographical distribution of the indices. The nature of the variation of the indices in relation to elevation above sea level, degree of continentality of the climate, slope exposure and grade, etc. is considered in the process. As already pointed out, average perennial characteristics of climatic elements, as well as the recurrence interval and modulus of agroclimatic indices, are used in zoning. Specific agroclimatic zoning is carried out sucessfully in 80–90% of the years.

The findings of the above-mentioned studies are being charted, and regions with varying agroclimatic conditions or simply with a varying measure of climatic factors necessary for the growth and productivity of a given crop (strain, hybrid) are being mapped with isolines. At the present time, agroclimatic zoning operations for major crops are being conducted in most of the countries of the world.

Subject Index

Advanced Series in Agricultural Sciences

Co-ordinating Editor: B. Yaron
Editors: G. W. Thomas, B. R. Sabey, Y. Vaadia,
L. D. Van Vleck

Volume 1: A. P. A. Vink

Land Use in Advancing Agriculture

1975. 94 figures, 115 tables. X, 394 pages
ISBN 3-540-07091-5

Contents: Land Use Surveys. – Land Utilization Types. – Land Resources. – Landscape Ecology and Land Conditions. – Land Evaluation. – Development of Land Use in Advancing Agriculture.

Volume 2: H. Wheeler

Plant Pathogenesis

1975. 19 figures, 5 tables. X, 106 pages
ISBN 3-540-07358-2

Contents: Concepts and Definitions. – Mechanisms of Pathogenesis. – Responses of Plants to Pathogens. – Disease-Resistance Mechanisms. – Genetics of Pathogenesis. – Nature of the Physiological Syndrome.

Volume 3: R. A. Robinson

Plant Pathosystems

1976. 15 figures, 2 tables. X, 184 pages
ISBN 3-540-07712-X

Contents: Systems. – Plant Pathosystems. – Vertical Pathosystem Analysis. – Vertical Pathosystem Management. – Horizontal Pathosystem Analysis. – Horizontal Pathosystem Management. – Polyphyletic Pathosystems. – Crop Vulnerability. – Conclusions. – Terminology.

Volume 4: H. C. Coppel, J. W. Mertins

Biological Insect Pest Suppression

1977. 46 figures, 1 table. XIII, 314 pages
ISBN 3-540-07931-9

Contents: Glossary. – Historical, Theoretical, and Philosophical Bases of Biological Insect Pest Suppression. – Organisms used in Classical Biological Insect Pest Suppression. – Manipulation of the Biological Environment for Insect Pest Suppression. – A Fusion of Ideas. – Index.

Volume 5: J. J. Hanan, W. D. Holley, K. L. Goldsberry

Greenhouse Management

1978. 283 figures, 117 tables, XIV, 530 pages
ISBN 3-540-08478-9

Contents: Introduction. – Light. – Greenhouse Construction. – Temperature. – Water. – Soils and Soil Mixtures. – Nutrition. – Carbon Dioxide and Pollution. – Insect and Disease Control. – Chemical Growth Regulation. – Business Management. – Marketing. – Appendices: Conversion Tables. Symbolism. Definitions.

Volume 6: J. E. Vanderplank

Genetic and Molecular Basis of Plant Pathogenesis

1978. 3 figures, 36 tables. XI, 167 pages
ISBN 3-540-08788-5

Contents: Variation in the Resistance of the Host and in the Pathogenicity of the Parasite. – The Gene-for-Gene and the Protein-for-Protein Hypotheses. – The Protein-for-Protein Hypothesis: Temperature Effects and Other Matters. – Common Antigenic Surfaces in Host and Pathogen. – Other Large Molecules in Relation to Gene-for-Gene Disease. – Population Genetics of the Pathogen. – Horizontal Resistance to Disease. – Selective Pathotoxins in Host-Pathogen Specificity. – A Molecular Hypothesis of Vertical and Horizontal Resistance. – Biotrophy, Necrotrophy, and the Lineage of Symbiosis.

Volume 7: J. K. Matsushima

Feeding Beef Cattle

1979. 31 figures, 23 tables. Approx. 150 pages
ISBN 3-540-09198-X

Contents: Nutrients. – Classification of Feeds. – Procedures in Ration Formulation. – Processing Feeds for Beef Cattle. – Systems of Feeding. – Feed Additives. – Growth Stimulants.

Springer-Verlag
Berlin
Heidelberg
New York

Irrigation Science

ISSN 0342-7188   Title No. 271

Editor in Chief: Gerald Stanhil,
Agricultural Research Organisation, The Volcani Center,
Institute of Soils and Water, P.O.B., Bet Dagan, Israel

Editors: J.F.Bierhuizen, Department of Horticulture,
Agricultural University, Haagsteeg 3, Wageningen, The
Netherlands; **A.E.Hall,** University of California, Depart-
ment of Botany & Plant Sciences, Riverside, CA 92521,
USA; **Th.C.Hsiao,** Department of Water Science, Univer-
sity of California, Davis, CA 95616, USA; **M.E.Jensen,**
Soil and Water Conservation Laboratory, A.R.S.,
U.S.D.A.,Kimberly, ID 83341, USA; **St.L.Rawlins,** U.S.
Salinity Laboratory, P.O.B. 672, Riverside, CA 92501,
USA; **C.W.Rose,** School of Australian Environmental
Studies, Griffith University, Nathan, Queensland 4111,
Australia; **D.Shimshi,** Agricultural Research Organisa-
tion, Regional Experiment Station, Mobile Post Negev 2,
Gilat, Israel; **B.Slavik,** Czechoslovak Academy of
Sciences, Institute of Experimental Botany, Department
of Plant Physiology, Flemingovo Nám. 2, Praha 6,
Czechoslovakia; **N.C.Turner,** CSIRO Division of Plant
Industry, P.O.Box 1600, Canberra City, A.C.T.2601,
Australia

Irrigation Science publishes original contributions and
short communications reporting the results of irrigation
research, including relevant contributions from the plant,
soil and atmospheric sciences as well as the analysis of
field experimentation. Special emphasis is given to multi-
disciplinary studies dealing with the problems involved in
maintaining the long term productivity of irrigated lands
and in increasing the efficiency of agricultural water use.
Aspects of particular interest are:

- Physical and chemical aspects of water status and movement in the plant-soil-atmosphere system.
- Physiology of plant growth and yield response to water
  status.
- Plant-soil-atmosphere and water management in irrigation practice.
- Measurement, modification and control of crop water
  requirements.
- Salinity and alkalinity control by
  soil and water management.
- Ecological aspects of irrigated agriculture.

Springer-Verlag
Berlin
Heidelberg
New York

Subscription information and sample copy upon request.